INTRODUCTION TO ABSTRACT ALGEBRA

TEXTBOOKS in MATHEMATICS

Series Editor: Denny Gulick

PUBLISHED TITLES

COMPLEX VARIABLES: A PHYSICAL APPROACH WITH APPLICATIONS AND MATLAB®
Steven G. Krantz

INTRODUCTION TO ABSTRACT ALGEBRA
Jonathan D. H. Smith

LINEAR ALBEBRA: A FIRST COURSE WITH APPLICATIONS
Larry E. Knop

FORTHCOMING TITLES

ENCOUNTERS WITH CHAOS AND FRACTALS
Denny Gulick

TEXTBOOKS in MATHEMATICS

INTRODUCTION TO ABSTRACT ALGEBRA

Jonathan D. H. Smith

Iowa State University
Ames, Iowa, U.S.A.

CRC Press
Taylor & Francis Group
Boca Raton   London   New York

CRC Press is an imprint of the
Taylor & Francis Group, an **informa** business

A CHAPMAN & HALL BOOK

Chapman & Hall/CRC
Taylor & Francis Group
6000 Broken Sound Parkway NW, Suite 300
Boca Raton, FL 33487-2742

© 2009 by Taylor & Francis Group, LLC
Chapman & Hall/CRC is an imprint of Taylor & Francis Group, an Informa business

Library of Congress Cataloging-in-Publication Data

Smith, Jonathan D. H., 1949-
  Introduction to abstract algebra / Jonathan D.H. Smith.
    p. cm. -- (Textbooks in mathematics ; 3)
  Includes bibliographical references and index.
  ISBN 978-1-4200-6371-4 (hardback : alk. paper)
  1. Algebra, Abstract. I. Title.

QA162.S62 2008
512'.02--dc22   2008027689

**Visit the Taylor & Francis Web site at
http://www.taylorandfrancis.com**

**and the CRC Press Web site at
http://www.crcpress.com**

Contents

Preface

This book is designed as an introduction to "abstract" algebra, particularly for students who have already seen a little calculus, as well as vectors and matrices in 2 or 3 dimensions. The emphasis is not placed on abstraction for its own sake, or on the axiomatic method. Rather, the intention is to present algebra as the main tool underlying discrete mathematics and the digital world, much as calculus was accepted as the main tool for continuous mathematics and the analog world.

Traditionally, treatments of algebra at this level have faced a dilemma: groups first or rings first? Presenting rings first immediately offers familiar concepts such as polynomials, and builds on intuition gained from working with the integers. On the other hand, the axioms for groups are less complex than the axioms for rings. Moreover, group techniques, such as quotients by normal subgroups, underlie ring techniques such as quotients by ideals. The dilemma is resolved by emphasizing semigroups and monoids along with groups. Semigroups and monoids are steps up to groups, while rings have both a group structure and a semigroup or monoid structure.

The first three chapters work at the concrete level: numbers, functions, and equivalence. Semigroups of functions and groups of permutations appear early. Functional composition, cycle notation for permutations, and matrix notation for linear functions provide techniques for practical computation, avoiding less direct methods such as generators and relations or table look-up. Equivalence relations are used to introduce rational numbers and modular arithmetic. They also enable the First Isomorphism Theorem to be presented at the set level, without the requirement for any group structure. If time is short (say just one quarter), the first three chapters alone may be used as a quick introduction to algebra, sufficient to exhibit irrational numbers or to gain a taste of cryptography.

Abstract groups and monoids are presented in the fourth chapter. The examples include orthogonal groups and stochastic matrices, while concepts such as Lagrange's Theorem and groups of units of monoids are covered. The fifth chapter then deals with homomorphisms, leading to Cayley's Theorem reducing abstract groups to concrete groups of permutations. Rings form the topic of the sixth chapter, while integral domains and fields follow in the seventh. The first six or seven chapters provide basic coverage of abstract algebra, suitable for a one-semester or two-quarter course.

Subsequent chapters deal with slightly more advanced topics, suitable for a second semester or third quarter. Chapter 8 delves deeper into the theory of rings and fields, while modules — particularly vector spaces and abelian groups — form the subject of Chapter 9. Chapter 10 is devoted to group theory, and Chapter 11 gives an introduction to quasigroups.

The final four chapters are essentially independent of each other, so that instructors have the freedom to choose which topics they wish to emphasize. In particular, the treatment of fields in Chapter 8 does not make use of any of the concepts of linear algebra, such as vector space, basis, or dimension, which are covered in Chapter 9. For a one-semester introduction to groups, one could replace Chapter 6 with Chapter 10, using the field of integers modulo a prime in the examples that call for a finite field.

Each chapter includes a range of exercises, of varying difficulty. Chapter notes point out variations in notation and approach, or list the names of mathematicians that are used in the terminology. No biographical sketches are given, since libraries and the Internet can offer much more detail as required.

A special feature of the book is the inclusion of the "Study Projects" at the end of each chapter. The use of these projects is at the instructor's discretion. Some of them may be incorporated into the main presentation, offering typical applications or extensions of the algebraic topics. Some are coherent series of exercises, that could be assigned along with the other problems, or used for extra credit. Some projects are suitable for group study by students, occasionally involving some outside research.

I have benefited from many discussions with my students and colleagues about algebra, its presentation and application. Specific acknowledgments are due to Mark Ciecior, Dan Nguyen, Jessica Schuring, Dr. Sungyell Song, Shibi Vasudevan, and anonymous referees for helpful comments on a preliminary version of the book. The original impetus for the project came from Bob Stern at Taylor & Francis. I am grateful to him, and the publishing staff, for bringing it to fruition.

Chapter 1

NUMBERS

Algebra begins as the art of working with numbers. The *integers* are the whole numbers, positive, negative, and zero. Put together, they form the set

$$\mathbb{Z} = \{\ldots, -2, -1, 0, 1, 2, 3, \ldots\} \tag{1.1}$$

(the letter $\mathbb{Z}$ coming from the German word $\mathfrak{Zahlen}$, meaning "numbers"). The *natural numbers* are the nonnegative integers, including zero. They are "natural" because they are the possible numbers of elements in a finite set. For example, 4 is the number of elements of the set

$$\{\spadesuit, \heartsuit, \diamondsuit, \clubsuit\} \tag{1.2}$$

of suits in a deck of cards, while 13 is the number of elements of the set

$$\{A\heartsuit, K\heartsuit, Q\heartsuit, J\heartsuit, 10\heartsuit, 9\heartsuit, 8\heartsuit, 7\heartsuit, 6\heartsuit, 5\heartsuit, 4\heartsuit, 3\heartsuit, 2\heartsuit\} \tag{1.3}$$

of cards in the suit $\heartsuit$ of hearts. Note that 0 is the number of elements in the empty set $\varnothing$ or $\{\ \}$. The natural numbers form the set

$$\mathbb{N} = \{0, 1, 2, 3, \ldots\}. \tag{1.4}$$

Another set of numbers familiar from calculus is the set $\mathbb{R}$ of *real numbers*, like -17, $\sqrt{2} = 1.41421\ldots$, $e = 2.71828\ldots$, $\pi = 3.14159\ldots$, and so on. It is hard to display the set of real numbers as a list of elements between braces, like the sets (1.1)–(1.4) above. Instead, the set $\mathbb{R}$ is pictured as the *real line*

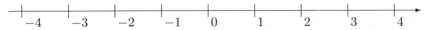

(like an axis in the graph of a function). Pictures like this are useful as geometric visualizations of real numbers. At times similar pictures can even be useful for natural numbers or integers, since these numbers also happen to be real numbers.

1.1 Ordering numbers

In calculus, order relations between real numbers are crucial, for instance when we want to find the maximum value of a function over a certain range.

Recall that $x < y$ (read "x less than y") means $y - x$ is positive, while $x \leq y$ (read "x less than or equal to y") means that $y - x$ is nonnegative. We can also write $y > x$ ("x greater than y") instead of $x < y$, or $y \geq x$ ("x greater than or equal to y") instead of $x \leq y$. In the real line picture, with the positive numbers going off to the right, the relation $x < y$ becomes an arrow $x \longrightarrow y$. It is often helpful to signify the relation $x \leq y$ with an arrow from x to y, without requiring the arrow to go horizontally from left to right.

Since algebra also needs to work with order relations between numbers, it is important to know the rules for manipulating them. The first rule is called *reflexivity*:

$$x \leq x \tag{1.5}$$

for any real (or integral, or natural) number x. This particular rule doesn't seem to be saying very much, but it often serves as a place-holder. The second rule is *transitivity*:

$$(\ x \leq y \quad \text{and} \quad y \leq z \ ) \quad \text{implies} \quad x \leq z \tag{1.6}$$

for any real (or integral, or natural) numbers x, y, and z. If Xavier can't beat Yerkes, and Yerkes can't beat Zandor, then Xavier can't beat Zandor either. Why does (1.6) hold? Well, if $x \leq y$ and $y \leq z$, the quantities $y - x$ and $z - y$ are nonnegative. In that case, so is their sum $z - x$, meaning that $x \leq z$. Transitivity makes a natural arrow picture:

... "completing the triangle." The final rule for the order relation is the one that yields conclusions of proofs, when you want to show that two numbers are actually equal:

$$(\ x \leq y \quad \text{and} \quad y \leq x \ ) \quad \text{implies} \quad x = y \tag{1.7}$$

for real numbers x and y. This rule is called *antisymmetry*. If Xavier can't beat Yerkes, and Yerkes can't beat Xavier either, then Xavier and Yerkes will tie.

Rules for an order relation

(R)	**Reflexivity:**	$x \leq x$
(T)	**Transitivity:**	$x \leq y$ and $y \leq z$ imply $x \leq z$
(A)	**Antisymmetry:**	$x \leq y$ and $y \leq x$ imply $x = y$

As an illustration of the use of the rules, here's a proposition with its proof.

PROPOSITION 1.1 (Squeezing.)
Suppose x, y, and z are real numbers. If $x \le y \le z \le x$, then $x = z$.

PROOF   Since $x \le y \le z$, transitivity shows that $x \le z$. But also $z \le x$, so antisymmetry gives $x = z$. ⬚

1.2 The Well-Ordering Principle

Compare (1.1) with (1.4). The elements of $\mathbb{Z}$ in (1.1) stretch off arbitrarily far to the left inside the braces: There is no smallest integer. In a version of the schoolyard game "My Dad earns more than your Dad," consider two players trying to name the smaller integer. Whatever number the first player names, say $-10,000,000$, the second player can always choose $-10,000,001$ or something even more negative. With the natural numbers, the situation is different. It is summarized by the following statement, the so-called

Well-Ordering Principle:

Each nonempty subset S of $\mathbb{N}$ has a least element inf S.

(Compare Exercise 7. The mathematical notation inf S stands for the *infimum* of S.) Of course, the principle is only required for infinite subsets S. For finite nonempty subsets S, the least element inf S, in this case often denoted as the *minimum* min S, can be located easily (Project 2).

Example 1.2 (**An application of the Well-Ordering Principle.**)
Suppose $S = \{n \in \mathbb{N} \mid 10^n < \frac{1}{2}n^n\}$, the set of natural numbers n for which the power 10^n is less than half the power n^n. The set S is nonempty, indeed infinite, since as n increases beyond 10, the power n^n grows faster than 10^n. (Formally, $\lim_{n \to \infty}(\frac{1}{2}n^n/10^n) = \infty$.) The Well-Ordering Principle guarantees that S has a least element inf S. You are invited to find it in Exercise 5. ⬚

In one of its main applications, the Well-Ordering Principle underwrites the techniques known as *recursion* and *mathematical induction*. For example, consider the definition of the *factorial* $n!$ of a natural number n. This quantity is usually defined recursively as follows:

$$0! = 1, \qquad (n+1)! = (n+1) \cdot n!$$

How can we be sure that the definition is complete, that it will not leave a quantity such as 50001200! undefined?

For generality, consider a property $P(n)$ of a natural number n, say the property that $n!$ is defined by the given recursive procedure.

- The **Induction Basis** is the statement that the property $P(0)$ holds.

- The **Induction Step** is the statement that truth of the property $P(n)$ implies the truth of the property $P(n+1)$.

- The **Principle of Induction** states: The Induction Basis and Induction Step together guarantee that $P(n)$ holds for all natural numbers n.

To justify the Principle of Induction, suppose that it goes wrong. In other words, the set

$$S = \{n \mid P(n) \text{ is false }\}$$

is nonempty. By the Well-Ordering Principle, the set S has a least element s. The Induction Basis shows that s cannot be 0. Thus $s > 0$, and $s - 1$ is a natural number. Since $s - 1$ does not lie in S, the property $P(s-1)$ holds. The Induction Step then gives the contradiction that $P(s)$ is true. Thus the Principle of Induction cannot go wrong.

Example 1.3 (A model proof by induction.)
Let $P(n)$ be the statement that the identity

$$1^2 + 2^2 + 3^2 + \ldots + (n-1)^2 + n^2 = \frac{n(n+1)(2n+1)}{6} \qquad (1.8)$$

holds for a natural number n. As Induction Basis, note that (1.8) reduces to the triviality $0 = 0$ for $n = 0$, so $P(0)$ is true. For the Induction Step, suppose that $P(n)$ is true, so that (1.8) holds as written. Then

$$1^2 + 2^2 + 3^2 + \ldots + (n-1)^2 + n^2 + (n+1)^2$$
$$= \frac{n(n+1)(2n+1)}{6} + (n+1)^2$$
$$= \frac{n(n+1)(2n+1) + 6(n+1)^2}{6}$$
$$= \frac{(n+1)(2n^2 + 7n + 6)}{6}$$
$$= \frac{(n+1)(n+2)\big(2(n+1)+1\big)}{6},$$

so that $P(n+1)$ is true. This proves (1.8) by induction.

1.3 Divisibility

The set $\mathbb{Z}$ of integers is a subset of the set $\mathbb{R}$ of real numbers; so integers can certainly be compared using the order relation $\leq$ for real numbers. However, in many cases a different relation between integers is more relevant. This is the relation of *divisibility*. Given two integers m and n, the integer m is said to be a *multiple* of n if there is an integer r such that $m = r \cdot n$. For example, 946 is a multiple of 11, since $946 = 86 \cdot 11$. Even integers are the multiples of 2. Zero is a multiple of every integer. Turning the relationship around, an integer n is said to *divide* an integer m, or to be a *divisor* of m, if m is a multiple of n. Summarizing,

$$\boxed{n \text{ divides } m} \quad \text{is equivalent to} \quad \boxed{m \text{ is a multiple of } n}. \quad (1.9)$$

The statement "n divides m" is written symbolically as $n \mid m$.

It is useful to compare the two equivalent concepts of (1.9). Divisibility is most convenient for formulating mathematical claims. On the other hand, it is generally easier to prove those claims by working with the corresponding equation $m = r \cdot n$ from the relation of being a multiple. As an example, consider the proof that the divisibility relation $\mid$ on $\mathbb{Z}$ shares the reflexivity (R) and transitivity (T) properties of the relation $\leq$ on $\mathbb{R}$ (page 2).

PROPOSITION 1.4 (Divisibility on $\mathbb{Z}$ is reflexive and transitive.)
Let m, n, and p be integers. Then:

(R) $m \mid m$;

(T) $(m \mid n$ and $n \mid p)$ *implies* $m \mid p$.

PROOF (R) For each integer m, the equation $m = 1 \cdot m$ holds, so m is a multiple of m.

(T) Since $m \mid n$, there is an integer r with $n = rm$. Since $n \mid p$, there is an integer s with $p = sn$. Then

$$p = sn = s(rm) = (sr)m$$

is a multiple of m, so $m \mid p$. □

However, the relation $\mid$ on $\mathbb{Z}$ is not antisymmetric. For example, $5 \mid -5$ since $-5 = (-1) \cdot 5$, and $-5 \mid 5$ since $5 = (-1) \cdot (-5)$. Nevertheless, $5 \neq -5$. The situation changes when we restrict ourselves to natural numbers. We regain all three properties: reflexivity (R), transitivity (T), and antisymmetry (A).

PROPOSITION 1.5 (Divisibility on ℕ is an order relation.)
Let m, n, and p be natural numbers. Then:

(R) $m \mid m$;

(T) $(\, m \mid n \text{ and } n \mid p \,)$ *implies* $m \mid p$;

(A) $(\, m \mid n \text{ and } n \mid m \,)$ *implies* $m = n$.

The proof of Proposition 1.5 is assigned as Exercise 14. The proposition means that divisibility relations between natural numbers may be displayed with arrow diagrams, just like the order relations between real numbers. For example, the set

$$\{1, 2, 3, 4, 6, 12\}$$

of divisors of 12 is exhibited in Figure 1.1. The diagram explicitly displays divisibilities such as $3 \mid 6$ with arrows: $3 \longrightarrow 6$. Other relations, such as $3 \mid 12$ or $4 \mid 4$, are implicit from the transitivity and reflexivity guaranteed by Proposition 1.5.

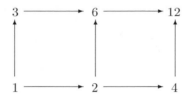

FIGURE 1.1:   The positive divisors of 12.

1.4   The Division Algorithm

To check whether a positive integer d divides a given integer a (positive, negative, or zero), a formal procedure known as the *Division Algorithm* is available. Given the

$$\text{input}: \quad \text{a } \texttt{positive integer } d \text{ (the } divisor\text{) and} \qquad (1.10)$$

$$\text{an } \texttt{integer } a \text{ (the } dividend\text{),} \qquad (1.11)$$

the Division Algorithm (Figure 1.2) produces the

$$\text{output}: \quad \text{an } \texttt{integer } q \text{ (the } quotient\text{) and} \qquad (1.12)$$

$$\text{an } \texttt{integer } r \text{ (the } remainder\text{),} \qquad (1.13)$$

satisfying the following:

$$a = dq + r\,; \tag{1.14}$$
$$0 \leq r < d\,. \tag{1.15}$$

For example, given the divisor 5 and dividend 37, the algorithm produces 7 as the quotient and 2 as the remainder: $37 = 5 \cdot 7 + 2$, with $0 \leq 2 < 5$. Given divisor 5 and dividend -42, it produces $-42 = 5 \cdot (-9) + 3$, with $0 \leq 3 < 5$. In general, the dividend a is a multiple of the divisor d if and only if the remainder r is zero.

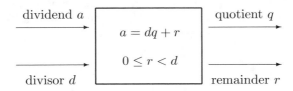

FIGURE 1.2: The Division Algorithm.

The word *dividend* in (1.11) means "the thing that is to be divided," like the profits of a company being divided among the shareholders. The word *quotient* in (1.12) is Latin for "How many times?" (the divisor d has to be added to itself to approach or equal the dividend). Then the remainder r is what is left after subtracting q times the divisor d from the dividend a.

The following proposition, with its proof, is a guarantee that the Division Algorithm will always perform as claimed. The proof relies on the use of the Well-Ordering Principle as presented in Section 1.2.

PROPOSITION 1.6
Given a dividend a as in (1.11), *and a divisor d as in* (1.10), *there is a unique quotient q as in* (1.12) *and a unique remainder r as in* (1.13), *such that the equation* (1.14) *and inequalities* (1.15) *hold.*

PROOF   Define a subset S of $\mathbb{N}$ by

$$S = \{a - dk \mid k \in \mathbb{Z},\ a - dk \geq 0\} \tag{1.16}$$

— the set all integers of the form $a - dk$ in which k is an element of the set $\mathbb{Z}$ of integers, and such that the inequality $a - dk \geq 0$ is satisfied.

Claim 1: The set S is nonempty.

If $a \geq 0$, then $a - d \cdot 0 = a$ is an element of S. Now d is a positive integer, so $d - 1 \geq 0$. Then if $a < 0$, we have $a - da = (-a)(d - 1) \geq 0$, as a product of two nonnegative integers. Thus $a - da$ is an element of S in this case.

With Claim 1 established, we can appeal to the Well-Ordering Principle. It tells us that the nonempty subset S of $\mathbb{N}$ has a least element $\inf S$. Set

$$r = \inf S \,. \tag{1.17}$$

Since r is an element of S, we have $0 \leq r$, the left-hand inequality in (1.15). And again since r is an element of S, we know that it is of the form $r = a - dk$ for some integer k. Set the quotient q to be the integer with

$$r = a - dq \,. \tag{1.18}$$

Adding dq to both sides of this equation yields (1.14).

Claim 2: $r < d$.

Could Claim 2 possibly be false? Could it happen that $r \geq d$? Well, if so, $r - d$ is still a natural number. But by (1.18),

$$r - d = a - d(q + 1) \,,$$

so $r - d$ would be a member of S strictly less than r. That would contradict (1.17), so the assumption that led to the contradiction, namely $r \geq d$, must be false. This shows that Claim 2 must be true, and verifies the right-hand inequality in (1.15).

Claim 3: The integers q and r satisfying (1.14) and (1.15) are unique.

Suppose $a = dq' + r'$ for integers q' and r' with $0 \leq r' < d$. Now $r' < r$ cannot be true, for otherwise we would have $0 \leq r' = a - dq'$ as an element of S less than r, the least element of S. Conversely, $r < r'$ cannot be true either, for then we would have $q > q'$, i.e., $(q - q') > 0$ and $(q - q') \geq 1$, with

$$r' = r + (r' - r) = r + \big((a - dq') - (a - dq)\big) = r + d(q - q') \geq d \,,$$

in contradiction to $r' < d$. Thus $r = r'$ and $q = q'$. $\square$

1.5   Greatest common divisors

Let a and b be nonzero integers. A positive integer c is said to be a *common divisor* of a and b if it divides both a and b:

$$c \mid a \text{ and } c \mid b.$$

For example, consider the divisors of 72 displayed in Figure 1.3. It is apparent that 4 is a common divisor of 24 and 36.

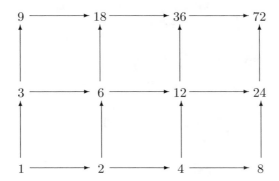

FIGURE 1.3:   The positive divisors of 72.

There are other common divisors of 24 and 36, such as 2 and 12.

DEFINITION 1.7 (Greatest common divisor, relatively prime.)
Let a and b be nonzero integers.

(a) *A positive integer d is the* greatest common divisor *(GCD) of a and b if*

- *d is a common divisor of a and b, and*
- *if c is a common divisor of a and b, then $c \leq d$.*

(b) *The integers a and b are said to be* relatively prime *or* coprime *if their greatest common divisor is 1.*

For instance, 12 is the greatest common divisor of 24 and 36. The numbers 8 and 9 are relatively prime. Note that 1 is coprime to every nonzero integer.

Why should the greatest common divisor of two nonzero integers a and b be guaranteed to exist? Well, the set of common divisors of a and b is a finite set

S, the intersection of the finite sets of positive divisors of a and b. (Compare Exercise 11.) The greatest common divisor is then just the maximum element of the finite set S. Since each pair a, b of nonzero integers has a uniquely defined greatest common divisor, we may use a functional notation

$$\gcd(a,b)$$

to denote that number. For example, $\gcd(24,36) = 12$. Note that

$$\gcd(a,a) = |a|, \tag{1.19}$$

$$\gcd(b,a) = \gcd(a,b), \tag{1.20}$$

and

$$\gcd(a,b) = \gcd(-a,b) = \gcd(a,-b) = \gcd(-a,-b) \tag{1.21}$$

for nonzero integers a and b (compare Exercise 26).

The defining properties of the greatest common divisor of a pair of nonzero integers a and b may be summarized as follows:

$d = \gcd(a,b)$ if and only if:

- $d \mid a$ and $d \mid b$; $\tag{1.22}$
- $(c \mid a$ and $c \mid b)$ implies $c \leq d$. $\tag{1.23}$

1.6   The Euclidean Algorithm

Given nonzero integers a and b, how can we compute $\gcd(a,b)$? By (1.21), it is sufficient to consider the case where a and b are both positive. By (1.19), it is sufficient to consider the case where a and b are distinct. And finally, by (1.20), it is sufficient to consider the case where $a > b$. Then for positive integers $a > b$, the positive integer $\gcd(a,b)$ is produced by the *Euclidean Algorithm*.

In fact, the Euclidean Algorithm is capable of more. Borrowing terminology from matrix theory or linear algebra, define a real number z to be an *integral linear combination* of real numbers x and y if it can be expressed in the form

$$z = lx + my \tag{1.24}$$

with integer coefficients l and m. Much of the significance of integral linear combinations resides in the following simple result, whose proof is assigned as Exercise 27.

PROPOSITION 1.8 (Common divisor divides linear combination.)
A common divisor c of integers n and p is a divisor of each integral linear combination $ln + mp$ of n and p.

The **Euclidean Algorithm** not only produces $\gcd(a, b)$, but if required may also be used to exhibit $\gcd(a, b)$ as an integral linear combination of a and b. Given integers $a > b > 0$, the algorithm works with a strictly decreasing sequence

$$r_{-1} > r_0 > r_1 > r_2 > \cdots > r_k > r_{k+1} = 0 \qquad (1.25)$$

of natural numbers. Following the initial specification

$$r_{-1} = a \quad \text{and} \quad r_0 = b,$$

the natural numbers (1.25) are produced by a series of steps. For $0 \le i \le k$, Step (i) applies the Division Algorithm with r_{i-1} as the dividend and r_i as the divisor:

$$r_{i-1} = q_{i+1} r_i + r_{i+1}, \qquad (1.26)$$

obtaining r_{i+1} as the remainder with $r_i > r_{i+1} \ge 0$ (and some integer q_{i+1} as the quotient). The Euclidean Algorithm makes its last call to the Division Algorithm in Step (k), obtaining the remainder $r_{k+1} = 0$. At that time the greatest common divisor $\gcd(a, b)$ is output as r_k, the last nonzero remainder in the list (1.25).

Why is $r_k = \gcd(a, b)$, and how is r_k produced as a linear combination of a and b.? To answer these questions, it is helpful to rewrite (1.26) as the matrix equation

$$\begin{bmatrix} r_{i-1} \\ r_i \end{bmatrix} = \begin{bmatrix} q_{i+1} & 1 \\ 1 & 0 \end{bmatrix} \begin{bmatrix} r_i \\ r_{i+1} \end{bmatrix} \qquad (1.27)$$

holding for $0 \le i \le k$. (Compare Section 2.3, page 28, for a review of matrix multiplication.) Note that (1.27) is an equality between 2-dimensional column vectors with integral entries. Equality of the bottom entries is trivial, while (1.26) is the equality between the top entries. Now

$$\begin{bmatrix} 0 & 1 \\ 1 & -q_{i+1} \end{bmatrix} \begin{bmatrix} q_{i+1} & 1 \\ 1 & 0 \end{bmatrix} = \begin{bmatrix} 1 & 0 \\ 0 & 1 \end{bmatrix} = \begin{bmatrix} q_{i+1} & 1 \\ 1 & 0 \end{bmatrix} \begin{bmatrix} 0 & 1 \\ 1 & -q_{i+1} \end{bmatrix},$$

so (1.27) is equivalent to the matrix equation

$$\begin{bmatrix} r_i \\ r_{i+1} \end{bmatrix} = \begin{bmatrix} 0 & 1 \\ 1 & -q_{i+1} \end{bmatrix} \begin{bmatrix} r_{i-1} \\ r_i \end{bmatrix} \qquad (1.28)$$

for $0 \le i \le k$. Repeated use of (1.28) gives

$$\begin{bmatrix} r_k \\ r_{k+1} \end{bmatrix} = \begin{bmatrix} 0 & 1 \\ 1 & -q_{k+1} \end{bmatrix} \cdots \begin{bmatrix} 0 & 1 \\ 1 & -q_1 \end{bmatrix} \begin{bmatrix} r_{-1} \\ r_0 \end{bmatrix} = \begin{bmatrix} s & t \\ u & v \end{bmatrix} \begin{bmatrix} r_{-1} \\ r_0 \end{bmatrix}$$

for integers s and t (computed by multiplying the 2×2 matrices in the middle term), so r_k is expressed as the integral linear combination

$$r_k = s r_{-1} + t r_0 = sa + tb \qquad (1.29)$$

of a and b. By Proposition 1.8, any common divisor c of a and b is a divisor of r_k, confirming that r_k satisfies the requirement (1.23) for the greatest common divisor of a and b. Finally, repeated use of (1.27) gives

$$\begin{bmatrix} a \\ b \end{bmatrix} = \begin{bmatrix} r_{-1} \\ r_0 \end{bmatrix} = \begin{bmatrix} q_1 & 1 \\ 1 & 0 \end{bmatrix} \cdots \begin{bmatrix} q_{k+1} & 1 \\ 1 & 0 \end{bmatrix} \begin{bmatrix} r_k \\ r_{k+1} \end{bmatrix} = \begin{bmatrix} s' & t' \\ u' & v' \end{bmatrix} \begin{bmatrix} r_k \\ 0 \end{bmatrix}$$

for integers s', t', u', and v', so that $a = s'r_k$ and $b = u'r_k$. This means that $r_k \mid a$ and $r_k \mid b$. Thus r_k satisfies the requirement (1.22) for the greatest common divisor of a and b.

Now we know that $r_k = \gcd(a,b)$, the import of the equation (1.29) may be recorded for future reference as follows. (Compare Exercise 28.)

PROPOSITION 1.9 (GCD as an integral linear combination.)
Let a and b be nonzero integers. Then the greatest common divisor $\gcd(a,b)$ may be expressed as an integral linear combination of a and b.

Example 1.10 (A run of the Euclidean Algorithm.)
Consider the determination of $\gcd(7,5)$ with the Euclidean Algorithm. The calls to the Division Algorithm are as follows:

$$\begin{aligned} \text{Step } (0): \quad & 7 = 1 \cdot 5 \;+\; 2 \\ \text{Step } (1): \quad & 5 = 2 \cdot 2 \;+\; 1 \\ \text{Step } (2): \quad & 2 = 2 \cdot 1 \;+\; 0 \end{aligned}$$

Thus $\gcd(7,5)$ emerges as 1, the remainder from the penultimate Step (1). The matrix equations (1.27) become

$$\begin{bmatrix} 7 \\ 5 \end{bmatrix} = \begin{bmatrix} 1 & 1 \\ 1 & 0 \end{bmatrix} \begin{bmatrix} 5 \\ 2 \end{bmatrix}, \qquad \begin{bmatrix} 5 \\ 2 \end{bmatrix} = \begin{bmatrix} 2 & 1 \\ 1 & 0 \end{bmatrix} \begin{bmatrix} 2 \\ 1 \end{bmatrix}, \qquad \begin{bmatrix} 2 \\ 1 \end{bmatrix} = \begin{bmatrix} 2 & 1 \\ 1 & 0 \end{bmatrix} \begin{bmatrix} 1 \\ 0 \end{bmatrix}.$$

The matrix equations (1.28) become

$$\begin{bmatrix} 1 \\ 0 \end{bmatrix} = \begin{bmatrix} 0 & 1 \\ 1 & -2 \end{bmatrix} \begin{bmatrix} 2 \\ 1 \end{bmatrix}, \qquad \begin{bmatrix} 2 \\ 1 \end{bmatrix} = \begin{bmatrix} 0 & 1 \\ 1 & -2 \end{bmatrix} \begin{bmatrix} 5 \\ 2 \end{bmatrix}, \qquad \begin{bmatrix} 5 \\ 2 \end{bmatrix} = \begin{bmatrix} 0 & 1 \\ 1 & -1 \end{bmatrix} \begin{bmatrix} 7 \\ 5 \end{bmatrix}.$$

Thus

$$\begin{bmatrix} 1 \\ 0 \end{bmatrix} = \begin{bmatrix} 0 & 1 \\ 1 & -2 \end{bmatrix} \begin{bmatrix} 0 & 1 \\ 1 & -2 \end{bmatrix} \begin{bmatrix} 0 & 1 \\ 1 & -1 \end{bmatrix} \begin{bmatrix} 7 \\ 5 \end{bmatrix} = \begin{bmatrix} -2 & 3 \\ 5 & -7 \end{bmatrix} \begin{bmatrix} 7 \\ 5 \end{bmatrix},$$

whence $\gcd(7,5) = 1 = (-2) \cdot 7 + 3 \cdot 5$.   ⬛

1.7   Primes and irreducibles

The positive number 35 can be reduced to a product $5 \cdot 7$ of smaller positive numbers 5 and 7. On the other hand, neither 5 nor 7 can be reduced further. In fact, if $5 = a \cdot b$ for positive integers a and b, then $a = 1$ and $b = 5$ or $a = 5$ and $b = 1$. We define a positive integer p to be *irreducible* if $p > 1$ and

$$0 < d \mid p \quad \text{implies} \quad (d = 1 \text{ or } d = p) \tag{1.30}$$

for integers d. Irreducibility is an "internal" or "local" property of a positive integer p, only involving the finite set of positive divisors of p.

Now look outwards rather than inwards. The positive number 35 may divide a product, without necessarily dividing any of the factors in that product. For example, 35 divides $7 \cdot 10$, but 35 does not divide 7 or 10. On the other hand, 5 divides the product $7 \cdot 10$, and then 5 divides the factor 10 in the product. We define a positive integer p to be *prime* if $p > 1$ and

$$p \mid a \cdot b \quad \text{implies} \quad (p \mid a \text{ or } p \mid b) \tag{1.31}$$

for any integers a and b. Primality may be considered as an "external" or "global" property of a positive integer p, since it involves arbitrary integers a and b. The two properties are summarized as follows:

Properties of an integer $p > 1$:

(internal)   **irreducible:** $0 < d \mid p$ implies $(d = 1 \text{ or } d = p)$

(external)   **prime:** $p \mid a \cdot b$ implies $(p \mid a \text{ or } p \mid b)$

It is a feature of the integers that the internal concept of irreducibility agrees with the external concept of primality.

PROPOSITION 1.11 ("Prime" ≡ "irreducible" for integers.)
Let $p > 1$ be an integer.

(a) *If p is prime, then it is irreducible.*

(b) *If p is irreducible, then it is prime.*

PROOF   (a):   Suppose p is prime and $0 < d \mid p$, say $p = d'd$ for some positive integer d'. Then $p \mid d'd$. Since p is prime, it follows that $p \mid d'$ or $p \mid d$. In the latter case, $d \mid p$ and $p \mid d$, so $d = p$ by antisymmetry. In the former case, the same argument (replacing d by d') shows $d' = p$. Then $d = 1$.

(b): Suppose p is irreducible and $p \mid a \cdot b$, say $ab = pk$ for some integer k. Suppose p does not divide a. It will be shown that $p \mid b$. Since p is irreducible, its only positive divisors are 1 and p. Thus $\gcd(p, a) = 1$, for $\gcd(p, a) = p$ would mean $p \mid a$. Using Proposition 1.9, write $\gcd(p, a)$ as an integral linear combination

$$1 = lp + ma$$

of p and a. Postmultiplying by b gives

$$b = lpb + mab$$
$$= lpb + mpk = p(lb + mk),$$

so that $p \mid b$ as required. $\square$

With Proposition 1.11 proved, prime numbers (as in Figure 1.4) may be characterized equally well by either the irreducibility (1.30) or the primality (1.31). (See the Notes to this section on page 23.)

2	3	5	7	11	13	17	19	23	29
31	37	41	43	47	53	59	61	67	71
73	79	83	89	97	101	103	107	109	113
127	131	137	139	149	151	157	163	167	173
179	161	191	193	197	199	211	223	227	229

FIGURE 1.4: The first 50 prime numbers.

There is a traditional adjective for numbers which are not prime:

DEFINITION 1.12 (Composite numbers.) *An integer n is said to be* composite *if $n > 1$, but n is not prime.*

Thus a number $n > 1$ is composite if it is not irreducible, i.e., if it has a nontrivial factorization $n = a \cdot b$ with integers $1 < a < n$ and $1 < b < n$.

1.8 The Fundamental Theorem of Arithmetic

In Figure 1.3, the number 72 is displayed as the product $72 = 8 \cdot 9 = 2^3 \cdot 3^2 = 2 \cdot 2 \cdot 2 \cdot 3 \cdot 3$ of prime numbers. The latter product may be written with the factors in various orders, such as $72 = 2 \cdot 3 \cdot 2 \cdot 2 \cdot 3$ or $72 = 2 \cdot 3 \cdot 2 \cdot 3 \cdot 2$. But to within such reorderings of the prime factors, the factorization is unique. The *Fundamental Theorem of Arithmetic* states that every integer greater than 1 has a factorization as a product of primes, unique up to reordering of the factors.

The existence part of the theorem is stated and proved as follows.

THEOREM 1.13 (Existence of factorizations.)
Each integer $n > 1$ may be expressed as a product of prime numbers.

PROOF Let B be the set of integers $n > 1$ which cannot be expressed as a product of primes. If the theorem is false, then B is nonempty. In that case, the Well-Ordering Principle says that B has a least element b. Since the integer b lies in the set B, it is not itself prime (or irreducible), so it has divisors g_1 and g_2 with

$$b = g_1 g_2 \qquad (1.32)$$

and $1 < g_1, g_2 < b$. Since the divisors g_1 and g_2 are strictly less than b, the least element of B, they are expressible as products of primes. But then (1.32) expresses the integer b as a product of primes, contradicting its status as a member of B. Since falsehood of the theorem leads to an inevitable contradiction, we conclude that the theorem is true. ∎

Implicit in the proof of Theorem 1.13 is a method, however slow, to produce the factorization of a given integer larger than 1 as a product of primes. For example, consider $b = 500$, which factorizes as $b = g_1 g_2$ with $g_1 = 50$ and $g_2 = 10$. Then $g_1 = 5 \cdot 10 = 5 \cdot 2 \cdot 5$ and $g_2 = 2 \cdot 5$, so $500 = 5 \cdot 2 \cdot 5 \cdot 2 \cdot 5$. If b is less friendly, e.g., $b = 281957$, then one has to try dividing b in turn by successive primes $p = 2, 3, 5, 7, 11, \ldots$ up to $\sqrt{b}$ (compare Exercise 36).

We now state the uniqueness half of the fundamental theorem.

THEOREM 1.14 (Uniqueness of factorization.)
Suppose that $p_1, p_2, \ldots, p_r$ and $q_1, q_2 \ldots, q_s$ are primes. Then if

$$p_1 \cdot p_2 \cdot \ldots \cdot p_r = q_1 \cdot q_2 \cdot \ldots \cdot q_s, \qquad (1.33)$$

$r = s$, and each p_i on the left hand side of (1.33) appears as a q_j on the right hand side of (1.33).

To prove Theorem 1.14, we will use a subsidiary result, a "lemma."

LEMMA 1.15
Suppose that $p_1, q_1, q_2, \ldots, q_s$ are primes. Then if

$$p_1 \mid q_1 \cdot q_2 \cdot \ldots \cdot q_s, \qquad (1.34)$$

there is some $1 \leq j \leq s$ such that $p_1 = q_j$.

PROOF Suppose that the lemma is false. Let S be the set of natural numbers s for which there are primes $p_1, q_1, q_2, \ldots, q_s$ with (1.34) holding, but where p_1 does not appear as any q_j with $1 \leq j \leq s$. Since the lemma is false, the set S is nonempty, and thus has a least element s. Consider $p_1, q_1, q_2, \ldots, q_s$ as in (1.34) for this integer s. Now p_1 does not divide the product $q_1 \cdot q_2 \cdot \ldots \cdot q_{s-1}$, for then the minimality of s in S would mean that p_1 shows up among $q_1, q_2, \ldots, q_{s-1}$. Since p_1 is prime, and (1.34) holds, it follows that $p_1 \mid q_s$. Since $1 < p_1$ and q_s is irreducible, $p_1 = q_s$, in contradiction to the assumption. Thus the lemma is true after all. $\square$

To complete the proof of Theorem 1.14, suppose (1.33) holds. Then

$$p_1 \mid q_1 \cdot q_2 \cdot \ldots \cdot q_s \,.$$

By Lemma 1.15, there is some $1 \leq j \leq s$ such that $p_1 = q_j$. Then

$$p_2 \cdot p_3 \cdot \ldots \cdot p_r = q_1 \cdot \ldots \cdot q_{j-1} \cdot q_{j+1} \cdot \ldots \cdot q_s \,.$$

By Lemma 1.15, p_2 cancels with some q_k from the right-hand side. Continuing in this fashion, the p_i on the left of (1.33) are paired off with the q_j on the right. In particular, the number r of factors on the left-hand side of (1.33) agrees with the number s of factors on the right.

The Fundamental Theorem of Arithmetic makes a connection between the two order relations $\leq$ and $\mid$ on the set $\mathbb{N}$ of natural numbers. Specifically, for distinct primes $p_1, p_2, \ldots, p_r$ and natural numbers $e_1, f_1, e_2, f_2, \ldots, e_r, f_r$,

$$p_1^{e_1} \cdot p_2^{e_2} \cdot \ldots \cdot p_r^{e_r} \mid p_1^{f_1} \cdot p_2^{f_2} \cdot \ldots \cdot p_r^{f_r} \text{ if and only if } e_1 \leq f_1, \ \ldots, \ e_r \leq f_r.$$

We conclude with an application of this idea.

DEFINITION 1.16 (Least common multiple.) *Let a and b be nonzero integers. The* least common multiple *$\mathrm{lcm}(a, b)$ of a and b is the minimum element of the set $S = \{m \mid m > 0, \ a \mid m, \ b \mid m\}$ of positive common multiples of a and b.*

Write $\max\{e, f\}$ for the maximum of integers e and f. The Fundamental Theorem of Arithmetic yields the following result. Its proof is assigned as Exercise 39.

PROPOSITION 1.17 (Computing the least common multiple.)
Let $a = p_1^{e_1} \cdot p_2^{e_2} \cdot \ldots \cdot p_r^{e_r}$ and $b = p_1^{f_1} \cdot p_2^{f_2} \cdot \ldots \cdot p_r^{f_r}$ for distinct primes $p_1, p_2, \ldots, p_r$, and natural numbers $e_1, f_1, e_2, f_2, \ldots, e_r, f_r$. Then

$$\mathrm{lcm}(a, b) = p_1^{\max\{e_1, f_1\}} \cdot p_2^{\max\{e_2, f_2\}} \cdot \ldots \cdot p_r^{\max\{e_r, f_r\}} \,.$$

1.9   Exercises

1. Suppose x, y, and z are real numbers. If $x \leq y \leq z \leq x$, give a formal proof that $y = z$ by use of transitivity and antisymmetry.

2. Suppose that $x_0, x_1, \ldots, x_n$ are real numbers, with $x_0 \leq x_1 \leq \cdots \leq x_n$. If $x_n \leq x_0$, show that $x_0 = x_r$ for $1 \leq r \leq n$.

3. Why is (1.7) true?

4. Why is (1.5) true?

5. Find the least element inf S of the set S from Example 1.2.

6. Find the smallest integer n for which $2^n < n!$.

7. Let S be a nonempty subset of $\mathbb{N}$. Let s be an element of S. The intersection $\{0, 1, \ldots, s - 1\} \cap S$ denotes the set of elements of S less than s.

   (a) If the intersection $\{0, 1, \ldots, s-1\} \cap S$ is empty, show that inf $S = s$.

   (b) If the intersection $\{0, 1, \ldots, s - 1\} \cap S$ is nonempty, show that inf $S = \min(\{0, 1, \ldots, s - 1\} \cap S)$.

8. Prove
$$1^3 + 2^3 + 3^3 + \ldots + n^3 = \left(\frac{n(n+1)}{2} \right)^2$$
for natural numbers n.

9. (a) Prove
$$1 + 2 + 3 + \ldots + n = \frac{n(n+1)}{2} \tag{1.35}$$
for natural numbers n by induction.

   (b) Can you prove (1.35) directly, without using induction?

10. Prove $n < 2^n$ for natural numbers n.

11. Let m be a nonzero integer.

   (a) Show that $n \mid m$ implies $|n| \leq |m|$. (In words: each divisor of a nonzero integer is no greater than that integer in absolute value.)

   (b) If you are uncomfortable with absolute values, show instead that $n \mid m$ implies $n^2 \leq m^2$.

   (c) Conclude that the set of divisors of m is finite.

12. Show that every integer divides zero.

13. There are 36 inches in a yard, and 100 centimeters in a meter.

   (a) In how many ways can a piece of wood a yard long be divided into equal pieces whose length is an integral number of inches?

   (b) In how many ways can a piece of wood a meter long be divided into equal pieces whose length is an integral number of centimeters?

14. Prove Proposition 1.5. [Hint: To prove the antisymmetry (A) that does not hold for divisibility on $\mathbb{Z}$, consider the solutions x of the equation $x^2 = 1$ in $\mathbb{Z}$ and $\mathbb{N}$.]

15. Describe the divisibility relation $\mid$ on the set $\mathbb{R}$ of real numbers.

16. Consider running the Division Algorithm on the inputs $a = 1$ and $d = 0$.

   (a) For the set S of (1.16), what is $\inf S$?

   (b) Show that a unique remainder r is obtained, but that the quotient q is not unique.

   (c) Is Proposition 1.6 contradicted?

17. Let d be a positive odd number. Show that for each integer a, there are unique integers q and r such that $a = dq + r$ with $|r| < d/2$. In other words, each integer a can be approximated by a multiple of d to within an error of less than $d/2$.

18. Consider the 16×11 rectangular array of 176 pixels in a display.

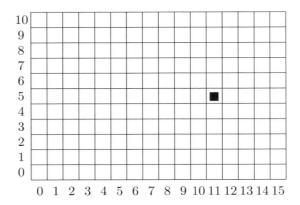

The pixels are located by their coordinates in the array, so that the bottom left pixel has coordinates $(0,0)$, and the top right pixel has coordinates $(15, 10)$. The pixels are addressed by the numbers from 0 to 175. The address of the pixel with coordinates (q, r) is

$$a = 11q + r \, .$$

(a) What is the address of the pixel with the black square?

(b) What are the coordinates of the pixel with address 106?

19. Let $d > 1$ be a fixed integer, known as the *base*. To represent a given positive integer n as a sequence $n = n_k n_{k-1} \ldots n_2 n_1$ of digits in base d, with $0 \le n_i < d$ for $1 \le i \le k$, consider the following algorithm:

(a) Initialize with $q_0 = n$ and $i = 1$;

(b) At Step (i), obtain $q_{i-1} = q_i d + n_i$ with the Division Algorithm;

(c) Stop at Step (k) when $q_k = 0$;

(d) Otherwise, replace i by $i + 1$ and return to (b).

Show that $n = n_k d^{k-1} + n_{k-1} d^{k-2} + \ldots + n_2 d + n_1$.

20. Express the base 10 number 3817 as a *hexadecimal* (base 16) number. Use $A = 10$, $B = 11$, $C = 12$, $D = 13$, $E = 14$, $F = 15$ for the digits above 9.

21. In a certain state, persons under age 21 are not allowed into bars that serve intoxicating beverages. If 21 were read as an *octal* number (to base 8), what would be the minimum age (to the usual base 10) of persons allowed into bars?

22. In Figure 1.3:

(a) Identify the set of positive divisors of 18.

(b) Identify the set of positive divisors of 24.

(c) Identify the set S of common divisors of 18 and 24.

(d) Identify $\gcd(18, 24)$ as the largest element of the set S.

23. Find all pairs of relatively prime positive integers less than 10.

24. Show that 1 is the only positive integer that is relatively prime to every positive integer.

25. In a gearbox, gear wheel A meshes with gear wheel B. The two rotate together many times. Gear wheel A has a teeth, and gear wheel B has b teeth. Show that each tooth of A meshes with each tooth of B at some time if and only if a and b are relatively prime.

26. Prove the equalities (1.19), (1.20), and (1.21).

27. Prove Proposition 1.8.

28. Without appealing to the discussion of the Euclidean Algorithm, give a direct proof of Proposition 1.9. [Hint: Applying the Well-Ordering Principle, show that $\gcd(a, b)$ is the smallest member of the set S of positive, integral linear combinations of a and b.]

29. Let c be a positive common divisor of two nonzero integers a and b.

    (a) Show that c divides $\gcd(a, b)$.

    (b) Show that $\gcd(a, b)/c = \gcd(a/c, b/c)$.

30. For nonzero integers a, b, and c, with $c > 0$, show that $\gcd(ac, bc) = \gcd(a, b) \cdot c$.

31. Let a and b be distinct nonzero integers. Show that the greatest common divisor $\gcd(a, b)$ can be expressed in infinitely many distinct ways as an integral linear combination $\gcd(a, b) = la + mb$ of a and b.

32. Use the Euclidean Algorithm to determine $\gcd(109, 60)$, and to express it as an integral linear combination of 109 and 60.

33. Show that $2n + 1$ and $3n + 1$ are coprime for all natural numbers n.

34. Show that the Euclidean Algorithm will make at most b calls to the Division Algorithm when it computes $\gcd(a, b)$ with $a > b > 0$.

35. (a) In how many ways can 72 be expressed as an *ordered* product of three twos and two threes?

    (b) Interpret each such expression $72 = p_1 p_2 p_3 p_4 p_5$ (with $p_i = 2$ or 3) as a walk from 1 to 72 along the path

    $$1 \rightarrow p_1 \rightarrow p_1 p_2 \rightarrow p_1 p_2 p_3 \rightarrow p_1 p_2 p_3 p_4 \rightarrow p_1 p_2 p_3 p_4 p_5 = 72$$

    in Figure 1.3.

    (c) Conversely, show that each path from 1 to 72, following the arrows at each step, determines an ordered factorization.

36. (a) Show that a composite number b has a prime divisor p with $p \le \sqrt{b}$.

    (b) Conclude that an integer n is prime if it is not divisible by any prime less than $\sqrt{n}$.

37. Factorize $b = 281957$ as a product of primes.

38. Can you prove that $n^2 - n + 41$ is prime for each natural number n?

39. Prove Proposition 1.17.

40. For positive integers a and b, show that an integer is a multiple of both a and b if and only if it is a multiple of $\text{lcm}(a, b)$.

41. Use the Fundamental Theorem of Arithmetic to obtain a formula for $\gcd(a, d)$, similar to the formula for $\text{lcm}(a, b)$ given in Proposition 1.17.

42. For nonzero integers a, b, and c, show that $\gcd(a, bc) = 1$ if and only if both $\gcd(a, b) = 1$ and $\gcd(a, c) = 1$.

43. For positive integers a and b, prove $a \cdot b = \gcd(a,b) \cdot \operatorname{lcm}(a,b)$. [Hint: For natural numbers e and f, prove $e + f = \min\{e,f\} + \max\{e,f\}$.]

44. (a) Give an example of prime numbers p_1, p_2 and natural numbers e_1, f_1, e_2, f_2 such that

$$\operatorname{lcm}\left(p_1^{e_1} p_2^{e_2}, p_1^{f_1} p_2^{f_2}\right) \neq p_1^{\max\{e_1,f_1\}} \cdot p_2^{\max\{e_2,f_2\}}.$$

(b) Why does this not contradict Proposition 1.17?

45. (a) Let $p_1 = 2, p_2 = 3, p_3 = 5, \ldots, p_r$ be the first r primes. Show that

$$n = p_1 \cdot p_2 \cdot \ldots \cdot p_r + 1$$

is not divisible by any of $p_1, p_2, \ldots, p_r$.

(b) Applying Theorem 1.13 to n, deduce that there is a prime number p_s with $p_r < p_s \leq n$.

(c) Conclude that there is an infinite number of primes.

46. Let n be a positive integer. A positive integer d is said to be a *unitary divisor* of n if d divides n, and $\gcd(d, n/d) = 1$. In this case, n is said to be a unitary multiple of d.

(a) Determine the unitary divisors of 72 and 1200.

(b) Determine the least common unitary multiple of 18 and 45.

(c) Show that there is no least common unitary multiple of 3 and 9.

47. Consider a world in which the only positive numbers are the numbers

$$1, 5, 9, 13, 17, 21, 25, 29, \ldots \qquad (1.36)$$

of the form $4r + 1$ for r in $\mathbb{N}$. Suppose that the numbers are only multiplied, not added.

(a) Show that the product of two numbers from the list (1.36) also appears in the list.

(b) Show that the numbers below 25 in the list (1.36) are irreducible within this alternative world.

(c) Show that 9 divides $21 \cdot 21$, but 9 does not divide 21.

(d) Conclude that in this world, the property of being prime is distinct from the property of being irreducible.

1.10   Study projects

1. For a sport competition of your choice (say one season of a particular league), determine whether the transitivity rule (1.6) and antisymmetry rule (1.7) apply.

2. Consider the problem of finding the minimum $\min S$ of a finite set S of natural numbers with n elements. Design a procedure to do this with just $n-1$ comparisons between pairs of elements from S. As inspiration, look at the brackets for a single-elimination sport competition in a league with n members. (Compare Figure 1.5 for the case $n = 6$.)

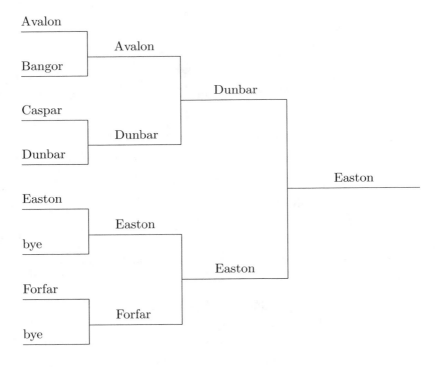

FIGURE 1.5:   Brackets for a competition.

3. The number 946 is a multiple of 11. Also, the difference between the respective sums $9 + 6$ and 4 of the odd-placed and even-placed digits of 946 is (a multiple of) 11. Is this just a coincidence, or can you extend the observation to derive a quick way of recognizing multiples of 11?

4. Discuss why Proposition 1.6 is needed. Why is it not enough to claim that your computer or calculator can produce a unique quotient q and remainder r if you give it a dividend a and (positive) divisor d? [Hint: Can your computer accept a very large integer?]

5. **Speed of the Euclidean Algorithm.** Exercise 34 gives a crude bound for the number of steps required by the Euclidean Algorithm. Can you improve on this bound? Or for any positive integer k, can you always find integers $a > b > k$ for which the Euclidean Algorithm requires approximately b steps?

6. **Greatest common divisors.** Consider the following method to find the greatest common divisor of positive integers a and b:

   (a) If a and b are even, remember $\gcd(a, b) = 2 \cdot \gcd(a/2, b/2)$ and compute $\gcd(a/2, b/2)$ instead. (Compare Exercise 29.)

   (b) If say a is even and b is odd, remember $\gcd(a, b) = \gcd(a/2, b)$ and compute $\gcd(a/2, b)$ instead.

   (c) If a and b are odd, say $a > b$, remember $\gcd(a, b) = \gcd(a - b, b)$ and compute $\gcd(a - b, b)$ instead.

   Use this method to compute greatest common divisors of pairs of large integers. How does this method compare with the Euclidean Algorithm? Can you adapt this new method to express $\gcd(a, b)$ as a linear combination of a and b?

1.11   Notes

Section 1.6

Euclid ($Ευκλειδης$) was a Greek mathematician living in the third century B.C.

Section 1.7

When discussing integers, it has been traditional to define a number p to be "prime" if it is irreducible. The proof of primality as we have defined it — Proposition 1.11(b) — is then known as *Euclid's Lemma*. Historically, the distinct terminology for the internal and external properties emerged in the 19th century, as mathematicians started to consider other systems of factorization (for example the system of Exercise 47). In these cases, the two properties may no longer coincide.

Chapter 2

FUNCTIONS

Algebra, just like calculus, works with many different kinds of functions. In this chapter, we will learn how to specify functions, and how to compose them. We will also see how functions form mathematical structures: semigroups, monoids, and groups.

2.1 Specifying functions

Let X and Y be sets. Then a *map* or *function* $f : X \to Y$ or $X \xrightarrow{f} Y$ is a rule that assigns a unique element $f(x)$ of Y to each element x of X. In this context, the elements x of X are called the *arguments* of the function f, while the elements $f(x)$ of Y are called the *values* of the function. As examples, consider the *squaring* function

$$\mathrm{sq} : \mathbb{Z} \to \mathbb{N} \tag{2.1}$$

defined by $\mathrm{sq}(n) = n^2$ for each integer n, or the *absolute value* function

$$\mathrm{abs} : \mathbb{Z} \to \mathbb{N} \tag{2.2}$$

defined by $\mathrm{abs}(n) = |n|$ for each integer n. In a function $f : X \to Y$, the set X is called the *domain*, while the set Y is called the *codomain*. Thus the domain of (2.1) is $\mathbb{Z}$, while the codomain of (2.1) is $\mathbb{N}$. Note that (2.1) is considered as different from the function

$$\mathrm{sq} : \mathbb{Z} \to \mathbb{Z} \tag{2.3}$$

with $\mathrm{sq}(n) = n^2$, since the two functions have different codomains. In general, two functions $f : X \to Y$ and $g : Z \to T$ are equal if and only if all three of the following conditions are satisfied:

- The domain X of f equals the domain Z of g ;

- The codomain Y of f equals the codomain T of g ;

- The function values $f(x)$ and $g(x)$ agree on each argument x in X.

The reason for including the domain and codomain in the specification of a function will become apparent in Section 2.5.

A function f must be able to assign a function value $f(x)$ to each argument x in its domain. For instance, we cannot have a function

$$\text{inv} : \mathbb{R} \to \mathbb{R}$$

with $\text{inv}(x) = x^{-1}$, since this rule does not work for the element 0 of the domain $\mathbb{R}$. On the other hand, elements y of the codomain of a function $f : X \to Y$ are not required to show up as actual function values $f(x)$. While each natural number does occur as the absolute value of some integer, there are many natural numbers (such as 3) which are not the square of any integer. The only demand placed on the codomain is that it be big enough to contain all the function values that are generated. For example, we cannot set up a function

$$\text{sqrt} : \mathbb{N} \to \mathbb{N}$$

with $\text{sqrt}(n) = \sqrt{n}$ for natural numbers n, since the function value $\sqrt{3}$ does not lie in $\mathbb{N}$. But setting

$$\text{sqrt} : \mathbb{N} \to \mathbb{R}$$

would be fine, since square roots of natural numbers are always real numbers.

In summary, **the domain should always be small enough** to guarantee that the function rule will work on each element of the domain. On the other hand, **the codomain should always be large enough** to contain all the function values that occur. In a function $f : X \to Y$, the set

$$f(X) = \{f(x) \mid x \in X\} \tag{2.4}$$

of function values of domain elements is called the *image* of the function. For example, the image of the squaring function (2.1) is the set

$$\{0, 1, 4, 9, 16, 25, 36, \dots\}$$

of *perfect squares*.

It is sometimes helpful to be able to specify a function without naming it explicitly. To this end, we will denote the action of a function at the element level using barred arrows, e.g., $\text{sq} : n \mapsto n^2$. Thus the squaring function (2.1) might have been specified as

$$\mathbb{Z} \to \mathbb{N}; n \mapsto n^2$$

without having to receive the (rather artificial) name sq. The barred arrow notation is especially helpful when we examine functions whose arguments or values are themselves sets (as in Section 3.2, for example). Suppose that A and B are sets. Then $f : A \to B$ will denote a function with domain A and codomain B, while $f : A \mapsto B$ means that a certain function f takes the argument A to the value B.

Warning: In calculus, the notation "$f(x)$" is sometimes used to denote a function, for example when we speak of "the function $\sin(x)$." In algebra, the notation "$f(x)$" is reserved for the value of a function f at an argument x. Do not confuse functions with elements of their domains or codomains.

2.2   Composite functions

Consider two functions $f : X \to Y$ and $g : Y \to Z$, where the codomain Y of f is also the domain of g. Then there is a *composite function*

$$g \circ f : X \to Z; x \mapsto g(f(x))$$

whose domain is the domain of f and whose codomain is the codomain of g. For example, the squaring function $\text{sq} : \mathbb{Z} \to \mathbb{Z}$ of (2.3) may be composed with the absolute value function $\text{abs} : \mathbb{Z} \to \mathbb{N}$ of (2.2) to yield the function

$$\text{abs} \circ \text{sq} : \mathbb{Z} \to \mathbb{N}; n \mapsto |n^2|.$$

In fact, since $|n^2| = n^2$ for any natural number n, this composite function $\text{abs} \circ \text{sq}$ is the same as the original squaring function (2.1).

Composition of functions $f : X \to Y$ and $g : Y \to Z$ may be illustrated by an arrow picture, strongly reminiscent of the picture of transitivity on page 2:

If you find yourself getting confused by a profusion of functions, it can be helpful to draw such pictures.

Suppose that there are functions $f : X \to Y$, $g : Y \to Z$, and $h : Z \to T$. These functions may be composed in two different ways:

$$h \circ (g \circ f) : X \to T; x \mapsto h(g \circ f(x))$$

and

$$(h \circ g) \circ f : X \to T; x \mapsto (h \circ g)(f(x)).$$

However,

$$h(g \circ f(x)) = h(g(f(x))) = (h \circ g)(f(x))$$

for all x in X, so in fact we have the *associative law*

$$h \circ (g \circ f) = (h \circ g) \circ f \tag{2.5}$$

for $X \xrightarrow{f} Y \xrightarrow{g} Z \xrightarrow{h} T$.

2.3   Linear functions

Linear functions form one of the most important classes of functions. For positive integers m and n, consider the set

$$\mathbb{R}_m^n = \left\{ \begin{bmatrix} a_{11} & \cdots & a_{1n} \\ \vdots & & \vdots \\ a_{m1} & \cdots & a_{mn} \end{bmatrix} \Bigg| \ a_{ij} \ \text{real} \right\}$$

of $m \times n$ real matrices. In particular, $\mathbb{R}_2^1$ is the set of 2-dimensional real column vectors

$$\mathbf{x} = \begin{bmatrix} x_1 \\ x_2 \end{bmatrix}$$

with x_1, x_2 in $\mathbb{R}$. Each 2×2 real matrix

$$A = \begin{bmatrix} a_{11} & a_{12} \\ a_{21} & a_{22} \end{bmatrix}$$

gives a *linear* function

$$L_A : \mathbb{R}_2^1 \to \mathbb{R}_2^1; \ \begin{bmatrix} x_1 \\ x_2 \end{bmatrix} \mapsto \begin{bmatrix} a_{11}x_1 + a_{12}x_2 \\ a_{21}x_1 + a_{22}x_2 \end{bmatrix} \tag{2.6}$$

or

$$L_A(\mathbf{x}) = A\mathbf{x}$$

using matrix multiplication. Note that

$$L_A \begin{bmatrix} 1 \\ 0 \end{bmatrix} = \begin{bmatrix} a_{11} \\ a_{21} \end{bmatrix} \quad \text{and} \quad L_A \begin{bmatrix} 0 \\ 1 \end{bmatrix} = \begin{bmatrix} a_{12} \\ a_{22} \end{bmatrix},$$

so the linear function L_A determines the matrix A.

Given a second matrix

$$B = \begin{bmatrix} b_{11} & b_{12} \\ b_{21} & b_{22} \end{bmatrix}$$

with a corresponding linear function $L_B : \mathbf{x} \mapsto B\mathbf{x}$, the matrix product BA is defined by

$$\begin{bmatrix} b_{11} & b_{12} \\ b_{21} & b_{22} \end{bmatrix} \begin{bmatrix} a_{11} & a_{12} \\ a_{21} & a_{22} \end{bmatrix} = \begin{bmatrix} b_{11}a_{11} + b_{12}a_{21} & b_{11}a_{12} + b_{12}a_{22} \\ b_{21}a_{11} + b_{22}a_{21} & b_{21}a_{12} + b_{22}a_{22} \end{bmatrix}. \tag{2.7}$$

This apparently complicated formula is designed to make the equation

$$L_{BA}(\mathbf{x}) = L_B \circ L_A(\mathbf{x}) \tag{2.8}$$

true for all $\mathbf{x}$ in $\mathbb{R}_2^1$ (Exercise 3): Matrix multiplication tracks the composition of the corresponding linear functions. In particular, the associativity of matrix multiplication is a direct consequence of the associativity (2.5) of function composition.

2.4 Semigroups of functions

Let X be a set. A map or function $f : X \to X$ from X to itself is often described as a *self-map* of the set X. In this context, the set X is sometimes called the *base set* for the function $f : X \to X$.

DEFINITION 2.1 (Semigroup of functions.) *A set S of functions* $f : X \to X$ *with domain X and codomain X is said to be a* semigroup of functions *on the base set X if*

$$f \quad and \quad g \quad in \quad S \quad imply \quad g \circ f \quad in \quad S\,. \tag{2.9}$$

We also say that the set S is closed under composition.

If f is an element of a semigroup S of functions, the powers f^n for positive integers n are defined recursively by $f^1 = f$ and $f^{n+1} = f^n \circ f$.

Here are some important examples of semigroups of functions.

Example 2.2 (Self-maps.)
For a base set X, define X^X to be the set of all functions from X to X. Then X^X forms a semigroup of functions on X. (For a justification of the notation, see Exercise 5.) □

Example 2.3 (Constant functions.)
Let X be a set, and let Y be a subset of X. For each element y of Y, define a *constant function*

$$c_y : X \to X; x \mapsto y\,.$$

Note that for each element x of X, and for y, z in the subset Y, we have

$$c_z \circ c_y(x) = c_z(c_y(x)) = c_z(y) = z = c_z(x)\,,$$

so that $c_z \circ c_y = c_z$. Thus the set

$$C_Y = \{c_y \mid y \text{ in } Y\} \tag{2.10}$$

forms a semigroup of functions on X. □

Example 2.4 (Nondecreasing functions.)
Recall that in calculus, a function $f : \mathbb{R} \to \mathbb{R}$ is *nondecreasing* if $x \leq y$ implies $f(x) \leq f(y)$. Then the set of nondecreasing functions forms a semigroup of functions on $\mathbb{R}$ (Exercise 6). □

Example 2.5 (Real shifts.)

For each real number r, define the *shift* by r to be the map

$$\sigma_r : \mathbb{R} \to \mathbb{R}; x \mapsto r + x \,.$$

Note that for real numbers r, s, and x, we have

$$\sigma_r \circ \sigma_s(x) = r + (s + x) = (r + s) + x = \sigma_{r+s}(x) \,,$$

so $\sigma_r \circ \sigma_s = \sigma_{r+s}$. Thus the set Σ of shifts forms a semigroup of functions on $\mathbb{R}$. For a positive integer n and real number r, the equation

$$\sigma_r^n = \sigma_{nr} \tag{2.11}$$

holds (Exercise 9). □

Example 2.6 (Computable functions.)

Define a function $f : \mathbb{N} \to \mathbb{N}$ to be *computable* if there is a computer program producing $f(n)$ as output whenever a natural number n is given as the input. Then if f and g are computable, so is their composite $g \circ f$. In fact, given the input n, a program for $g \circ f$ could just take the output $f(n)$ of a program for f, and feed this number $f(n)$ as input to the program for g, obtaining the required output $g \circ f(n)$. Thus the set of computable functions forms a semigroup of functions on $\mathbb{N}$. □

DEFINITION 2.7 (Identity function.) *For any set X, the* identity *function id_X is defined by $\mathrm{id}_X : X \to X; x \mapsto x$.*

Note that for sets X, Y and $f : X \to Y$, we have

$$\mathrm{id}_Y \circ f = f = f \circ \mathrm{id}_X \,. \tag{2.12}$$

DEFINITION 2.8 (Monoid of functions.) *A set S of self-maps on a base set X is said to be a* monoid of functions *on X if it forms a semigroup, and if the identity function id_X is an element of S.*

Clearly X^X is a monoid on X. For a slightly less trivial example, note that the identity function $\mathrm{id}_{\mathbb{N}}$ on the set $\mathbb{N}$ of natural numbers is computable. Given a natural number n as input, consider the "lazy" program which immediately returns n as output. Thus the set of computable functions in Example 2.6 is a monoid of functions on $\mathbb{N}$.

For an element f of a monoid of functions on a set X, the power notation may be extended by setting $f^0 = \mathrm{id}_X$.

Example 2.9 (Linear functions.)
By (2.8), the set $L(2,\mathbb{R})$ of linear functions from $\mathbb{R}_2^1$ to itself forms a semigroup of functions on $\mathbb{R}_2^1$. Now for the 2×2 *identity matrix*

$$I_2 = \begin{bmatrix} 1 & 0 \\ 0 & 1 \end{bmatrix}, \tag{2.13}$$

the linear function L_{I_2} is the identity function $\mathrm{id}_{\mathbb{R}_2^1}$, so $L(2,\mathbb{R})$ forms a monoid of functions on $\mathbb{R}_2^1$. □

2.5 Injectivity and surjectivity

A function $f : X \to Y$ is required to assign a unique function value $f(x)$ in the codomain Y to each argument x from the domain X. On the other hand, it may happen that different arguments are assigned the same function value. For instance, with the squaring function $\mathrm{sq} : \mathbb{Z} \to \mathbb{N}$ of (2.1), we have

$$\mathrm{sq}(-5) = (-5)^2 = 25 = 5^2 = \mathrm{sq}(5).$$

DEFINITION 2.10 (Injective function.) *A function $f : X \to Y$ is said to be* injective, *or an* injection, *or "one-to-one," if*

$$f(x) = f(x') \quad implies \quad x = x' \tag{2.14}$$

for all elements x and x' of the domain X.

Expressing the injectivity condition (2.14) another way, the equation

$$f(x) = y \tag{2.15}$$

is required to have a unique solution x in X for each element y of the image (2.4) of f. By default, any function with empty domain is injective. The restricted squaring function

$$\mathrm{sq} : \mathbb{N} \to \mathbb{N}; n \mapsto n^2 \tag{2.16}$$

is injective, while the original squaring function $\mathrm{sq} : \mathbb{Z} \to \mathbb{N}$ is not. This shows why the domain is an integral part of the specification of a function.

PROPOSITION 2.11 (Retracts of injective functions.)
Let $f : X \to Y$ be injective, with nonempty domain. Then there is a function

$$r : Y \to X \tag{2.17}$$

such that

$$r \circ f = \mathrm{id}_X \,. \qquad (2.18)$$

PROOF   Pick an element x_0 of X. For a codomain element y that does not lie in the image $f(X)$, define $r(y) = x_0$. Now consider an element y of the image $f(X)$. By the definition (2.4) of the image, the equation

$$f(x) = y$$

has a solution. Since f is injective, the solution is unique. Define $r(y)$ to be this unique solution element x_y. We obtain a function $r : Y \to X$. Now $r \circ f : X \to X$. Then for each element x of X, we have

$$r \circ f(x) = r\big(f(x)\big) = x_{f(x)} = x = \mathrm{id}_X(x)\,,$$

verifying (2.18).   □

DEFINITION 2.12 (Retracts.)   *A function $r : Y \to X$ is called a* retract *of a function $f : X \to Y$ if $r \circ f = \mathrm{id}_X$.*

PROPOSITION 2.13 (Functions with retracts are injective.)
If a function $f : X \to Y$ has a retract, then it is injective.

PROOF   Let $r : Y \to X$ be a retract for f. Then

$$f(x) = f(x') \ \text{ implies } \ x = r \circ f(x) = r \circ f(x') = x'$$

for x, x' in X.   □

Proposition 2.11 shows that each injection with nonempty domain has a retract. Note that an injection f might have many retracts, because of the arbitrary choice of the element x_0 in the proof of the existence of the retract (Exercise 22). Also, note that the identity function $\mathrm{id}_\varnothing$ on the empty set is its own retract.

DEFINITION 2.14 (Surjective function.)   *A function $f : X \to Y$ is said to be* surjective, *or a* surjection, *or to map* onto *its codomain, if the codomain and image coincide: $Y = f(X)$.*

Equivalently, a solution x to the equation $f(x) = y$ of (2.15) exists for each element y of the codomain. In yet another formulation, the *inverse image*

$$f^{-1}\{y\} = \{x \text{ in } X \mid f(x) = y\}$$

is required to be nonempty for each element y of Y. Note that the only surjective function with an empty domain is the identity function $\mathrm{id}_\varnothing$ on the empty set. The absolute value function abs : $\mathbb{Z} \to \mathbb{N}$ of (2.2) is surjective, while the absolute value function

$$\mathrm{abs} : \mathbb{Z} \to \mathbb{Z}; n \mapsto |n| \qquad (2.19)$$

is not, since the equation

$$|n| = -5$$

has no solution n. This shows why the codomain is an integral part of the specification of a function.

PROPOSITION 2.15 (Sections of surjective functions.)
Let $f : X \to Y$ be surjective. Then there is a function

$$s : Y \to X \qquad (2.20)$$

such that

$$f \circ s = \mathrm{id}_Y . \qquad (2.21)$$

PROOF If X is empty, then so is Y, and f is just the identity function $\mathrm{id}_\varnothing$. In this case, $s = \mathrm{id}_\varnothing$ makes (2.21) work. Now suppose that X is nonempty. For an element y of Y, there is an element x of X such that $f(x) = y$. Choose the function value $s(y)$ as one such element x_y. Then a function $s : Y \to X$ is defined. For each element y of Y, we have

$$f \circ s(y) = f\big(s(y)\big) = f(x_y) = y = \mathrm{id}_Y(y) ,$$

verifying (2.21). □

DEFINITION 2.16 (Sections.)   *A function $s : Y \to X$ is called a section of a function $f : X \to Y$ if $f \circ s = \mathrm{id}_Y$.*

PROPOSITION 2.17 (Functions with sections are surjective.)
If a function $f : X \to Y$ has a section, then it is surjective.

PROOF Let $s : Y \to X$ be a section for f. Then

$$f\big(s(y)\big) = f \circ s(y) = \mathrm{id}_Y(y) = y$$

for each element y of Y. □

Proposition 2.15 shows that each surjection has a section. Note that a surjection f might have many sections, because of the arbitrary choice of the elements x_y in the proof of the existence of the section (Exercise 23).

2.6   Isomorphisms

DEFINITION 2.18 (Bijective function, isomorphism of sets.)   *A function $f : X \to Y$ is said to be* bijective, *or an* isomorphism *(of sets), or a* bijection, *if it is both injective and surjective.*

PROPOSITION 2.19 (Inverses of isomorphisms.)
Let $f : X \to Y$ be a bijection. Then there is a function

$$g : Y \to X \qquad (2.22)$$

such that

$$g \circ f = \mathrm{id}_X \quad and \quad f \circ g = \mathrm{id}_Y . \qquad (2.23)$$

PROOF   By Propositions 2.11 and 2.15, we have

$$r \circ f = \mathrm{id}_X \quad and \quad f \circ s = \mathrm{id}_Y . \qquad (2.24)$$

By the associativity of function composition, we have

$$r = r \circ \mathrm{id}_Y = r \circ f \circ s = \mathrm{id}_X \circ s = s .$$

Define $g = r = s$. Then (2.23) follows from (2.24).   ⬚

DEFINITION 2.20 (Inverse, invertible function.)   *For a function $f : X \to Y$, a function $g : Y \to X$ satisfying $g \circ f = \mathrm{id}_X$ and $f \circ g = \mathrm{id}_Y$ is called an* inverse *of f. A function is described as* invertible *if it has an inverse.*

Example 2.21 **(Natural logarithms and the exponential function.)**
Let $(0, \infty)$ denote the set (open interval) of positive real numbers. Then the exponential function $\exp : \mathbb{R} \to (0, \infty); x \mapsto e^x$ is invertible, with the natural logarithm function $\log : (0, \infty) \to \mathbb{R}$ as an inverse. The equations (2.23) are the familiar relationships

$$\log e^x = x$$

for real numbers x and

$$e^{\log y} = y$$

for positive real numbers y.   ⬚

Proposition 2.19 shows that bijections are invertible. Conversely, an inverse g of an invertible function f is both a retraction and a section for f. Thus an invertible function f is both injective and surjective. In other words, invertible functions are bijective.

Example 2.22 (Inverses of real shifts.)
For each real number r, the shift σ_r of Example 2.5 has the shift σ_{-r} as an inverse. Thus the shifts are bijective. □

Recall that sections and retractions need not be unique. With inverses, the situation is different.

PROPOSITION 2.23 (Uniqueness of inverses.)
The inverse of an invertible function is unique.

PROOF Let $g : Y \rightarrow X$ be an inverse of a function $f : X \rightarrow Y$. If a function $h : Y \rightarrow X$ satisfies

$$h \circ f = \mathrm{id}_X \quad \text{or} \quad f \circ h = \mathrm{id}_Y ,$$

then

$$h = h \circ \mathrm{id}_Y = h \circ f \circ g = \mathrm{id}_X \circ g = g$$

or

$$g = g \circ \mathrm{id}_Y = g \circ f \circ h = \mathrm{id}_X \circ h = h$$

respectively. In particular, $g : Y \rightarrow X$ is uniquely specified by (2.23). □

In view of Propositon 2.23, we can speak of *the* inverse f^{-1} of an invertible function f. Note that

$$(f^{-1})^{-1} = f$$

for an invertible function f, so that inverses of invertible functions are again invertible.

If there is an isomorphism $f : X \rightarrow Y$ from a set X to a set Y, we often write

$$X \cong Y \tag{2.25}$$

and say that the sets X and Y are *isomorphic*. In this case $Y \cong X$, by virtue of the isomorphism f^{-1}. We will often want to show that two sets X and Y are isomorphic. The standard technique for this is to exhibit two mutually inverse functions $f : X \rightarrow Y$ and $g : Y \rightarrow X$.

Example 2.24
The set $\mathbb{Z}$ of integers is isomorphic with the subset $\mathbb{N}$ of natural numbers. Consider the function $f : \mathbb{N} \rightarrow \mathbb{Z}$ defined by setting $f(0) = 0$ and $f(2r) = r$, $f(2r - 1) = -r$ for positive integers r. Consider the function $g : \mathbb{Z} \rightarrow \mathbb{N}$

defined by $g(n) = 2n$ for $n \geq 0$ and $g(n) = 2|n| - 1$ for $n < 0$. Then f and g are mutually inverse.   ⬚

The proof of the following proposition is left as Exercise 34.

PROPOSITION 2.25
Consider functions $f : X \to Y$ and $g : Y \to Z$.

(a) *If f and g are injective, then so is $g \circ f$.*

(b) *If f and g are surjective, then so is $g \circ f$.*

(c) *If f and g are bijective, then so is $g \circ f$. Moreover, $(g \circ f)^{-1} = f^{-1} \circ g^{-1}$.*

Example 2.26 (Counting.)
For each natural number n, consider the finite set

$$\mathbf{n} = \{0, 1, 2, \ldots, n-1\} \tag{2.26}$$

of natural numbers less than n. Note that the set $\mathbf{n}$ has n elements. In particular, $\mathbf{0}$ is the empty set. Now if a finite set X has n elements, say $X = \{x_0, x_1, \ldots, x_{n-1}\}$, then there is a bijection

$$k : \mathbf{n} \to X; i \mapsto x_i. \tag{2.27}$$

Indeed, a set X has n elements if and only if there is a bijection $k : \mathbf{n} \to X$. We may say that k *counts* the elements of X. The number of elements in a finite set X is called the *size* or *order* of X. It is written as $|X|$. Proposition 2.25(c) may be used to show that two finite sets X and Y are isomorphic if and only if $|X| = |Y|$ (Exercise 35).   ⬚

2.7   Groups of permutations

DEFINITION 2.27 (Groups of permutations.)   *Let X be a set.*

(a) *A bijective function $f : X \to X$ is called a* permutation *of the set X.*

(b) *A set G of permutations on X is said to be a* group of permutations *of X or a* permutation group *on the set X if it is a monoid of functions satisfying the additional property*

$$f \text{ in } G \text{ implies } f^{-1} \text{ in } G. \tag{2.28}$$

The property (2.28) is known as closure under inversion.

Let $X!$ be the set of all permutations of a given set X. According to Proposition 2.25(c), $X!$ forms a semigroup of functions on X. Since id_X is a permutation, $X!$ forms a monoid of functions on X. Finally, the fact that inverses of invertible functions are invertible means that $X!$ forms a group of permutations. This group $X!$ is known as the *symmetric group* on X. (For a justification of the notation, see Exercise 44.)

Example 2.28 (The group of real shifts.)

The monoid $\Sigma_{\mathbb{R}}$ of shifts on $\mathbb{R}$ (compare Example 2.5 and Exercise 8) forms a group of permutations of $\mathbb{R}$, since $(\sigma_r)^{-1} = \sigma_{-r}$ as noted in Example 2.22. On the other hand, consider the set

$$\Sigma_{\mathbb{R}}^+ = \{\sigma_r \mid r \geq 0\}$$

of shifts by nonnegative real numbers r. This set does form a monoid of permutations on $\mathbb{R}$. However, it does not form a permutation group on $\mathbb{R}$, since it does not satisfy the property (2.28) of closure under inversion.   ▯

Example 2.29 (The symmetric groups S_n.)

For each natural number n, write S_n for the symmetric group $\underline{n}!$ on the set (2.26) of natural numbers less than n. The group S_n is called the *symmetric group on n symbols*. For a set $\{a_1, a_2, \ldots, a_r\}$ of distinct elements of $\underline{n}$, the *cycle*

$$(a_1\ a_2\ \ldots\ a_r)$$

denotes the bijection

$$\underline{n} \to \underline{n}\,;\, a_1 \mapsto a_2, a_2 \mapsto a_3, \ldots, a_{r-1} \mapsto a_r, a_r \mapsto a_1$$

with $x \mapsto x$ for elements x of $\underline{n}$ not included in the set $\{a_1, a_2, \ldots, a_r\}$. It is conventional to write $\mathrm{id}_{\underline{n}}$ as the cycle (0). Two cycles $(a_1\ a_2\ \ldots\ a_r)$ and $(b_1\ b_2\ \ldots\ b_s)$ are said to be *disjoint* if the corresponding sets $\{a_1, a_2, \ldots, a_r\}$ and $\{b_1, b_2, \ldots, a_s\}$ are disjoint (have no common element). Each permutation may be written as a product of mutually disjoint cycles. For example,

$$0 \mapsto 3, 1 \mapsto 1, 2 \mapsto 7, 3 \mapsto 6, 4 \mapsto 8, 6 \mapsto 0, 7 \mapsto 2, 8 \mapsto 4$$

in S_9 may be written as the product $(0\ 3\ 6) \circ (2\ 7) \circ (4\ 8)$ of disjoint cycles. By following the effect of these functions, it is easy to express products of permutations as products of disjoint cycles. For example,

$$(0\ 3\ 6) \circ (2\ 7) \circ (4\ 8) \circ (0\ 4\ 7\ 2\ 6) \circ (1\ 8) = (0\ 8\ 1\ 4\ 2) \circ (3\ 6)$$

since $(0\ 3\ 6) \circ (2\ 7) \circ (4\ 8) \circ (0\ 4\ 7\ 2\ 6) \circ (1\ 8)(0)$

$$= (0\ 3\ 6) \circ (2\ 7) \circ (4\ 8) \circ (0\ 4\ 7\ 2\ 6)(0)$$
$$= (0\ 3\ 6) \circ (2\ 7) \circ (4\ 8)(4) = (0\ 3\ 6) \circ (2\ 7)(8) = 8$$

and so on.   ▯

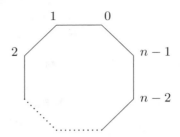

FIGURE 2.1:   A regular n-gon.

Example 2.30 (The cyclic groups C_n.)
For each positive integer n, the *cyclic group C_n* consists of the n permutations

$$\left(0 \ 1 \ 2 \ 3 \ \ldots \ (n-2) \ (n-1)\right),$$
$$\left(0 \ 1 \ 2 \ 3 \ \ldots \ (n-2) \ (n-1)\right)^{2},$$
$$\left(0 \ 1 \ 2 \ 3 \ \ldots \ (n-2) \ (n-1)\right)^{3}, \ \ldots$$
$$\ldots, \ \left(0 \ (n-1) \ (n-2) \ \ldots \ 3 \ 2 \ 1\right),$$

and (0) from S_n. These permutations correspond to the respective counter-clockwise rotations of a regular n-gon by the angles

$$\frac{2\pi}{n}, \ 2\frac{2\pi}{n}, \ 3\frac{2\pi}{n}, \ \ldots, \ (n-1)\frac{2\pi}{n}, \ 0$$

radians (Figure 2.1). ⬜

Example 2.31 (The Klein 4-group.)
The *Klein 4-group V_4* is the set

$$\{ \ (0), \ (0 \ 1)(2 \ 3), \ (0 \ 2)(1 \ 3), \ (0 \ 3)(1 \ 2) \ \}$$

of permutations. It forms a group of permutations of the set $\underline{\mathbf{n}}$ for each natural number $n \geq 4$ (Exercise 40). ⬜

The cycle notation is extended to denote permutations of arbitrary sets. For example,

$$(A\heartsuit \ 3\heartsuit \ J\heartsuit \ 7\heartsuit) \circ (K\heartsuit \ 4\heartsuit) \circ (10\heartsuit \ 9\heartsuit \ 2\heartsuit)$$

might denote a shuffle of the suit $\heartsuit$ of hearts from (1.3). For an application of permutations to elementary cryptography, see Study Project 3 at the end of the chapter.

2.8 Exercises

1. Show that the empty set $\varnothing = \{\ \}$ cannot be the codomain of a function $f : X \to \varnothing$ with nonempty domain X.

2. Draw an arrow picture to illustrate all the functions

$$f,\ g,\ h,\ g \circ f,\ h \circ (g \circ f),\ h \circ g,\ (h \circ g) \circ f$$

involved in the associative law (2.5).

3. Verify that (2.8) holds for all 2-dimensional real column vectors $\mathbf{x}$.

4. Let m, n, and p be positive integers. Show that for an $m \times n$ real matrix A, there is a function $\mathbb{R}^p_n \to \mathbb{R}^p_m; X \mapsto AX$.

5. Let X be a finite set with n elements. Show that the semigroup X^X has n^n elements.

6. Show that the set of nondecreasing functions forms a monoid of functions on $\mathbb{R}$.

7. A function $f : \mathbb{R} \to \mathbb{R}$ is said to be *strictly increasing* if $x < y$ implies $f(x) < f(y)$. Show that the set of strictly increasing functions forms a monoid of functions on $\mathbb{R}$.

8. Show that the set Σ of shifts in Example 2.5 forms a monoid of functions on $\mathbb{R}$.

9. Verify the equation (2.11), and show that it also holds for $n = 0$. (Hint: Consider using induction with $n = 0$ as the basis.)

10. For each natural number n, define the *power map*

$$p_n : \mathbb{R} \to \mathbb{R}; x \mapsto x^n. \tag{2.29}$$

Show that the set P of all power maps forms a monoid of functions on the set $\mathbb{R}$.

11. Let n be an integer such that $nx = c_n(x)$ for all integers x. Show that $n = 0$.

12. Let Y be a subset of a set X. Under what conditions on X and Y does the set (2.10) of constant functions form a monoid of functions on X?

13. A function $f : \mathbb{R} \to \mathbb{R}$ is said to be *affine* if there are real numbers m and c such that

$$f : x \mapsto m \cdot x + c \tag{2.30}$$

Show that the set A of all affine functions forms a monoid of functions on $\mathbb{R}$. (In calculus, affine functions are often called "linear," but in algebra it is best to reserve this term for the case $c = 0$.)

14. A function $f : \mathbb{R} \to \mathbb{R}$ is said to be a *polynomial function* if there is a natural number n and real numbers $f_0, f_1, \ldots, f_n$ such that

$$f(x) = f_n x^n + \ldots + f_1 x + f_0$$

for x in $\mathbb{R}$.

(a) Show that the set of all polynomial functions forms a monoid of functions on $\mathbb{R}$.

(b) Show that there is a function $f : \mathbb{R} \to \mathbb{R}$ which is not a polynomial function.

15. Show that the set $C(\mathbb{R})$ of all continuous functions $f : \mathbb{R} \to \mathbb{R}$ forms a monoid of functions on $\mathbb{R}$.

16. Let r be a positive integer. Let $C^r(\mathbb{R})$ denote the set of all functions $f : \mathbb{R} \to \mathbb{R}$ for which the r-th derivative $f^r(x)$ exists at each real number x. Show that $C^r(\mathbb{R})$ forms a monoid of functions on $\mathbb{R}$.

17. Let X be an infinite set. A function $f : X \to X$ is said to be *almost identical* if the set

$$\{x \in X \mid x \neq f(x)\}$$

of elements x of X, differing from their image $f(x)$ under f, is finite. Let F be the subset of X^X consisting of the almost identical functions. Show that F is a monoid of functions.

18. Show that the power map p_n of (2.29) is injective if and only if n is odd.

19. Show that the power map p_n of (2.29) is surjective if and only if n is odd.

20. Show that sections are injective.

21. Show that retracts are surjective.

22. Show that the injection (2.16) has infinitely many retracts.

23. Show that the surjection (2.2) has infinitely many sections.

24. Consider the function

$$f : \mathbb{R} \to \mathbb{R}; x \mapsto x(x-1)(x+1).$$

(a) Show that f is not injective.

(b) Using the Intermediate Value Theorem or otherwise, show that f is surjective.

25. Consider the function

$$f : \mathbb{R} \to \mathbb{R}; x \mapsto e^x .$$

(a) Show that f is injective.

(b) Is the natural logarithm function a retraction for f?

(c) Show that f is not surjective.

26. Let $f : X \to X$ be a function with finite domain X. Show that the following three conditions are equivalent:

(a) f is injective;

(b) f is surjective;

(c) f is bijective.

27. Consider the 2×2 real matrix

$$A = \begin{bmatrix} a & b \\ c & d \end{bmatrix} .$$

Show that the following three conditions are equivalent:

(a) The linear function L_A is injective;

(b) $ad - bc \neq 0$;

(c) The linear function L_A is surjective.

28. Let $f : X \to Y$ be a function from a finite set X with m elements to a finite set Y with n elements. If $m > n$, show that f is not injective. (This is known as the *Pigeonhole Principle*: If m pigeons occupy n holes, and $m > n$, then at least two pigeons have to share.)

29. Show that the set $\mathbb{R}$ of real numbers is isomorphic to its proper subset $(0, \infty)$ of positive real numbers. (Compare Example 2.21.)

30. Show that a finite set X cannot be isomorphic to a proper subset Y of X. (Recall that a subset Y of a set X is *proper* if there is an element x of X that does not lie in Y.)

31. Let $f : X \to Y; x \mapsto f(x)$ be a function.

(a) Show that there is a unique subset Y' of Y such that the *minimal corestriction*

$$X \to Y'; x \mapsto f(x) \tag{2.31}$$

is a surjective function.

(b) What is the minimal corestriction of the absolute value function (2.19)?

32. Let $f : X \to Y; x \mapsto f(x)$ be a function.

   (a) Show that there is a subset X' of X such that the function

$$X' \to Y; x \mapsto f(x) \qquad (2.32)$$

   is an injective function.

   (b) Give an example to show that the subset X' need not be unique.

33. Let $f : X \to Y$ be a function with nonempty domain X. Show that there is a function $g : Y \to X$ such that $f = f \circ g \circ f$.

34. Prove Proposition 2.25.

35. Let X and Y be finite sets. Show that $X \cong Y$ if and only if $|X| = |Y|$.

36. Let X be a set. Suppose that a nonempty semigroup G of permutations of X satisfies the property (2.28) of closure under inversion.

   (a) Show that id_X lies in G.

   (b) Conclude that G is a permutation group on X.

37. Express

$$(0\ 7\ 2\ 1) \circ (3\ 4\ 5\ 6) \circ (0\ 6\ 2\ 4) \circ (3\ 1\ 5\ 7)$$

   as a product of disjoint cycles.

38. Let β and

$$\alpha = (x_1\ x_2\ \ldots\ x_{r-1}\ x_r)$$

   be permutations of a finite set X. Show that

$$\beta \circ \alpha \circ \beta^{-1} = \left(\beta(x_1)\ \beta(x_2)\ \ldots\ \beta(x_{r-1})\ \beta(x_r) \right).$$

39. Let n be a positive integer. Show that S_n has $n!$ elements: To specify a permutation α of $\mathbf{n}$, there are n choices for $\alpha(0)$, then $n - 1$ choices for $\alpha(1)$ (avoiding $\alpha(0)$), then $n - 2$ choices for $\alpha(2)$ (avoiding $\alpha(0)$ and $\alpha(1)$), and so on.

40. Show that V_4 forms a group of permutations of each set $\mathbf{n}$ with $n \geq 4$.

41. Show that for distinct elements $a_1, a_2, \ldots, a_{r-1}, a_r$ of the set $\mathbf{n}$,

$$(a_1\ a_2\ \ldots\ a_{r-1}\ a_r) = (a_2\ a_3\ \ldots\ a_r\ a_1).$$

42. Show that for distinct elements $a_1, a_2, \ldots, a_{r-1}, a_r$ of the set $\underline{\mathbf{n}}$,

$$(a_1 \ a_2 \ \ldots \ a_{r-1} \ a_r)^{-1} = (a_r \ a_{r-1} \ \ldots \ a_2 \ a_1).$$

43. Show that

$$(a_1 \ a_2 \ \ldots \ a_{r-1} \ a_r) \circ (b_1 \ b_2 \ \ldots \ b_{s-1} \ b_s)$$
$$= (b_1 \ b_2 \ \ldots \ b_{s-1} \ b_s) \circ (a_1 \ a_2 \ \ldots \ a_{r-1} \ a_r)$$

for disjoint cycles $(a_1 \ a_2 \ \ldots \ a_r)$ and $(b_1 \ b_2 \ \ldots \ b_s)$.

44. Let X be a finite set with n elements. Show that the symmetric group $X!$ has $n!$ elements.

45. Let $\mathrm{Aff}(\mathbb{R})$ be the set of all affine functions (2.30) with $m \neq 0$ (compare Exercise 13). Show that $\mathrm{Aff}(\mathbb{R})$ forms a group of permutations of $\mathbb{R}$.

46. Suppose that a group G of permutations of $\mathbb{R}$ contains the real shifts σ_a and σ_b for real numbers a and b.

   (a) Show that G contains σ_{ma} for each positive integer m.
   (b) Show that G contains σ_{ma} for each integer m.
   (c) Show that G contains σ_{ma+nb} for each integral linear combination $ma + nb$ of a and b.

47. Suppose that a group G of permutations of $\mathbb{R}$ contains the real shifts σ_2 and σ_5. Show that G contains σ_n for each integer n.

2.9 Study projects

1. **Not all functions are computable.** Consider a program to compute a certain function $\mathbb{N} \to \mathbb{N}$. This program, as a list of instructions in a certain programming language, can ultimately be written out as a long but finite string of binary (base 2) digits:

   10001100101110000111100010010010010101000000 . . . 011001001

   Add a 1 to the left of each such string. (Why do we need to do this?) The number that is represented to base 2 by the augmented string — compare Exercise 19 of Chapter 1 — is called the *Gödel number* of the program. List all possible such programs, for all computable functions $\mathbb{N} \to \mathbb{N}$, in increasing order of Gödel numbers. Suppose that the list is $P_0, P_1, P_2, \ldots$. Suppose that for natural numbers m and n, the program

P_m outputs the function value p_{mn} when given input n. Now show that the function

$$f : \mathbb{N} \to \mathbb{N}; x \mapsto p_{xx} + 1$$

is not computable, since for each natural number n, the function value $f(n) = p_{nn} + 1$ differs from the function value p_{nn} computed by P_n when applied to the argument n. (The technique used here to construct f is known as *Cantor diagonalization*.)

2. **Symmetries of the regular tetrahedron.** Consider the solid regular tetrahedron:

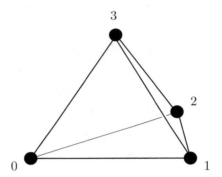

(a) Show that the rotations by $120°$ and $240°$ about an axis through the vertex 0 and the midpoint of the triangular face 123 implement the permutations (123) and (132) of the vertices.

(b) Show that the symmetries of the regular tetrahedron include all the permutations

$$\left\{ \begin{matrix} (123), & (023), & (013), & (012), \\ (132), & (032), & (031), & (021) \end{matrix} \right\} \qquad (2.33)$$

(c) Show that the full set of symmetries of the tetrahedron consists of the union of the set (2.33) and the Klein 4-group V_4.

(d) Determine a geometric or combinatorial rule that can decide when a product of two elements of (2.33) will lie in V_4.

(e) Look up the structure of a methane molecule CH_4. How do its symmetries relate to the symmetries of the regular tetrahedron?

3. **Cryptography** is the art of designing secret codes. An original text, the *cleartext*, is transformed by a bijection c into an encoded text for transmission across some channel that may be prone to eavesdropping. The person for whom the message is intended then applies the inverse bijection $d = c^{-1}$ to decode the encoded text.

Among the most elementary codes are those given by a permutation of the alphabet, for example the "keyboard permutation"

$$c = (QWERTYUIOP) \circ (ASDFGHJKL) \circ (ZXCVBNM). \quad (2.34)$$

(a) Apply the keyboard permutation (2.34) to encode the cleartext

$$SEND\ MORE\ MONEY.$$

(b) Suppose that the keyboard permutation was used to produce the coded message

$$JPDYSHRD\ GTRRF.$$

What was the original cleartext?

4. **Cryptanalysis** is the eavesdropper's art of breaking secret codes, of reading encoded messages without explicitly being given details of the bijection c used to encode them. Codes given by a permutation c such as the keyboard permutation (2.34) are quite easy to break, using the fact that various letters and letter combinations in the English language have different frequencies. For example, the commonest letters in decreasing order of frequency are

$$E,\ T,\ A,\ O,\ I,\ N,\ \dots, \quad (2.35)$$

while combinations such as TH arise often, and Q is almost always followed by U. Sometimes, knowledge of the context of the message can be used. For example, if the message concerns Persian Gulf countries, like Iraq or Qatar, then the letter following a Q might not be a U.

Suppose that an eavesdropper intercepts the message

$$CZYSY\quad ABEE\quad FY\quad UKCDZYQ\quad TO\quad SKBR$$
$$TS\quad OTL\quad BR\quad UKSCQ\quad TO\quad CZY\quad YKQC \quad (2.36)$$

encoded by a permutation c.

(a) Order the letters in the message (2.36) according to their relative frequency.

(b) Compare this ordered list of letters with (2.35), and make an initial partial guess at the decoding bijection $d = c^{-1}$.

(c) Now, assuming that the message (2.36) consists of ordinary English words, try to decode the message as completely as you can.

(d) If you cannot make sense of the message, try a different assignment of the commonest letters. Also, be aware of the few possibilities for two-letter words in the English language. Try using the fact that TH is a common pair of letters.

In order to remove the extra information given by the spaces between words, a secret message such as (2.36) might be transmitted in a series of 5-letter groups such as

$$CZYSY \quad ABEEF \quad YUKCD \quad ZYQTO \quad SKBRT$$
$$SOTLB \quad RUKSC \quad QTOCZ \quad YYKQC$$

instead.

2.10 Notes

Section 2.1

Various sources use various names in place of our domain/codomain/image terminology. For instance, the terms "target" and "range" appear frequently, the latter in all three possible roles:

- The range is the place where one goes to shoot bullets, so the "range" is the domain, the place from which the arguments shoot out;

- The codomain is the set over which the function values "range";

- The image is the exact set over which the function values "range."

We hope that our terms are unambiguous.

Section 2.5

Strictly speaking, the proof of Proposition 2.15 uses the Axiom of Choice to select the elements x_y. In fact, the existence of a section to each surjective function is one of the many equivalent formulations of the Axiom.

Section 2.7

For infinite sets X, some authors reserve the term "permutation" to denote bijections $f : X \to X$ that are almost identical in the sense of Exercise 17.

C.F. Klein was a German mathematician who lived from 1849 to 1925.

Section 2.9

K. Gödel was an Austrian (Moravian) logician and mathematician, later emigrating to the United States, who lived from 1906 to 1978. G. Cantor was a German mathematician who lived from 1845 to 1918.

2.11 Summary

In algebra, the notation for a function is more precise than the notation commonly used in calculus. Associated with a function

$$f : X \to Y; x \mapsto f(x)$$

are a number of key terms:

- The function assigns a unique **value** $f(x)$ ...

- ... to each **argument** x;

- The **domain** X is the set from which the arguments are taken;

- The **codomain** Y is the set in which the function values are expected to lie;

- The **image** is the set

$$f(X) = \{f(x) \mid x \text{ in } X\}$$

  of actual values which occur;

- For each element y of the codomain Y, the **inverse image** of y is the set

$$f^{-1}\{y\} = \{x \text{ in } X \mid f(x) = y\} \tag{2.37}$$

  of arguments x that are assigned the function value y.

Note that the inverse image sets (2.37) exist for all functions f, regardless of whether f is invertible or not. If f does happen to be invertible, then $f^{-1}\{y\} = \{f^{-1}(y)\}$, so the notation is consistent in that case.

At a more basic level, it is important to distinguish between elements, sets, and functions. An equation may say that two elements are equal:

$$\mathrm{id}_X(x) = x$$

or that two sets are equal:

$$\mathrm{sq}(\mathbb{R}) = \{s \in \mathbb{R} \mid s \geq 0\}$$

or that two functions are equal:

$$\mathrm{abs} = \mathrm{sqrt} \circ \mathrm{sq}. \tag{2.38}$$

Do not write equations with a function on one side, and an element on the other. For example, " $\mathrm{id}_X = x$ " would be meaningless, since the left-hand side is a function, while the right-hand side is an element. In calculus, one might see an equation of the form

$$|x| = \sqrt{x^2}\,,$$

and this equation could be ambiguous. It could be expressing the equation (2.38) between functions, or it could be an expression of the equality between two real numbers, say if x had been specified as a certain real number. The notation in algebra is carefully designed to avoid this kind of ambiguity, and to make mathematical reasoning more transparent.

Added care is required when discussing a semigroup G of functions or group G of permutations on a set X. In this case, we must distinguish between the elements of the base set X, and the elements of the semigroup G. The semigroup G is a set whose elements are functions, of the form $f : X \to X$.

Chapter 3

EQUIVALENCE

When we do mathematics, we study the structures that underlie the various phenomena encountered in the world. For this to work, mathematics has to be able to filter out all the detail that is not relevant to the particular structure being studied. Equivalence is the filter.

The most basic example is the concept of number. What does the number 3 stand for? A set X has 3 elements if and only if there is a set isomorphism

$$f : \{1, 2, 3\} \to X \tag{3.1}$$

counting off the elements of X as $f(1)$, $f(2)$, and $f(3)$. The function f has to be injective, so that no element of X gets counted twice. The function f has to be surjective, to make sure that each element of X gets counted.

The only problem here is the circularity. To characterize the number 3, we have used that number in the domain of the function (3.1). To escape the circularity, we can decide to consider two sets as equivalent for the purposes of counting whenever they are isomorphic. The number 3 then emerges as the property which is common to each of the sets that are isomorphic to some given 3-element set (for instance $\{1, 2, 3\}$ or $\{\varnothing, \{\varnothing\}, \{\{\varnothing\}\}\}$). The particular details of the elements in the sets are not relevant to the problem of counting. They are filtered out by the equivalence.

Equivalence relations play a key role in the analysis of general functions. Each function determines an equivalence relation on its domain, identifying two elements whenever they have the same function value. Conversely, it transpires that every equivalence relation is of this type.

3.1 Kernel and equivalence relations

Consider the squaring function $\mathrm{sq} : \mathbb{Z} \to \mathbb{Z}; n \mapsto n^2$ of (2.3). For two integers n and n',

$$\mathrm{sq}\,(n) = \mathrm{sq}\,(n') \quad \text{if and only if} \quad n' = \pm n\,.$$

In other words, the integers n and n' are assigned the same function value if and only if they both lie in the same *equivalence class* $\{r, -r\}$. These equivalence classes *partition* the domain set $\mathbb{Z}$ of integers, meaning that $\mathbb{Z}$ decomposes as the union

$$\mathbb{Z} = \{0\} \cup \{\pm 1\} \cup \{\pm 2\} \cup \{\pm 3\} \cup \dots \qquad (3.2)$$

of mutually disjoint subsets, the equivalence classes.

DEFINITION 3.1 (Kernel relation of a function.)   *Consider a function $f : X \to Y$. A pair (x, x') of elements of X is said to be in the kernel relation* **ker** *f, and we write x* **ker** *f x' or x (**ker** f) x', if and only if x and x' are assigned the same function value by f. Formally,*

$$x \ (\textbf{ker} \ f) \ x' \qquad \text{if and only if} \qquad f(x) = f(x'). \qquad (3.3)$$

Previously, we studied order relations such as $\leq$ on $\mathbb{R}$ or the divisibility relation $|$ on $\mathbb{N}$. These order relations had the properties of being reflexive, transitive, and antisymmetric. Now the kernel relation **ker** f of a function $f : X \to Y$ is certainly reflexive:

$$x \ (\textbf{ker} \ f) \ x$$

for all x in X. It is also transitive:

$$(\ x \ (\textbf{ker} \ f) \ x' \quad \text{and} \quad x' \ (\textbf{ker} \ f) \ x'' \ ) \qquad \text{implies} \qquad x \ (\textbf{ker} \ f) \ x'',$$

since $f(x) = f(x')$ and $f(x') = f(x'')$ imply $f(x) = f(x'')$. The third property of the kernel relation is called *symmetry*:

$$x \ (\textbf{ker} \ f) \ x' \qquad \text{implies} \qquad x' \ (\textbf{ker} \ f) \ x.$$

These properties of kernel relations are formalized in the important concept of an equivalence relation. (Remember RST in alphabetical order!)

DEFINITION 3.2 (Equivalence.)   *Let R be a relation on a set X.*

(R) *The relation R is* reflexive *if $x \ R \ x$ for each element x of X.*

(S) *The relation R is* symmetric *if for elements x and x' of X, the relation $x \ R \ x'$ implies $x' \ R \ x$.*

(T) *The relation R is* transitive *if for elements x, x', and x'' of X, the relations $x \ R \ x'$ and $x' \ R \ x''$ imply $x \ R \ x''$.*

Finally, the relation R is an equivalence relation *on X if it satisfies all three conditions (R), (S), and (T).*

PROPOSITION 3.3 (Kernels are equivalence relations.)
Let $f : X \to Y$ be a function. Then the kernel relation **ker** *f of f, specified by (3.3), is an equivalence relation on the domain X of the function f.*

3.2 Equivalence classes

The kernel of the squaring function sq : $\mathbb{Z} \to \mathbb{Z}$ yielded the partition (3.2) of $\mathbb{Z}$. Each equivalence relation on a set yields a partition of the set.

DEFINITION 3.4 (Equivalence class.) *If R is an equivalence relation on a set X, define the* equivalence class *of x under R to be the set*

$$[x]_R = \{x' \text{ in } X \mid x \; R \; x'\} \tag{3.4}$$

of all elements x' of X that are related to x by R.

Note that each class $[x]_R$ is nonempty, since by reflexivity it at least contains the element x itself. For the kernel relation **ker** f of a function $f : X \to Y$, and for an element x of the domain X, the equivalence classes are given by the inverse image sets

$$[x]_{\mathbf{ker} \, f} = f^{-1}\{f(x)\}. \tag{3.5}$$

Here is the key partitioning property of equivalence relations.

PROPOSITION 3.5 (Equivalence classes are disjoint or equal.)
Let R be an equivalence relation on a set X. Let x_1 and x_2 be elements of X. Then the two equivalence classes $[x_1]_R$, $[x_2]_R$ are either disjoint:

$$[x_1]_R \cap [x_2]_R = \varnothing$$

or else equal: $[x_1]_R = [x_2]_R$. In the latter case, $x_1 \; R \; x_2$.

PROOF Suppose that $[x_1]_R$ and $[x_2]_R$ are not disjoint, so they have a common element x'. Then $x_1 \; R \; x'$ and $x_2 \; R \; x'$ by the definition (3.4) of the equivalence classes. By symmetry, $x' \; R \; x_2$. Then $x_1 \; R \; x'$ and $x' \; R \; x_2$ imply $x_1 \; R \; x_2$ by transitivity.
Now suppose that x'' is an element of $[x_1]_R$, so that $x_1 \; R \; x''$. Then

$$x_2 \; R \; x_1 \; R \; x''$$

implies $x_2 \; R \; x''$ by transitivity, so that x'' is an element of $[x_2]_R$. Similarly, each element of $[x_2]_R$ is an element of $[x_1]_R$. It follows that the two classes $[x_1]_R$ and $[x_2]_R$ are equal. $\quad\square$

To conclude this section, we will show that each equivalence relation R on a set X is the kernel relation of a suitable function with X as domain. Let X_R denote the set

$$\{[x]_R \mid x \text{ in } X\}$$

of all equivalence classes under R. It is very important to note that X_R is a set of sets: The elements C of the set X_R are themselves sets (the equivalence classes). One of the main difficulties in understanding algebra arises if the different levels of the hierarchy

$$\text{elements} - \text{sets} - \text{sets of sets}$$

are confused.

PROPOSITION 3.6 (Equivalence relations are kernels.)
Let R be an equivalence relation on a set X.

(a) *There is a surjective function*

$$\mathbf{nat}\, R : X \to X_R; x \mapsto [x]_R \,.$$

(b) *The kernel relation of the function* $\mathbf{nat}\, R$ *is R itself.*

PROOF   Part (a) is immediate. For part (b), note that two elements x_1, x_2 of X are related by the kernel relation of $\mathbf{nat}\, R$ if and only if $[x_1]_R = [x_2]_R$. By Proposition 3.5, the latter condition holds if and only if x_1 and x_2 are related by R. □

By Proposition 2.15, the surjective function $\mathbf{nat}\, R : X \to X_R$ has a section $\mathbf{rep} : X_R \to X$, with $\mathbf{rep}(C)$ as an element of C for each equivalence class C in X_R. The element $\mathbf{rep}(C)$ is called a *representative* for the equivalence class C. Each equivalence class C may be written as the class

$$C = [\mathbf{rep}(C)]_R$$

of its chosen representative. Sometimes, to avoid having to consider the set X_R of sets, it is convenient to consider the image set $\mathbf{rep}(X_R)$ instead, the set of representative elements. Note that the sets X_R and $\mathbf{rep}(X_R)$ are isomorphic, by virtue of the mutually inverse functions $\mathbf{rep} : X_R \to \mathbf{rep}(X_R)$ and $\mathbf{rep}(X_R) \to X_R; x \mapsto [x]_R$.

Example 3.7 (Choosing representatives.)
Let X be the set of citizens. Suppose that two citizens are related by the equivalence relation R if and only if they belong to the same congressional district (riding, parliamentary constituency, …). Then as a representative $\mathbf{rep}(C)$ for an equivalence class C, one may choose the congressional representative (Member of Parliament, …) for that district. Of course this choice of representative is not unique, and may change after an election! □

3.3 Rational numbers

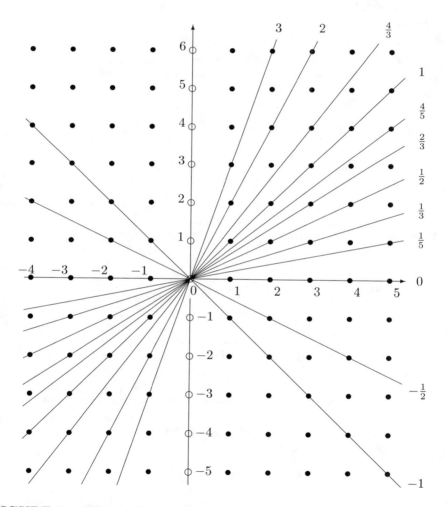

FIGURE 3.1:   Rational numbers.

Rational numbers provide a good example of the use of equivalence classes in algebra. Consider the set

$$X = \{(n, m) \mid n, m \text{ in } \mathbb{Z}, \ n \neq 0\}$$

of pairs of integers, the first integer of each pair being nonzero. The set X is illustrated by the solid dots in Figure 3.1. Define a relation R on X by

$$(n_1, m_1) \ R \ (n_2, m_2) \qquad \text{if and only if} \qquad m_1 n_2 = m_2 n_1 . \tag{3.6}$$

PROPOSITION 3.8
The relation R is an equivalence relation on X.

PROOF   The conditions required by Definition 3.2 have to be verified. The reflexivity (R) and symmetry (S) are immediate. To verify the transitivity, suppose

$$(n_1, m_1) \ R \ (n_2, m_2) \ R \ (n_3, m_3) \tag{3.7}$$

for elements (n_i, m_i) of X (with $1 \leq i \leq 3$). Then

$$m_1 n_2 = m_2 n_1 \quad \text{and} \quad m_2 n_3 = m_3 n_2 .$$

Multiplying these two equations together yields

$$m_1 n_2 m_2 n_3 = m_3 n_2 m_2 n_1 .$$

Since n_2 is a nonzero integer, it can be canceled to yield

$$m_1 m_2 n_3 = m_3 m_2 n_1 .$$

If m_2 is nonzero, it can also be canceled to give the equation

$$m_1 n_3 = m_3 n_1 \tag{3.8}$$

showing that $(n_1, m_1) \ R \ (n_3, m_3)$. If $m_2 = 0$, then (3.7) shows $m_1 n_2 = 0$ and $m_3 n_2 = 0$. Since n_2 is nonzero, these latter equations give $m_1 = 0 = m_3$. The equation (3.8) then holds trivially, so again $(n_1, m_1) \ R \ (n_3, m_3)$. □

In Figure 3.1, the equivalence classes for R are the sets of solid dots lying on the same line through the origin. (A few of these lines have been drawn.) The set $\mathbb{Q}$ of *rational numbers* is defined to be the set X_R of equivalence classes. For an element (n, m) of X, write m/n or

$$\frac{m}{n} = [(n, m)]_R$$

for the corresponding equivalence class. If $n = 1$, the class is often written just as m instead of $m/1$. If $m = 0$, the class is often written just as 0 instead of $0/n$. For nonzero rationals m/n, a preferred representative is the one in *lowest terms*, meaning $n > 0$ and $\gcd(n, m) = 1$. Thinking of the rational as a line through the origin in Figure 3.1, this representative is the first dot encountered on proceeding along the line to the right from the origin (Exercise 9).

The set $\mathbb{Q}$ of rational numbers has a well defined multiplication given by

$$\frac{m_1}{n_1} \cdot \frac{m_2}{n_2} = \frac{m_1 m_2}{n_1 n_2}. \tag{3.9}$$

It has a well-defined addition given by

$$\frac{m_1}{n_1} + \frac{m_2}{n_2} = \frac{m_1 n_2 + m_2 n_1}{n_1 n_2}. \tag{3.10}$$

The multiplication (3.9) and addition (3.10) appear to depend on the choice of the particular representatives (n_1, m_1) and (n_2, m_2). However, saying that these operations are "well defined" means that the same answer is obtained in each case, regardless of the particular choice of representative. For example, if $m_1/n_1 = m_1'/n_1'$ and $m_2/n_2 = m_2'/n_2'$, the definition (3.6) of R gives $m_1 n_1' = m_1' n_1$ and $m_2 n_2' = m_2' n_2$. Then $m_1 m_2 n_1' n_2' = m_1' m_2' n_1 n_2$, so

$$\frac{m_1 m_2}{n_1 n_2} = \frac{m_1' m_2'}{n_1' n_2'},$$

showing that the multiplication (3.9) is well defined. Verification that the addition (3.10) is well defined is assigned as Exercise 7.

The equivalence relation R on the set X actually arises as the kernel relation of the division function

$$\backslash : X \to \mathbb{R}; (n, m) \mapsto n^{-1} m \tag{3.11}$$

(Exercise 8). The set $\mathbb{Q}$ of rationals is usually identified with the image of the function (3.11), embedded in the codomain $\mathbb{R}$. This is possible because of the isomorphism between the set X_R of kernel classes and the image of (3.11). In the next section, the First Isomorphism Theorem will provide a comparable isomorphism for every function. To conclude this section, we show that the function (3.11) is not surjective, proving the existence of *irrational* real numbers, real numbers that are not expressible in the form $n^{-1} m$ with (n, m) in X.

THEOREM 3.9
Irrational numbers exist. In particular, $\sqrt{2}$ is not rational.

PROOF Suppose that $\sqrt{2}$ is rational, say m/n. Then $(m/n)^2 = 2$, so we obtain the equation

$$m^2 = 2n^2$$

between positive integers. However, there can be no such equation, since it would violate Theorem 1.14 (page 15). Indeed, Theorem 1.13 yields a factorization

$$m = 2^e p_2^{e_2} \dots p_r^{e_r}$$

of m and a factorization

$$n = 2^f p_2^{f_2} \ldots p_r^{f_r}$$

of n. Then the factorization of m^2 contains the even number $2e$ of prime factors 2, while the factorization of the same number m^2 in the form $2n^2$ contains the odd number $1 + 2f$ of prime factors 2. This contradicts the uniqueness part of the Fundamental Theorem of Arithmetic. ∎

3.4   The First Isomorphism Theorem for Sets

The division function (3.11) decomposes as a composite of the surjection $X \to X_R$, the isomorphism $X_R \cong \mathbb{Q}$, and the injection $\mathbb{Q} \hookrightarrow \mathbb{R}$. The topic of this section, the First Isomorphism Theorem for Sets, shows that every function can be written as a composition

$$\langle \text{injection} \rangle \circ \langle \text{isomorphism} \rangle \circ \langle \text{surjection} \rangle .$$

Consider a function $f : X \to Y$. Since the kernel relation **ker** f is an equivalence relation, Proposition 3.6(a) shows that there is a surjective function

$$s : X \to X_{\mathbf{ker}\,f}; x \mapsto [x]_{\mathbf{ker}\,f} . \tag{3.12}$$

On the other hand, there is an injection

$$j : f(X) \hookrightarrow Y; y \mapsto y \tag{3.13}$$

inserting the image $f(X)$ as a subset into the codomain Y. The remaining ingredient is an isomorphism between the set $X_{\mathbf{ker}\,f}$ of kernel classes and the image $f(X)$.

PROPOSITION 3.10
Let $f : X \to Y$ be a function. Then there is a well-defined bijection

$$b : X_{\mathbf{ker}\,f} \to f(X); [x]_{\mathbf{ker}\,f} \mapsto f(x) . \tag{3.14}$$

PROOF   It will first be shown that b is a well-defined injection. Note that well-definedness is an issue, since the specification (3.14) apparently depends on the choice of the representative x for the kernel class $[x]_{\mathbf{ker}\,f}$. However, for elements x and x' of X, we have

$$[x]_{\mathbf{ker}\,f} = [x']_{\mathbf{ker}\,f} \quad \text{if and only if} \quad x\,\mathbf{ker}\,f\,x' \quad \text{if and only if} \quad f(x) = f(x')$$

by Proposition 3.5 and the definition (3.3) of the kernel relation. Reading in the "only if" direction shows that b is well defined. Reading in the "if" direction shows that b is injective. Finally, it is immediate that b is surjective, by the definition (2.4) of the image $f(X)$. □

We now obtain the First Isomorphism Theorem for Sets. (Later, it will be embellished with additional algebraic structure.)

THEOREM 3.11 (First Isomorphism Theorem for Sets.)
Let $f : X \to Y$ be a function. Then f decomposes as the composite

$$f = j \circ b \circ s$$

of the surjection s of (3.12), the bijection b of (3.14), and the injection j of (3.13).

The theorem is summarized in the following diagram, of the kind proposed in Section 2.2:

$$
\begin{array}{ccc}
X & \xrightarrow{\;f\;} & Y \\
{\scriptstyle s}\downarrow & & \uparrow{\scriptstyle j} \\
X_{\mathbf{ker}\,f} & \xrightarrow[b]{} & f(X)
\end{array}
$$

Figure 3.2 presents a more naive illustration that may nevertheless be helpful.

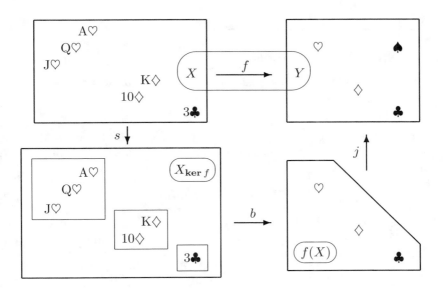

FIGURE 3.2: An illustration of the First Isomorphism Theorem.

The domain X is the set of cards in a hand. The codomain Y is the full set of suits. The function f maps each card in the hand to its suit, so two cards are in the relation $\mathbf{ker}\, f$ if and only if they lie in the same suit. The equivalence class

$$[Q\heartsuit]_{\mathbf{ker}\, f} = \{J\heartsuit, Q\heartsuit, A\heartsuit\}$$

consists of all the hearts in the hand, the class

$$[K\diamondsuit]_{\mathbf{ker}\, f} = \{10\diamondsuit, K\diamondsuit\}$$

consists of all the diamonds in the hand, and the class $[3\clubsuit]_{\mathbf{ker}\, f}$ contains the unique club in the hand. The image

$$f(X) = \{\heartsuit, \diamondsuit, \clubsuit\}$$

is the set of suits appearing in the hand. The First Isomorphism Theorem exhibits this set as isomorphic to the set

$$X_{\mathbf{ker}\, f} = \{[Q\heartsuit]_{\mathbf{ker}\, f}, [K\diamondsuit]_{\mathbf{ker}\, f}, [3\clubsuit]_{\mathbf{ker}\, f}\}$$

of equivalence classes. Indeed, both $f(X)$ and $X_{\mathbf{ker}\, f}$ each have 3 elements. The fact that the 3 elements of the set $X_{\mathbf{ker}\, f}$ are each themselves sets is irrelevant here. When dealing with sets of equivalence classes, you disregard the internal details of the classes for a moment, and just consider each class as an element.

3.5　Modular arithmetic

Fix a positive integer d. For each integer a, define $a \bmod d$ by

$$a = qd + (a \bmod d). \tag{3.15}$$

In other words, $a \bmod d$ is the remainder given by the Division Algorithm with a as dividend and d as divisor. (Compare Proposition 1.6, page 7.) Consider the function

$$f : \mathbb{Z} \to \mathbb{N}; a \mapsto a \bmod d. \tag{3.16}$$

The kernel classes $[a]_{\bmod d}$ of this function are known as *congruence classes modulo d*. Two integers a and b are said to be *congruent* modulo d, written

$$a \equiv b \mod d, \tag{3.17}$$

if they are related by the kernel relation $\mathbf{ker}\, f$, or (equivalently) if they lie in the same congruence class, or if they leave the same remainder after division by d. To facilitate working with (3.17), it is helpful to summarize yet more equivalent forms of the relation.

PROPOSITION 3.12 (Characterizations of congruence.)
Let d be a positive integer. For integers a and b, the following are equivalent:

(a) $a \equiv b \mod d$;

(b) d *divides* $a - b$;

(c) $a - b$ *is a multiple of* d.

PROOF The equivalence of (b) and (c) is (1.9). Now if (c) holds, say $a - b = rd$ for some integer r, we have

$$b = a - rd = qd + (a \bmod d) - rd = (q - r)d + (a \bmod d),$$

using (3.15). It follows that $(b \bmod d) = (a \bmod d)$, so (a) holds. Conversely, suppose that (a) holds, say $a = qd + (a \bmod d)$ and $b = q'd + (a \bmod d)$. Then $a - b = qd - q'd = (q - q')d$, so (c) holds.   □

The bijection b of the First Isomorphism Theorem provides an isomorphism

$$\mathbb{Z}_{\bmod d} \to \mathbb{Z}/d; [a]_{\bmod d} \mapsto a \bmod d \qquad (3.18)$$

between the set of congruence classes modulo d and the set

$$\mathbb{Z}/d = \{0, 1, 2, \ldots, d - 1\}$$

of remainders or *integers modulo* d, the image of the function (3.16). The isomorphism is often used to identify a congruence class with its representative remainder, so the set of congruence classes is then written as $\mathbb{Z}/d$.

For $d = 2$, the set $\mathbb{Z}/2 = \{0, 1\}$ consists of the two *bits* or *binary digits* 0 and 1. The remainder 0 stands for the set of even integers, while 1 stands for the set of odd integers. Now these sets of odd and even integers follow simple arithmetical rules, summarized by the tables in Figure 3.3. The left-hand table means that the sum of two even integers is even, the sum of an odd and

+	0	1		·	0	1
0	0	1		0	0	0
1	1	0		1	0	1

FIGURE 3.3: Addition and multiplication modulo 2.

an even integer is odd, and the sum of two odd integers is even. The right-hand table means that the product of two even integers is even, the product of an odd and an even integer is even, and the product of two odd integers is odd. Similar *modular arithmetic* holds for general positive divisors d.

PROPOSITION 3.13

Let d be a positive integer. Suppose that for integers a_i and b_i (with $i = 1, 2$),

$$a_1 \equiv b_1 \mod d \quad and \quad a_2 \equiv b_2 \mod d. \tag{3.19}$$

Then

$$a_1 + a_2 \equiv b_1 + b_2 \mod d \quad and \quad a_1 \cdot a_2 \equiv b_1 \cdot b_2 \mod d. \tag{3.20}$$

PROOF   By Proposition 3.12 and (3.19), there are integers r_1 and r_2 with

$$a_1 - b_1 = r_1 d \quad and \quad a_2 - b_2 = r_2 d.$$

Then

$$(a_1 + a_2) - (b_1 + b_2) = (a_1 - b_1) + (a_2 - b_2) = r_1 d + r_2 d = (r_1 + r_2)d$$

and

$$a_1 a_2 = (b_1 + r_1 d)(b_2 + r_2 d)$$
$$= b_1 b_2 + (r_1 b_2 + b_1 r_2 + r_1 r_2)d,$$

so Proposition 3.12 yields the required relations (3.20).   ☐

COROLLARY 3.14

There are well-defined operations

$$[a]_{\mathbf{mod}\, d} + [b]_{\mathbf{mod}\, d} = [a + b]_{\mathbf{mod}\, d} \tag{3.21}$$

and

$$[a]_{\mathbf{mod}\, d} \cdot [b]_{\mathbf{mod}\, d} = [a \cdot b]_{\mathbf{mod}\, d} \tag{3.22}$$

on the sets $\mathbb{Z}_{\mathbf{mod}\, d}$ and $\mathbb{Z}/_d$.

Note that for each element a of $\mathbb{Z}/_d$, the function

$$\mathbb{Z}/_d \to \mathbb{Z}/_d; x \mapsto a + x$$

is the permutation

$$\big((0 \bmod d)\ (1 \bmod d)\ (2 \bmod d)\ \ldots\ (-1 \bmod d)\big)^a$$

of the cyclic group C_d (compare Example 2.30).

3.6   Exercises

1. Let X be a set of sets. Show that isomorphism between members of X is an equivalence relation on X.

2. For the functions sq of (2.1) and abs of (2.2), show that

$$\mathbf{ker}\,\mathrm{sq} = \mathbf{ker}\,\mathrm{abs}.$$

3. Verify (3.5).

4. Let $f : X \to Y$ be a function. Show that the following two conditions are equivalent:

   (a) f is injective;

   (b) For each element x of X, the equivalence class $[x]_{\mathbf{ker}\,f}$ has only one element.

5. Show that the set of natural numbers may be chosen as a particular set of representatives for the equivalence classes of the equal kernel relations of Exercise 2.

6. Show that the set of nonpositive numbers may be chosen as a set of representatives for the equivalence classes of the equal kernel relations of Exercise 2.

7. Verify that the addition (3.10) of rationals is well defined.

8. (a) Show that the relation R of (3.6) on the set X is the kernel relation of the division function (3.11).

   (b) Conclude directly that R is an equivalence relation, without using Proposition 3.8.

9. Suppose that n is a positive integer and m is a nonzero integer. Show that $\gcd(n, m) = 1$ if and only if there is no element of the set X lying on the interior of the line segment from the origin to (n, m).

10. Show that $\sqrt{3}$ is irrational.

11. Show that $\sqrt[3]{2}$ is irrational.

12. Deal yourself a hand of cards, and use the First Isomorphism Theorem to analyze the function mapping each card in the hand to its suit. Identify the set of equivalence classes for the kernel relation of the function.

13. Let a, b, and d be positive integers. Show that

$$a \equiv b \mod d$$

if and only if a and b have the same rightmost digit in their base d representation (compare Exercise 19 in Chapter 1).

14. Show that a positive integer n cannot be a perfect square if its decimal representation ends in one of the digits 2, 3, 7, or 8.

15. Consider the numbers

$$1 \quad 2 \quad 3 \quad 4 \quad 5 \quad 6 \quad 7 \quad 8 \quad 9 \quad 10 \,.$$

Is it possible to put positive or negative signs in front of each, so that the total sum of the signed numbers is zero?

16. Repeat Exercise 15, this time with the numbers

$$1 \quad 2 \quad 3 \quad 4 \quad 5 \quad 6 \quad 7 \quad 8 \quad 9 \quad 10 \quad 11 \,.$$

17. Show that $\log_2 3$ is irrational.

18. Consider an angle θ with

$$0 < \theta < \frac{\pi}{2} \,.$$

Suppose that

$$\cos \theta = \frac{l}{n} \quad \text{and} \quad \sin \theta = \frac{m}{n}$$

are rational numbers, with positive integers l, m, and n. Show that l and m cannot both be odd numbers.

19. Which of the following three conditions determines the kernel relation R of the cosine function $\cos : \mathbb{R} \to \mathbb{R}; x \mapsto \cos x$?

    (a) $x \, R \, y$ if and only if $x = \pm y$.
    (b) $x \, R \, y$ if and only if $x = 2\pi n \pm y$ for some integer n .
    (c) $x \, R \, y$ if and only if $x - y = 2\pi n$ for some integer n .

20. Define a relation P on the set $\mathbb{R}$ of real numbers by

$$x \, P \, y \quad \text{if and only if} \quad x - y = 2n\pi \quad \text{for some integer } n \,.$$

Show directly that P is an equivalence relation on $\mathbb{R}$.

21. Show that the relation P of Exercise 20 is the kernel relation of the column vector-valued function

$$f : \mathbb{R} \to \mathbb{R}_2^1; \theta \mapsto \begin{bmatrix} \cos \theta \\ \sin \theta \end{bmatrix} \,.$$

Conclude that P is an equivalence relation on $\mathbb{R}$.

3.7 Study projects

1. **Tonal music.** Consider the set X of frequencies of audible sounds, measured in Hertz (or cycles per second). Define a relation R on X by $f_1 \ R \ f_2$ if and only if

   f_1/f_2 is an integral power $\ldots 2^{-3}, 2^{-2}, 2^{-1}, 2^0, 2^1, 2^2, 2^3 \ldots$ of 2.

   (a) Show that R is an equivalence relation on the set X.

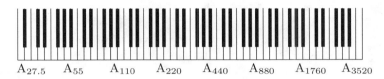

$A_{27.5}$ $\quad A_{55}$ $\quad A_{110}$ $\quad A_{220}$ $\quad A_{440}$ $\quad A_{880}$ $\quad A_{1760}$ $\quad A_{3520}$

FIGURE 3.4: A piano keyboard.

   (b) If the note A (compare Figure 3.4) labels the equivalence class

   $$\ldots, 55, 110, 220, 440, 880, 1760, \ldots ,$$

   describe how other notes also correspond to equivalence classes.

   (c) Explain why the equivalence classes are more important in music than the actual frequencies themselves.

2. **Continued fractions.** Theorem 3.9 shows that there are real numbers s, for example $\sqrt{2}$, that are not rational. In Figure 3.1, this means that the line of slope s through the origin moves off to infinity on the right without ever exactly hitting a solid dot (n, m). Nevertheless, the line does come very close to various dots along the way. For example, the line of slope $\sqrt{2}$ only narrowly misses $(2, 3)$, $(5, 7)$, $(12, 17)$, $(29, 41)$, etc. Which dots are close to a line of irrational slope s? How can we find good rational approximations to real numbers like e and π?

   The answer is given by *continued fractions*. A continued fraction is an expression of the form

$$x_0 + \cfrac{1}{x_1 + \cfrac{1}{x_2 + \cfrac{1}{\cdots + \cfrac{1}{x_{n-1} + \cfrac{1}{x_n}}}}}$$

or, equivalently,

$$x_0 + 1 \Big/ \left(x_1 + 1 \Big/ \left(x_2 + 1 \Big/ \left(\cdots + 1/(x_{n-1} + 1/x_n) \cdots \right) \right) \right).$$

Since these expressions are extremely unwieldy, they are rewritten as

$$[[x_0, x_1, \ldots, x_{n-1}, x_n]]. \tag{3.23}$$

In other words, $[[x_0]] = x_0$ and

$$[[x_0, x_1, \ldots, x_i, x_{i+1}]] = \left[\left[x_0, x_1, \ldots, x_i + \frac{1}{x_{i+1}} \right]\right] \tag{3.24}$$

for natural numbers i. In order to recover a fraction from (3.23), define p_i and q_i for $i \geq -2$ by the initial setting

$$\begin{bmatrix} p_{-1} & p_{-2} \\ q_{-1} & q_{-2} \end{bmatrix} = \begin{bmatrix} 1 & 0 \\ 0 & 1 \end{bmatrix} \tag{3.25}$$

and the recurrence

$$p_i = x_i p_{i-1} + p_{i-2} \tag{3.26}$$
$$q_i = x_i q_{i-1} + q_{i-2} \tag{3.27}$$

for all natural numbers i.

(a) Use induction to prove that

$$\begin{vmatrix} p_{k-1} & p_{k-2} \\ q_{k-1} & q_{k-2} \end{vmatrix} = (-1)^k \tag{3.28}$$

for all natural numbers k.

(b) Consider the equation

$$[[x_0, x_1, \ldots, x_{k-1}, x_k]] = \frac{p_k}{q_k} \tag{3.29}$$

for natural numbers k. Show that it holds for $k = 0$. If it holds for $k = n$, use (3.26) and (3.27) to deduce that

$$[[x_0, x_1, \ldots, x_{n-1}, x_n]] = \frac{x_n p_{n-1} + p_{n-2}}{x_n q_{n-1} + q_{n-2}}. \tag{3.30}$$

Replace x_n by $x_n + 1/x_{n+1}$ on both sides of (3.30). Use (3.24) to rewrite the left-hand side, and simplify the right-hand side to obtain (3.29) with $k = n + 1$. Conclude that (3.29) holds for all natural numbers k by induction. As a corollary, deduce that (3.30) holds for all natural numbers n.

(c) Use (3.29) to express $[[1,2,2]]$ and $[[1,2,2,2]]$ as rational numbers.

3. **Approximating irrationals.** For a real number x, the *floor* $\lfloor x \rfloor$ is defined to be the largest integer l with $l \le x$. The *fractional part* of x is $x - \lfloor x \rfloor$.

(a) If x is an irrational number, show that the fractional part is an irrational number, with $0 < x - \lfloor x \rfloor < 1$.

(b) If x is an irrational number, show that the reciprocal $(x - \lfloor x \rfloor)^{-1}$ of the fractional part of x is an irrational number with $1 < (x - \lfloor x \rfloor)^{-1}$.

(c) Let s be an irrational number. Define a sequence of real numbers s_i by $s_0 = s$ and
$$s_{i+1} = (s_i - \lfloor s_i \rfloor)^{-1}$$
for i in $\mathbb{N}$. Show that s_k is irrational for all natural numbers k.

(d) Show that $1 \le \lfloor s_k \rfloor$ for all positive integers k.

(e) Show that $s_i = \lfloor s_i \rfloor + 1/s_{i+1}$ for all natural numbers i.

(f) Show that
$$s = [[\lfloor s_0 \rfloor, \lfloor s_1 \rfloor, \ldots, \lfloor s_{i-1} \rfloor, \lfloor s_i \rfloor, s_{i+1}]] \tag{3.31}$$
for all natural numbers i.

(g) Setting $x_i = \lfloor s_i \rfloor$ for all natural numbers i, consider the numbers p_i and q_i defined by (3.25) through (3.27) above. Show that p_k and q_k are integers, with $q_k \ge k$ for all natural numbers k.

(h) Use (3.31) (with $i = k$), and (3.30) (with $n = k+1$, and s_{k+1} in place of x_{k+1}), to show that
$$s = \frac{s_{k+1}p_k + p_{k-1}}{s_{k+1}q_k + q_{k-1}} \tag{3.32}$$

(i) Use (3.32) to show that
$$s - \frac{p_k}{q_k} = \frac{-1}{q_k(s_{k+1}q_k + q_{k-1})} \begin{vmatrix} p_k & p_{k-1} \\ q_k & q_{k-1} \end{vmatrix}.$$

(j) Using (3.28), along with the inequalities $s_{k+1} > 1$ from (b) and $q_k \ge k$ from (g) above, conclude that the irrational number s is approximated by the rational number p_k/q_k to within a tolerance given by
$$\left| s - \frac{p_k}{q_k} \right| \le \frac{1}{k^2}$$
for all positive integers k.

(k) Use continued fractions to compute some rational approximants to e and π. For example, with $s = s_0 = \pi = 3.14159\ldots$, we have $s_1 = (\pi - 3)^{-1} = 7.06251\ldots$, so $x_0 = 3$, $x_1 = 7$, and

$$[[x_0, x_1]] = [[3, 7]] = \frac{22}{7}.$$

(l) Compare your approximants from (k) with those that are given by truncating the series

$$\frac{1}{0!} + \frac{1}{1!} + \frac{1}{2!} + \frac{1}{3!} + \cdots$$

for e and

$$\frac{1}{2} + \frac{1}{2} \cdot \frac{1}{2^3 \cdot 3} + \frac{1 \cdot 3}{2 \cdot 4} \cdot \frac{1}{2^5 \cdot 5} + \frac{1 \cdot 3 \cdot 5}{2 \cdot 4 \cdot 6} \cdot \frac{1}{2^7 \cdot 7} + \cdots$$

for $\pi/6$.

3.8 Notes

Section 3.5

Many authors use the notation $\mathbb{Z}_2$ to denote the set of integers modulo 2 (and similar notation for other moduli). However, this notation gives no hint of the inherent quotient structure (Example 5.20, page 103). Furthermore, it clashes with the standard notation $\mathbb{Z}_2$ for the set of dyadic integers (Study Project 3 in Chapter 5).

Chapter 4

GROUPS AND MONOIDS

Chapter 2 showed how sets of functions could form semigroups, monoids, or groups. Many other sets have a similar structure, even though they do not consist of functions. In order to study the structure, the key properties of the sets of functions are abstracted and formulated in general terms.

4.1 Semigroups

If S is a semigroup of functions, then we may consider function composition as a map

$$S \times S \to S; (g, f) \mapsto g \circ f \qquad (4.1)$$

whose domain is the set $S \times S$ of ordered pairs (g, f) of elements of S. The closure (2.9) under composition guarantees that S may serve as the codomain of the map (4.1). Recall that function composition is always associative. The abstract properties of semigroups of functions are then captured by the following definition.

DEFINITION 4.1 (Semigroups.) *Let S be a set equipped with a map*

$$S \times S \to S; (x, y) \mapsto x \cdot y \qquad (4.2)$$

assigning an element $x \cdot y$ or xy of S to each ordered pair (x, y) of elements of S.

(a) *In general, the map (4.2) is known as a* multiplication *on S or (more formally) as a* binary operation *on S.*

(b) *The existence of such a map is described as the* closure *of the set S with respect to the multiplication.*

(c) *The pair $(S, \cdot)$ consisting of the set S with its multiplication $\cdot$ is called a* semigroup *(or an* abstract semigroup*) if the* associative law

$$x \cdot (y \cdot z) = (x \cdot y) \cdot z \qquad (4.3)$$

holds for all elements x, y, and z of the set S.

DEFINITION 4.2 (Commuting elements.)   *Two elements x and y of a semigroup $(S, \cdot)$ are said to* commute *if $x \cdot y = y \cdot x$. The semigroup $(S, \cdot)$ is said to be* commutative *if $x \cdot y = y \cdot x$ for all x, y in S.*

Example 4.3 (The real interval $(1, \infty)$.)

Let S be the set or interval $(1, \infty)$ of real numbers x with $x > 1$. Then S forms a semigroup under the usual (associative and commutative) multiplication of real numbers. ▢

Example 4.4 (Irrationals.)

Let S be the set of irrational real numbers. Then S does not form a semigroup under the usual associative multiplication of real numbers, since $\sqrt{2}$ is a member of S (compare Theorem 3.9, page 55), but $x \cdot x = 2$ is not a member of S. ▢

Although the binary operation (4.2) on a general semigroup S is called a "multiplication," it does not have to be an actual multiplication of numbers in the usual sense. Here is one example. For another, see Example 4.9 below.

Example 4.5 (Positive integers.)

Let S be the set of positive integers. Define a "multiplication" on S by

$$m \cdot n = \gcd(m, n) \,. \tag{4.4}$$

Then $(S, \cdot)$ forms a commutative semigroup. ▢

Function composition is always associative. But with general operations as in (4.2), you should be careful not to take associativity for granted, even when you are on familiar ground.

Example 4.6 (Integers under subtraction.)

Consider the set $\mathbb{Z}$ of integers. Then $\mathbb{Z}$ is closed under the operation of subtraction:

$$\mathbb{Z} \times \mathbb{Z} \to \mathbb{Z}; (x, y) \mapsto x - y \,.$$

However, $\mathbb{Z}$ does not form a semigroup under subtraction, since subtraction is not associative. Indeed,

$$3 - (5 - 4) = 3 - 1 = 2 \,,$$

while

$$(3 - 5) - 4 = (-2) - 4 = -6 \,.$$

▢

A semigroup S of functions always forms a semigroup $(S, \circ)$, with function composition as the "multiplication," in the sense of the abstract Definition 4.1. In general, this multiplication is not commutative (compare Example 2.3, page 29). However, Exercise 43 in Chapter 2 shows that disjoint cycles commute.

A semigroup S of functions may also provide the underlying set for an abstract semigroup structure with a multiplication which is different from the composition of functions. For example, the set P of power maps

$$p_n : \mathbb{R} \to \mathbb{R}; x \mapsto x^n$$

for natural numbers n (compare Exercise 10 in Chapter 2) certainly forms a semigroup of functions. On the other hand, it also forms a semigroup under the usual "componentwise" function multiplication of calculus:

$$p_m \cdot p_n = p_{m+n}$$

or

$$(p_m \cdot p_n)(x) = p_m(x) \cdot p_n(x) = x^m \cdot x^n = x^{m+n} = p_{m+n}(x)$$

for x in $\mathbb{R}$. Compare this with

$$p_m \circ p_n = p_{mn}$$

for the function composition.

4.2 Monoids

A monoid of functions on a set X is a semigroup of functions on X that contains the identity function id_X on X. As far as function composition is concerned, the key property of the identity function is (2.12). This property is abstracted by (4.5) below.

DEFINITION 4.7 (Abstract monoids.)   *Let $(M, \cdot)$ be a semigroup with $\cdot$ as multiplication. Then M is said to form a* monoid *(or an* abstract monoid*) $(M, \cdot, e)$ if it contains an element e satisfying*

$$e \cdot x = x = x \cdot e \tag{4.5}$$

for all x in M. The element e is known as the identity element *of the monoid M. (Proposition 4.13 below shows that the identity element is unique.)*

Example 4.8 (The real interval $(1, \infty)$.)
The semigroup $S = (1, \infty)$ of Example 4.3 does not form a monoid. Certainly S does not contain the usual identity element 1 for the multiplication of real

numbers. In fact, for each element e of S, we have $e \cdot x > x$ for all x in S. Thus no element e of S can satisfy the identity property (4.5). □

Example 4.9 (Natural numbers.)

The set $\mathbb{N}$ of natural numbers forms a commutative monoid under addition, with 0 as the identity element. On the other hand, the semigroup of positive integers under addition does not form a monoid (Exercise 4). □

Example 4.10 (Least common multiples.)

Let S be the set of positive integers. Define a "multiplication" on S by

$$m \cdot n = \operatorname{lcm}(m, n). \tag{4.6}$$

Then $(S, \cdot, 1)$ forms a commutative monoid. □

Example 4.11 (Intersection.)

Let n be a natural number. Let $\mathcal{P}(\underline{n})$ denote the set of subsets of the set $\underline{n}$ of natural numbers less than n — compare (2.26). Then $\mathcal{P}(\underline{n})$ forms a commutative monoid under the "multiplication" $\cap$ of *intersection*:

$$X \cap Y = \{z \mid z \text{ in both } X \text{ and } Y\}$$

with the full set $\underline{n}$ as the identity element. (See Exercise 6.) □

Example 4.12 (Matrices.)

The set $\mathbb{R}_2^2$ of all 2×2 real matrices forms a monoid under the operation of matrix multiplication. The identity element is the 2×2 identity matrix (2.13). Note that this monoid is not commutative (Exercise 9). □

Definition 4.7 refers to *the* identity element of a monoid. The definite article is justified by the following result.

PROPOSITION 4.13 (Uniqueness of the identity.)

Let M be a monoid. If e and f are identity elements of M, then $e = f$. Thus the identity element of a monoid is unique.

PROOF We have $e = e \cdot f = f$. The first equality holds since f is an identity element. The second equality holds since e is an identity element. □

4.3   Groups

Recall that functions are bijections if and only if they are invertible. Thus a monoid G of functions on a set X is a group of permutations of X if and only if each element f of G has an inverse element f^{-1} in G, with

$$f \circ f^{-1} = \mathrm{id}_X = f^{-1} \circ f.$$

This property is abstracted by (4.7) below.

DEFINITION 4.14 (Abstract groups.)   *A monoid $(G, \cdot, e)$ is a* group *(or an* abstract group*) if each element x of G has an* inverse x^{-1} *in G with*

$$x \cdot x^{-1} = e = x^{-1} \cdot x. \tag{4.7}$$

In other words, a group $(G, \cdot, e)$ is a set G with a multiplication $\cdot$ satisfying the following properties:

Closure: *$x \cdot y$ lies in G for all x, y in G;*

Associativity: *$x \cdot (y \cdot z) = (x \cdot y) \cdot z$ for all x, y, z in G;*

Identity: *There is an element e in G with $e \cdot x = x = x \cdot e$ for all x in G;*

Inverses: *For each x in G, there is x^{-1} in G with $x \cdot x^{-1} = e = x^{-1} \cdot x$.*

Commutative groups are often described as abelian.

Consider the four properties of Definition 4.14. Note that semigroups are required to satisfy the closure and associativity, while monoids are required to satisfy closure, associativity, and the identity property.

The uniqueness of inverses of invertible functions (Proposition 2.23, page 35) has its abstract counterpart in groups (see Exercise 11 for the proof):

PROPOSITION 4.15 (Uniqueness of inverses.)
In a group G, each element x has a unique inverse.

We now consider some examples of groups.

Example 4.16 (Real numbers under addition.)
The real numbers form a group $(\mathbb{R}, +, 0)$ with addition as the commutative "multiplication" operation. The "inverse" or *additive inverse* of a real number r is its negation $-r$.   □

In general, if the "multiplication" in a group G is denoted by an addition $+$ as in Example 4.16, the group G is described as *additive*. Its identity element is written as a *zero*, and the inversion is described as *negation*. These conventions are normally reserved for abelian groups.

Example 4.17 (Muliplication of nonzero reals.)
Under multiplication, the nonzero real numbers form a commutative group $(\mathbb{R}^*, \cdot, 1)$. $\square$

Example 4.18 (The general linear group.)
Let $\mathrm{GL}(2, \mathbb{R})$ be the set of invertible 2×2 matrices with real entries. Then $\mathrm{GL}(2, \mathbb{R})$ forms a nonabelian group under the usual matrix multiplication, having the identity matrix I_2 as the identity element of the group, and with the usual inversion of matrices. The group $\mathrm{GL}(2, \mathbb{R})$ is called the (*real*) *general linear group* of dimension 2. $\square$

Examples 4.17 and 4.18 are special cases of a general source of abstract groups: sets of invertible elements of monoids.

DEFINITION 4.19 (Invertible elements.)   *Let $(M, \cdot, e)$ be a monoid. An element u of M is said to be* invertible *or a* unit *if there is an element v of M such that $u \cdot v = e = v \cdot u$.*

PROPOSITION 4.20 (Invertible elements form a group.)
Let $(M, \cdot, e)$ be a monoid. Then the set M^* of invertible elements of M forms a group $(M^*, \cdot, e)$.

PROOF   We verify the four conditions listed in Definition 4.14.
Closure: Suppose that u_1 and u_2 are units of M, with $u_1 v_1 = e = v_1 u_1$ and $u_2 v_2 = e = v_2 u_2$. Then

$$(u_1 u_2)(v_2 v_1) = u_1 u_2 v_2 v_1 = u_1 e v_1 = u_1 v_1 = e \tag{4.8}$$

and

$$(v_1 v_2)(u_2 u_1) = v_1 v_2 u_2 u_1 = v_1 e u_1 = v_1 u_1 = e \,, \tag{4.9}$$

so $u_1 u_2$ is also a unit.
Associativity: The associativity of the multiplication in M^* is a special case of the associativity of the multiplication in the monoid M.
Identity: The identity element of M^* is the identity element e of M: The equations

$$e \cdot u = u = u \cdot e$$

in M^* are special cases of the equation (4.5) in M.

Inverses: By the definition of the set M^*, each element u of M^* has an inverse v. Note that v lies in M^*, since it has u as its inverse. ☐

DEFINITION 4.21 (The group of units.) *For a monoid* $(M, \cdot, 1)$, *the group* $(M^*, \cdot, 1)$ *is known as the* group of units *of the monoid* M.

Example 4.22 (Integers under multiplication.)
The integers form a commutative monoid $(\mathbb{Z}, \cdot, 1)$ under multiplication. The group of units of the monoid of integers is $\{\pm 1\}$. ☐

Example 4.23 (Invertible real numbers.)
The notation of Definition 4.21 is consistent with Example 4.17: the set of units of the monoid of real numbers under multiplication is the set $\mathbb{R}^*$ of nonzero real numbers. ☐

Example 4.24 (The group of units of a group.)
If $(G, \cdot, 1)$ is a group, then it is certainly a monoid: Just forget the inversion. Now $(G, \cdot, 1)$ is its own group of units. In particular, Proposition 4.15 and the proof of Proposition 4.20 — specifically (4.8) and (4.9) — yield the important formula

$$(u_1 u_2)^{-1} = u_2^{-1} u_1^{-1} \tag{4.10}$$

for elements u_1, u_2 of a group G. Note the generalization of (4.10) given in Exercise 16. ☐

4.4 Componentwise structure

Starting with given examples of semigroups, monoids, or groups, there are methods to obtain new semigroups, monoids, or groups from the given ones. One such method is the direct product construction. Recall that for sets X and Y, the (*external*) *direct product* or *product* of X and Y is the set

$$X \times Y = \{(x, y) \mid x \text{ in } X, y \text{ in } Y\} \tag{4.11}$$

of ordered pairs (x, y) of elements x from X and y from Y (Figure 4.1). In this context, the sets X and Y are known as the (*direct*) *factors* of the direct product. The set $X \times Y$ is sometimes called the *cartesian product* of X and Y, since it follows René Descartes' recipe for constructing the real plane as $\mathbb{R} \times \mathbb{R}$. Recall that two ordered pairs (x, y) and (x', y') are equal if and only if $x = x'$ and $y = y'$. We write X^2 for $X \times X$, dsecribing it as the *direct square* of the set X.

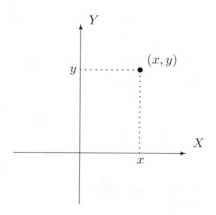

FIGURE 4.1: The product $X \times Y$.

Now suppose that X is a semigroup under the multiplication $\circ_X$, while Y is a semigroup under the multiplication $\circ_Y$. We may then define a multiplication on $X \times Y$ by

$$(x_1, y_1) \circ_{X \times Y} (x_2, y_2) = (x_1 \circ_X x_2, y_1 \circ_Y y_2) \tag{4.12}$$

for x_1, x_2 in X and y_1, y_2 in Y. The multiplication (4.12) is described as a *componentwise multiplication*, since it works individually on the respective x- and y-components of the ordered pairs. The following result is easily verified (Exercise 18).

PROPOSITION 4.25 (Direct product semigroup.)
Let $(X, \circ_X)$ and $(Y, \circ_Y)$ be semigroups. Then under the componentwise multiplication (4.12), the direct product $X \times Y$ forms a semigroup.

DEFINITION 4.26   *The semigroup $(X \times Y, \circ_{X \times Y})$ of Proposition 4.25 is called the* (external) direct product *of the semigroups $(X, \circ_X)$ and $(Y, \circ_Y)$.*

Example 4.27 (The real plane.)
The set $\mathbb{R}$ of real numbers forms a semigroup under multiplication. Then the real plane $\mathbb{R}^2$ forms a semigroup under componentwise multiplication.   ⧠

If the semigroups $(X, \circ_X)$ and $(Y, \circ_Y)$ are monoids, with respective identity elements e_X and e_Y, then the *componentwise identity element* is the element

$$e_{X \times Y} = (e_X, e_y) \tag{4.13}$$

of $X \times Y$. The following result, readily verified, is similar to Proposition 4.25.

PROPOSITION 4.28

Let $(X, \circ_X, e_X)$ and $(Y, \circ_Y, e_Y)$ be monoids. Then under the componentwise multiplication (4.12), the direct product $X \times Y$ forms a monoid

$$(X \times Y, \circ_{X \times Y}, e_{X \times Y})$$

with the componentwise identity element (4.13).

DEFINITION 4.29 (The direct product of two monoids.) The monoid $(X \times Y, \circ_{X \times Y}, e_{X \times Y})$ of Proposition 4.28 is called the (external) direct product of the two monoids $(X, \circ_X, e_X)$ and $(Y, \circ_Y, e_Y)$.

The final step in our examination of componentwise structure considers groups. Suppose that $(X, \circ_X, e_X)$ and $(Y, \circ_Y, e_Y)$ are groups. Then for an element (x, y) of $X \times Y$, define the *componentwise inverse*

$$(x, y)^{-1} = (x^{-1}, y^{-1}) \tag{4.14}$$

as an element of $X \times Y$.

PROPOSITION 4.30

Let $(X, \circ_X, e_X)$ and $(Y, \circ_Y, e_Y)$ be groups. Then the direct product $X \times Y$ forms a group

$$(X \times Y, \circ_{X \times Y}, e_{X \times Y})$$

under the componentwise multiplication (4.12), the componentwise identity element (4.13), and with componentwise inverses (4.14).

For the proofs of Propositions 4.28 and 4.30, see Exercise 19.

DEFINITION 4.31 (The direct product of two groups.) The group

$$(X \times Y, \circ_{X \times Y}, e_{X \times Y})$$

of Proposition 4.30 is called the (external) direct product of the two groups $(X, \circ_X, e_X)$ and $(Y, \circ_Y, e_Y)$.

Example 4.32 (The real plane.)
The set $\mathbb{R}$ of real numbers forms a group under addition. Then the real plane $\mathbb{R}^2$ forms a group under componentwise addition:

$$(x_1, y_1) + (x_2, y_2) = (x_1 + x_2, y_1 + y_2).$$

Note that this is just the usual addition for 2-dimensional real vectors. $\quad\Box$

The following theorem provides a good illustration of how componentwise structure is used.

THEOREM 4.33 (Groups of units of products.)

Let $(M_1, \cdot, e_1)$ and $(M_2, \cdot, e_2)$ be monoids. Then the group of units $(M_1 \times M_2)^*$ of the product monoid $M_1 \times M_2$ is the product $M_1^* \times M_2^*$ of the groups of units M_1^*, M_2^* of the respective factors M_1, M_2.

PROOF   The sets $(M_1 \times M_2)^*$ and $M_1^* \times M_2^*$ are both subsets of the product $M_1 \times M_2$. To prove the equality of the two subsets, it will be shown that each contains the other.

Let (u_1, u_2) be an element of $M_1^* \times M_2^*$. Thus there are elements v_1 of M_1 and v_2 of M_2 such that

$$u_1 v_1 = e_1 = v_1 u_1 \quad \text{and} \quad u_2 v_2 = e_2 = v_2 u_2 \,. \tag{4.15}$$

Then in the product monoid $\big(M_1 \times M_2, \cdot, (e_1, e_2)\big)$, we have

$$(u_1, u_2)(v_1, v_2) = (e_1, e_2) = (v_1, v_2)(u_1, u_2) \,, \tag{4.16}$$

so that (u_1, u_2) lies in $(M_1 \times M_2)^*$.

Conversely, suppose that (u_1, u_2) lies in $(M_1 \times M_2)^*$: There is an element (v_1, v_2) of $M_1 \times M_2$ such that (4.16) holds. In particular,

$$(u_1 v_1, u_2 v_2) = (e_1, e_2) = (v_1 u_1, v_2 u_2) \,.$$

The equality of the corresponding first components in this equation gives

$$u_1 v_1 = e_1 = v_1 u_1 \,,$$

so that u_1 lies in M_1^*. Examination of the second components shows that u_2 lies in M_2^*. It follows that (u_1, u_2) is an element of $M_1^* \times M_2^*$. □

It is relatively straightforward to extend the product constructions to larger numbers of factors. For example, a product $X \times Y \times Z$ of sets X, Y, and Z may be built recursively as $X \times (Y \times Z)$, or directly as the set

$$X \times Y \times Z = \{(x, y, z) \mid x \text{ in } X, \ y \text{ in } Y, \ z \text{ in } Z\}$$

of ordered triples. The product $X \times X \times X$ is known as the (*direct*) *cube* X^3 of the set X. For instance, the direct cube $\mathbb{R}^3$ of the additive group $(\mathbb{R}, +, 0)$ of real numbers, with componentwise structure, is the group of 3-dimensional vectors.

Componentwise structure is not limited to n-tuples. For example, the set $\mathbb{R}_2^2$ of 2×2 real matrices carries a componentwise additive group structure, with addition given by the usual addition

$$\begin{bmatrix} b_{11} & b_{12} \\ b_{21} & b_{22} \end{bmatrix} + \begin{bmatrix} a_{11} & a_{12} \\ a_{21} & a_{22} \end{bmatrix} = \begin{bmatrix} b_{11} + a_{11} & b_{12} + a_{12} \\ b_{21} + a_{21} & b_{22} + a_{22} \end{bmatrix} \tag{4.17}$$

of matrices. The same set carries a componentwise monoid structure, with multiplication given by the componentwise multiplication

$$\begin{bmatrix} b_{11} & b_{12} \\ b_{21} & b_{22} \end{bmatrix} \circ \begin{bmatrix} a_{11} & a_{12} \\ a_{21} & a_{22} \end{bmatrix} = \begin{bmatrix} b_{11}a_{11} & b_{12}a_{12} \\ b_{21}a_{21} & b_{22}a_{22} \end{bmatrix} \tag{4.18}$$

of matrices, and with the 2×2 *all ones matrix*

$$J_2 = \begin{bmatrix} 1 & 1 \\ 1 & 1 \end{bmatrix}$$

as the identity element. The matrix product (4.18) is called the *Hadamard product*. It is certainly different from the usual matrix multiplication (2.7). For an application, see (7.27), and also Study Project 5 in Chapter 5.

4.5 Powers

Another source of componentwise structure is found in sets of functions $f : X \to S$ from a certain domain X to a codomain S that carries algebraic structure. For example, in calculus the componentwise sum $f + g$ of two real-valued functions $f : \mathbb{R} \to \mathbb{R}$ and $g : \mathbb{R} \to \mathbb{R}$ is determined by the specification

$$(f + g)(x) = f(x) + g(x)$$

for all x in $\mathbb{R}$. Under this operation, the set $\mathbb{R}^{\mathbb{R}}$ of all real-valued functions forms an additive group, with the constant function zero as the zero (identity element), and with the inverse of a function f given by the negation $-f$, so that

$$(-f)(x) = -f(x)$$

for all real x. Here is the general definition. Verification of the claims embodied in the definition is deferred to Exercise 22.

DEFINITION 4.34 (Power structures.) *Let X and S be sets. Consider the set S^X of all functions $f : X \to S$ from X to S.*

(a) *If S carries a semigroup structure $(S, \cdot)$, then the X-th power $(S, \cdot)^X$ or S^X of the semigroup $(S, \cdot)$ is the set S^X equipped with the componentwise multiplication $f \cdot g$ given by*

$$(f \cdot g)(x) = f(x) \cdot g(x)$$

for x in X.

(b) *If S carries a monoid structure $(S, \cdot, e_S)$, then the X-th power $(S, \cdot, e_S)^X$ or S^X of the monoid $(S, \cdot, e_S)$ is the X-th power semigroup $(S^X, \cdot)$, with the constant function $E : X \to S; x \mapsto e_S$ as the componentwise identity element.*

(c) *If S carries a group structure $(S, \cdot, e_S)$, then the X-th power $(S, \cdot, e_S)^X$ or S^X of the group $(S, \cdot, e_S)$ is the X-th power monoid $(S^X, \cdot, E)$, with the componentwise inverse of a function $f : X \to S$ given by $f^{-1}(x) = f(x)^{-1}$ for each x in X.*

If X is the n-element set $\underline{\mathbf{n}} = \{0, 1, \ldots, n-1\}$ for a positive integer n, then the powers $S^{\underline{\mathbf{n}}}$ are known as the n-th powers S^n.

Example 4.35 (Bit strings.)

Let n be a positive integer. A *bit string* of *length n* is an element

$$b = b_{n-1}b_{n-2}\ldots b_2 b_1 b_0$$

of $\left(\mathbb{Z}/_2\right)^n$ (with b_i as the value $b(i)$ of i for $0 \le i < n$). For example, the bit string b might have been obtained as the *binary* (or base 2) expansion of the natural number

$$b_{n-1}2^{n-1} + b_{n-2}2^{n-2} + \cdots + b_2 2^2 + b_1 2^1 + b_0 2^0$$

(compare Exercise 19 in Chapter 1). According to Definition 4.34, the set $\left(\mathbb{Z}/_2\right)^n$ of bit strings of length n inherits respective group structures under $+$ and monoid structures under $\cdot$ from the addition and multiplication modulo 2 on $\mathbb{Z}/_2$ displayed in Figure 3.3. ⬚

Example 4.36 (Vectors.)

Let n be a positive integer. An *n-vector* or *n-dimensional real vector* is an element

$$(x_0, x_1, \ldots, x_{n-1})$$

of the power group $\mathbb{R}^n$. For example, in special relativity a 4-vector

$$(ct, x_1, x_2, x_3)$$

represents an event at time t and spatial location (x_1, x_2, x_3) in a certain frame of reference, c being the speed of light. ⬚

4.6   Submonoids and subgroups

Componentwise structure on product sets, as was studied in the preceding sections, is one rich source of new semigroups, monoids, and groups. Another is found from subsets that are closed under the given structure. Let $(S, \cdot)$ be a semigroup. Let X be a subset of S with the *closure property*

$$x, y \text{ in } X \qquad \text{implies} \qquad x \cdot y \text{ in } X. \qquad (4.19)$$

Then X forms a semigroup under the multiplication $\cdot$ inherited from S. The closure property is given by definition, and the associativity of $(X, \cdot)$ is just a special case of the associativity given in the semigroup $(S, \cdot)$.

DEFINITION 4.37 (Subsemigroups.) *Let S be a semigroup, and let X be a subset of S. Then X is described as a* subsemigroup *of the semigroup S if it satisfies the closure property* (4.19).

Trivially, the empty set is a subsemigroup of every semigroup.

Example 4.38 (Subsemigroups of the integers under addition.)
The set of negative integers forms a subsemigroup of the semigroup $(\mathbb{Z}, +)$ of integers under addition. The set of odd integers does not form a subsemigroup, since the closure property is violated by examples such as $1 + 3$. ⬛

DEFINITION 4.39 (Submonoids.) *A subset X of a monoid $(M, \cdot, e)$ is said to be a* submonoid *if it is a subsemigroup of the semigroup $(M, \cdot)$, and if it contains the identity element e of M.*

If $(X, \cdot, e)$ is a submonoid of a monoid $(M, \cdot, e)$, then $(X, \cdot, e)$ is itself a monoid: The identity property (4.5) for X is just a special case of the identity property (4.5) for M. Trivially, the set $\{e\}$ consisting only of the identity element is a submonoid of any monoid $(M, \cdot, e)$ with e as its identity element. Note that $\{e\}$ is a subsemigroup by the identity property: $e \cdot e = e$.

Example 4.40 (Submonoids of the integers under addition.)
The subsemigroup of negative integers does not form a submonoid of the monoid $(\mathbb{Z}, +, 0)$ of integers under addition, since it does not contain the identity element 0 of $\mathbb{Z}$. On the other hand, the monoid $(\mathbb{N}, +, 0)$ of natural numbers under addition (compare Example 4.9) does form a submonoid of $(\mathbb{Z}, +, 0)$. ⬛

Example 4.41 (Stochastic matrices.)
A 2×2 real matrix

$$A = \begin{bmatrix} p_1 & p_2 \\ q_1 & q_2 \end{bmatrix}$$

is said to be (*row*) *stochastic* if p_1, p_2, q_1, q_2 are all nonnegative,

$$p_1 + p_2 = 1, \qquad \text{and} \qquad q_1 + q_2 = 1.$$

Note that the identity matrix I_2 is stochastic. Let Π_2^2 be the set of 2×2 stochastic matrices. Then Π_2^2 forms a submonoid of the monoid $\mathbb{R}_2^2$ of all 2×2 matrices under matrix multiplication. (Compare Exercise 24.)   ▯

DEFINITION 4.42 (Subgroups.)   *A submonoid X of a group $(G, \cdot, e)$ is said to be a* subgroup *of G if it is closed under the inversion in G:*

$$x \quad in \ X \qquad implies \qquad x^{-1} \quad in \ X. \qquad (4.20)$$

Note that the set $\{e\}$ consisting only of the identity element is a subgroup of any group $(G, \cdot, e)$ with e as its identity element. Since a subgroup has to be a submonoid, with an identity element, it has to be nonempty. There is a quick way to check if a given nonempty subset X of a group G actually forms a subgroup of G.

PROPOSITION 4.43 (The subgroup test.)
Let X be a nonempty subset of a group $(G, \cdot, e)$. Then X is a subgroup of G if and only if it satisfies the closure property

$$x, y \quad in \ X \qquad implies \qquad x \cdot y^{-1} \quad in \ X. \qquad (4.21)$$

PROOF   First, suppose that X is a subgroup of G, and that x and y are elements of X. Then by the closure (4.20) under inversion, y^{-1} lies in X. Since x and y^{-1} lie in X, the closure property (4.19) then guarantees that $x \cdot y^{-1}$ lies in X.

Conversely, suppose that the nonempty subset X of the group G satisfies the closure property (4.21). Since X is nonempty, it contains an element a. Then the closure property (4.21) shows that the identity element $e = a \cdot a^{-1}$ lies in X. Again, for each element x of X, the closure property (4.21) shows that the inverse $x^{-1} = e \cdot x^{-1}$ lies in X. Finally, for x and y in X, the closure property (4.21) shows that the product $x \cdot y = x \cdot (y^{-1})^{-1}$ lies in X, so that X forms a subsemigroup of $(G, \cdot)$.   ▯

REMARK 4.44   In an additive group $(G, +, 0)$, the closure property (4.21) reduces to closure under the *subtraction*

$$x - y = x + (-y)$$

in G. If the operation of a group G is written as multiplication, it is sometimes convenient to define $x/y = x \cdot y^{-1}$, an operation known as *right division*.   ▯

Example 4.45 (Orthogonal matrices.)
A 2×2 (or larger square) matrix A is said to be *orthogonal* if the products AA^T and $A^T A$ of A with its transposed matrix A^T are the identity matrix I.

In particular, the inverse of an orthogonal matrix A is its transpose A^T, and the identity matrix I is orthogonal. Then by Proposition 4.43, the nonempty set $O_2(\mathbb{R})$ of orthogonal 2×2 real matrices forms a subgroup of the general linear group $GL(2, \mathbb{R})$ (compare Example 4.18). Indeed, if A and B are orthogonal, the computations

$$(AB^T)(AB^T)^T = AB^T(B^T)^T A^T = AB^T BA^T = AA^T = I$$

and

$$(AB^T)^T(AB^T) = (B^T)^T A^T AB^T = BB^T = I$$

show that $AB^T = AB^{-1}$ is orthogonal. The group $O_2(\mathbb{R})$ is called the (*real*) *orthogonal group* of dimension 2. ☐

We conclude this section with a classification of the subgroups of the group of integers under addition.

THEOREM 4.46 (Subgroups of the integers.)

Let J be a subgroup of the group $(\mathbb{Z}, +, 0)$ of integers under addition. Then there is a natural number d such that J consists of the set $d\mathbb{Z}$ of integral multiples of d.

PROOF Since J is a subgroup, it contains the identity element 0. If $J = \{0\}$, then $J = 0\mathbb{Z}$, the set of multiples of 0.

Otherwise, J contains a nonzero integer n. In this case it contains a positive integer (either n or $-n$). Consider the nonempty set S of positive elements of J. By the Well-Ordering Principle, the set S has a least element d. Then each integer multiple nd of d lies in J. Indeed if n is positive, nd is the sum

$$\overbrace{d + d + \cdots + d}^{n}$$

of n copies of d, which lies in J since J is closed under addition. If n is negative, then $nd = |n|(-d)$ is the sum

$$\overbrace{(-d) + (-d) + \cdots + (-d)}^{|n|}$$

of $|n|$ copies of $-d$, which lies in J since J is closed under negation and addition. Finally, $0d = 0$ lies in J.

Now suppose that J contains an element a which is not a multiple of d. Apply the Division Algorithm to express a as $a = dq + r$ with $0 < r < d$. Then $0 < r = a - dq = a + (-dq)$ lies in J by Proposition 4.43. This contradicts the choice of d as the smallest positive element of J. ☐

4.7   Cosets

A semigroup $(G, \cdot)$ carries an associative multiplication of its elements. It is very useful to extend this multiplication to subsets of G. Let X be a subset of a semigroup $(G, \cdot)$. If g is an element of G, define

$$Xg = \{xg \mid x \text{ in } X\} \qquad \text{and} \qquad gX = \{gx \mid x \text{ in } X\}. \qquad (4.22)$$

The sets of (4.22) are known respectively as the *right* and *left cosets* of the subset X with the element g. For example, the subgroup $d\mathbb{Z}$ of the group $(\mathbb{Z}, +, 0)$ in Theorem 4.46 is the coset of d in the semigroup $(\mathbb{Z}, \cdot)$. The notation (4.22) is extended by setting XY or

$$X \cdot Y = \{x \cdot y \mid x \text{ in } X, \ y \text{ in } Y\} \qquad (4.23)$$

for subsets X and Y of a semigroup $(G, \cdot)$. In particular, $Xg = X \cdot \{g\}$ and $\{g\} \cdot X = gX$ for an element g of G.

If X is a subset of a monoid G with identity element e, then the cosets eX and Xe both coincide with the subset X. There are further relations between the various cosets in a group $(G, \cdot)$.

PROPOSITION 4.47 (Group cosets are isomorphic as sets.)
Let X be a subset of a group G. Then for elements g_1, g_2 of G, the cosets Xg_1, Xg_2, and g_1X are all isomorphic as sets.

PROOF   The maps
$$X \to Xg_1; x \mapsto xg_1$$
and
$$Xg_1 \to X; y \mapsto yg_1^{-1}$$
are mutually inverse bijections, so $X \cong Xg_1$. Isomorphism is an equivalence relation (compare Exercise 1 in Chapter 2). It follows that Xg_1 and Xg_2 are isomorphic. Similarly, the maps
$$X \to g_1X; x \mapsto g_1x$$
and
$$g_1X \to X; y \mapsto g_1^{-1}y$$
are mutually inverse bijections, so $X \cong g_1X$. The rest of the proposition follows from the fact that isomorphism is an equivalence relation.   ▯

Since two finite sets are isomorphic if and only if they have the same number of elements (compare Example 2.26, page 36), we obtain the following.

COROLLARY 4.48 (Finite cosets are all the same size.)
Let X be a finite subset of a group G. Then for elements g_1, g_2 of G, the cosets Xg_1, Xg_2, and g_1X all have the same number of elements.

Cosets of subgroups are equivalence classes.

PROPOSITION 4.49
Let H be a subgroup of a group G.

(a) *Define a relation R on G by*

$$g_1 \; R \; g_2 \qquad \text{if and only if} \qquad hg_1 = g_2 \text{ for some } h \text{ in } H.$$

Then R is an equivalence relation on G.

(b) *The equivalence classes for R are the right cosets Hg.*

PROOF (a): **Reflexivity:** For g in G, we have $eg = g$ with e in H.
Symmetry: Suppose $g_1 \; R \; g_2$, say $hg_1 = g_2$ with h in H. Then $g_1 = h^{-1}g_2$, so $g_2 \; R \; g_1$.
Transitivity: Suppose $g_1 \; R \; g_2$ and $g_2 \; R \; g_3$, say $hg_1 = g_2$ and $h'g_2 = g_3$ with h, h' in H. Then $h'hg_1 = h'g_2 = g_3$, so $g_1 \; R \; g_3$.

(b) is immediate. □

In the proof of Proposition 4.49(a), it is worth noting the parallel between the three properties required for the equivalence relation R and the three closure properties of the subgroup H:

Equivalence relation is ...	Subgroup ...
... reflexive	... contains the identity
... symmetric	... is closed under inverses
... transitive	... is closed under multiplication

The Klein 4-group V_4 is a subgroup of the 24-element group S_4 (compare Exercise 39 in Chapter 2). In general, the number of elements of a subgroup of a finite group is always a divisor of the total number of elements in the group.

THEOREM 4.50 (Lagrange's Theorem.)
Let H be a subgroup of a finite group G. Then the number $|H|$ of elements of H divides the number $|G|$ of elements of G.

PROOF   By Propositions 4.49 and 3.5, two distinct right cosets of H are disjoint. Suppose that there are j right cosets altogether. By Corollary 4.48, each right coset has $|H|$ elements. Then

$$|G| = j|H|, \qquad\qquad (4.24)$$

so $|H|$ divides $|G|$.   □

The number $j = |G|/|H|$ in (4.24) is called the *index* of H in the group G. More generally, if G is an infinite group with a subgroup H, the *index* of H is the (possibly infinite) number of right cosets of H in G. (Compare Exercise 38.)

Lagrange's Theorem is useful for limiting the possible subgroups of a given finite group. For example, it shows that in the 24-element group S_4, a subgroup cannot be formed from the 9-element set consisting of the identity and the 8 permutations (2.33). In other words, without any calculation required, it shows that this 9-element set cannot be closed under multiplication.

In any group G with identity element e, the subgroup G is described as *improper*, while the smallest subgroup $\{e\}$ is described as *trivial*. A subgroup H is *proper* if it is not improper. Since prime numbers are irreducible (Proposition 1.11), Lagrange's Theorem yields the following result.

PROPOSITION 4.51 (Groups of prime order.)
A group with a prime number of elements can have no proper, nontrivial subgroups.

4.8   Multiplication tables

There are various ways to compute the product $x \cdot y$ of two elements x and y in a group G. If G is a group of matrices, we may use matrix multiplication. If G is a group of permutations, we may use function composition. A general method uses a table, the *multiplication table* of the group.

Consider the Klein 4-group V_4 of Example 2.31 (page 38). Writing the elements as $(0) = e$, $(0\ 1)(2\ 3) = a$, $(0\ 2)(1\ 3) = b$, and $(0\ 3)(1\ 2) = c$, the multiplication table appears as displayed in Figure 4.2. The table consists of four parts, separated by the lines. The top left-hand corner may contain the name of the group or its multiplication. The top right part consists of the *column labels*. The bottom left-hand part consists of the *row labels*. The bottom right-hand part is called the *body*. Note that in Figure 4.2, the group elements are presented in the same order as column and row labels, with the identity element appearing first. This is not necessary for a table to perform its function of specifying the group products, but it is a convention that is

V_4	e	a	b	c
e	e	a	b	c
a	a	e	c	b
b	b	c	e	a
c	c	b	a	e

FIGURE 4.2: Multiplication table of V_4.

usually followed. Putting the identity element first as a column and row label means that the first row and the first column of the body just repeat the respective column and row labels in order. For this reason, the body alone may be used to specify a group, the column labels being taken from the first row of the body, and the row labels from the first column of the body.

The body of the table in Figure 4.2 has a particular feature: each row and each column of the body contains each element of the group exactly once.

DEFINITION 4.52 (Latin squares.) *For a natural number n, let Q be a set with n elements. Then an $n \times n$ square containing each of the n elements of Q exactly once in each row and each column is called a* Latin square *(on the set Q).*

Theorem 4.55 below shows that the Latin square property of the body of the multiplication table of the Klein 4-group is actually typical of all finite groups. The theorem is preceded by a pair of results holding in general (not necessarily finite) groups.

PROPOSITION 4.53 (Cancellation in groups.)
Let G be a group, with elements x, y_1, y_2.

(a) *If*
$$x \cdot y_1 = x \cdot y_2 , \tag{4.25}$$
then $y_1 = y_2$.

(b) *If*
$$y_1 \cdot x = y_2 \cdot x ,$$
then $y_1 = y_2$.

PROOF (a): Multiplying both sides of the equation (4.25) on the left by

x^{-1}, we obtain

$$y_1 = e \cdot y_1 = x^{-1} \cdot x \cdot y_1 = x^{-1} \cdot x \cdot y_2 = e \cdot y_2 = y_2\,.$$

(b) is proved similarly (Exercise 41). □

COROLLARY 4.54 (Existence and uniqueness of solutions.)
Consider the equation

$$x \cdot y = z \tag{4.26}$$

in a group $(G, \cdot)$. If the equation (4.26) holds, knowledge of any two of the elements x, y, z specifies the third uniquely.

PROOF If x and y are given, then z is specified uniquely by the multiplication in G. If x and z are given, then a solution y to the equation (4.26) exists, namely $y = x^{-1} \cdot z$. Indeed:

$$x \cdot y = x \cdot (x^{-1} \cdot z) = (x \cdot x^{-1}) \cdot z = e \cdot z = z\,.$$

The solution is unique, by Proposition 4.53(a). The existence and uniqueness of a solution x to (4.26) given y and z follow similarly (Exercise 42). □

THEOREM 4.55 (Group tables are Latin squares.)
Let G be a finite group. Then the body of the multiplication table of G forms a Latin square on the set G.

PROOF Consider the row of the table labeled by an element x of G. Let z be an element of G. Now z appears in the row labeled x, namely in the column labeled y, if and only if the equation (4.26) holds. By Corollary 4.54, a solution y to this equation exists, so z indeed appears in the row labeled x. Moreover, the solution y is unique, so in the row labeled x, the element z only appears in the column labeled y, and not in any other column.

A similar argument shows that each element appears exactly once in each column of the body of the multiplication table (Exercise 43). □

Note that in Figure 3.3 (page 59), the body of the group table on the left is a Latin square, while the body of the monoid table on the right is not.

4.9 Exercises

1. Let S be the union $(-\infty, -1) \cup (1, \infty)$ of the real intervals $(-\infty, -1)$ and $(1, \infty)$. In other words, S is the set of real numbers x with $|x| > 1$. Show that S forms a semigroup under the usual multiplication of real numbers.

2. Verify the claims of Example 4.5.

3. Pick three integers l, m, and n at random (for example nonzero digits from the number of your telephone). See if there is a difference between $l - (m - n)$ and $(l - m) - n$.

4. (a) Show that the set of positive integers forms a semigroup under addition.

   (b) Show that for each positive integer e, the inequality $e + x > x$ holds for all positive integers x.

   (c) Conclude that the set of positive integers does not form a monoid under addition.

5. Verify the claims of Example 4.10.

6. Verify the claims of Example 4.11.

7. Show that the set $\mathcal{P}(\mathbf{n})$ (compare Example 4.11) forms a monoid under the operation $\cup$ of set union:

$$X \cup Y = \{x \mid x \text{ in } X \text{ or } x \text{ in } Y\}.$$

What is the identity element of this monoid?

8. Let $\mathcal{P}_{\mathbf{fin}}(\mathbb{N})$ denote the set of finite subsets of $\mathbb{N}$.

   (a) Show that $\mathcal{P}_{\mathbf{fin}}(\mathbb{N})$ forms a monoid under set union.

   (b) Show that $\mathcal{P}_{\mathbf{fin}}(\mathbb{N})$ forms a semigroup under set intersection.

   (c) Show that $\mathcal{P}_{\mathbf{fin}}(\mathbb{N})$ does not form a monoid under set intersection.

9. Exhibit two real 2×2 matrices X and Y such that $XY \neq YX$.

10. Consider the set $L(2, \mathbb{R})$ of linear functions from $\mathbb{R}_2^1$ to itself. (Compare Exercise 2.9 in Chapter 2.) Show that $L(2, \mathbb{R})$ forms a monoid under the *addition* defined componentwise by

$$(L_A + L_B)(\mathbf{x}) = L_A(\mathbf{x}) + L_B(\mathbf{x})$$

for $\mathbf{x}$ in $\mathbb{R}_2^1$ and A, B in $\mathbb{R}_2^2$.

11. Prove Proposition 4.15: Rewrite the proof of Proposition 2.23 in abstract terms.

12. Show that $\left(x^{-1}\right)^{-1} = x$ for each element of a group G.

13. Let X be a set. What is the group of units of the monoid X^X of functions?

14. What is the group of units of the monoid $(\mathbb{N}, +, 0)$ of Example 4.9?

15. Determine the group of units of the monoid A of affine functions (see Exercise 13 in Chapter 2).

16. Let u_1, u_2, $\ldots$, u_{r-1}, u_r be elements of a group G. Show that

$$\left(u_1 u_2 \ldots u_{r-1} u_r\right)^{-1} = u_r^{-1} u_{r-1}^{-1} \ldots u_2^{-1} u_1^{-1}.$$

17. Let a, b, and c be elements of a group $(G, \cdot, e)$, with $abc = e$. Give a careful proof that $cab = e$.

18. Prove Proposition 4.25:

    (a) Show that $X \times Y$ is closed under componentwise multiplication.

    (b) Show that the componentwise multiplication is associative.

19. Prove Propositions 4.28 and 4.30:

    (a) Show that

    $$(e_X, e_Y) \circ_{X \times Y} (x, y) = (x, y) = (x, y) \circ_{X \times Y} (e_X, e_Y)$$

    for each element (x, y) of $X \times Y$.

    (b) Show that

    $$(x^{-1}, y^{-1}) \circ_{X \times Y} (x, y) = (e_X, e_Y) = (x, y) \circ_{X \times Y} (x^{-1}, y^{-1})$$

    for each element (x, y) of $X \times Y$.

20. In the group $\mathbb{R} \times \mathbb{R}$ of Example 4.32, describe the identity element and inverses.

21. Consider the set $\mathbb{R}_2^2$ of real 2×2 matrices.

    (a) Determine inverses, and the identity element, in the group given by matrix addition.

    (b) Determine the group of units within the monoid of matrices under Hadamard multiplication.

22. In the respective parts (a), (b), and (c) of Definition 4.34, verify that S^X forms a semigroup, monoid, and group under the componentwise multiplication.

23. Let X be a set, and let S be a commutative semigroup. Show that the power S^X is commutative.

24. Show that the set Π_2^2 of all 2×2 row-stochastic matrices (as defined in Example 4.41) forms a submonoid of the monoid R_2^2 of all real 2×2 matrices under matrix multiplication.

25. A square matrix is said to be *column-stochastic* if its transpose is row-stochastic. Show that the set of 2×2 column-stochastic matrices forms a monoid under matrix multiplication.

26. A square matrix is said to be *doubly stochastic* if it is both row- and column-stochastic.

   (a) Give an example of a nonidentity 2×2 doubly stochastic matrix.

   (b) Show that the set of all 2×2 doubly stochastic matrices forms a monoid under matrix multiplication.

   (c) Determine the group of units within the monoid of 2×2 doubly stochastic matrices.

27. A 2×2 matrix

$$\begin{bmatrix} a & b \\ c & d \end{bmatrix}$$

   is said to be *upper triangular* if $c = 0$.

   (a) Show that the set of upper triangular real 2×2 matrices forms a monoid N under matrix multiplication.

   (b) Determine the group of units N^* of the monoid N.

28. Consider the set

$$K = \left\{ \begin{bmatrix} k & k \\ k & k \end{bmatrix} \ \middle| \ k \text{ real, nonzero} \right\}$$

   of all "constant" nonzero 2×2 real matrices.

   (a) Show that K forms a subsemigroup of the semigroup of all 2×2 real matrices under matrix multiplication.

   (b) Show that K does not form a submonoid of the monoid of all 2×2 real matrices under matrix multiplication.

   (c) Show that K forms a group under matrix multiplication.

29. Let H and K be subgroups of a group G. Show that the intersection $H \cap K$ of H and K (the set of elements of G common to both H and K) is also a subgroup of G.

30. Let G be a group. For each element i of an "index set" I, let H_i be a subgroup of G. Show that the intersection

$$\bigcap_{i \text{ in } I} H_i = \{g \mid g \text{ in } H_i \text{ for all } i \text{ in } I\}$$

of the subgroups H_i is also a subgroup of G.

31. Let $C(\mathbb{R})$ be the set of all continuous real-valued functions $f : \mathbb{R} \to \mathbb{R}$. Show that $C(\mathbb{R})$ forms a subgroup of the group $\mathbb{R}^{\mathbb{R}}$ of all real-valued functions with componentwise structure.

32. Let r be a positive integer. Show that the set $C^r(\mathbb{R})$ of all real-valued functions $f : \mathbb{R} \to \mathbb{R}$ with a continuous r-th derivative $f^{(r)}$ forms a subgroup of the group $\mathbb{R}^{\mathbb{R}}$ of all real-valued functions with componentwise structure.

33. For a real number θ (an angle in radians), show that the matrix

$$\begin{bmatrix} \cos\theta & -\sin\theta \\ \sin\theta & \cos\theta \end{bmatrix}$$

is orthogonal.

34. For real numbers θ and φ, show that the addition formulas for the cosine and sine are given by equating respective components on both sides of the product equation

$$\begin{bmatrix} \cos\theta & -\sin\theta \\ \sin\theta & \cos\theta \end{bmatrix} \cdot \begin{bmatrix} \cos\varphi & -\sin\varphi \\ \sin\varphi & \cos\varphi \end{bmatrix} = \begin{bmatrix} \cos(\theta+\varphi) & -\sin(\theta+\varphi) \\ \sin(\theta+\varphi) & \cos(\theta+\varphi) \end{bmatrix}$$

in the orthogonal group.

35. (a) Show that each subgroup of the group of integers under addition is a subsemigroup of the semigroup of integers under multiplication.

    (b) Exhibit an example of a subgroup of the group of real numbers under addition which is not a subsemigroup of the semigroup of real numbers under multiplication.

36. Let H be a subgroup of a group G. Show that two right cosets of H are either equal or disjoint.

37. Let H be a subgroup of a finite group G. Show that the number of left cosets of H in G is equal to the number of right cosets of H in G.

38. Show that the group of even integers has index 2 in the group $(\mathbb{Z}, +, 0)$ of integers under addition.

39. Let H and K be subgroups of a group G. Define a relation R on G by $g_1 \ R \ g_2$ if and only if $hg_1k = g_2$ for some h in H and k in K. Show that R is an equivalence relation on G. (The equivalence classes HgK are known as *double cosets*.)

40. Let H be a subgroup of a group G. Let x and y be elements of G. Show that the cosets Hx and Hy are equal if and only if xy^{-1} is an element of H.

41. Prove Proposition 4.53(b).

42. Let y and z be given elements of a group G. Show that there is a unique solution x in G to the equation $x \cdot y = z$.

43. Complete the proof of Theorem 4.55 by showing that each group element appears exactly once in each column of the body of the multiplication table.

4.10 Study projects

1. **Nim sums.** The game of *Nim* (see Figure 4.3) is played with a small collection of counters. The counters are arranged into several groups or *heaps*. There are two players, who take turns to play. When it is your

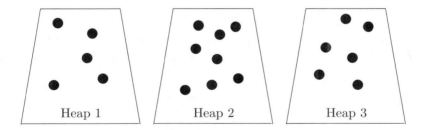

FIGURE 4.3: A Nim position with three heaps.

turn, you are allowed to remove counters from only one heap. You must remove at least one counter. The player removing the last counter is the winner.

How can you win at Nim? The game is analyzed using bit strings and the power group $(\mathbb{Z}/2, +)^n$ introduced in Example 4.35. Consider the position displayed in Figure 4.3. In the first heap, there are 5 counters, 101 or 0101 in binary. In the other two heaps, there are $8 = 1000$ and $6 = 110$ or 0110 counters respectively. Zeros are added as needed to make the lengths of all the strings equal.

Now the *Nim sum* $b +_2 c$ of two natural numbers b and c is defined by identifying natural numbers with their binary representations, and taking the sum of these bit strings in $(\mathbb{Z}/2, +)^n$ (for any positive integer n with $2^n > \max(b, c)$).

(a) Show that the Nim sum $b +_2 c$ of two natural numbers b and c is independent of the number of zeros appended in front of their binary representations.

(b) Show that the set of natural numbers forms a group $(\mathbb{N}, +_2, 0)$ under Nim sum.

(c) Compute the Nim sum $5 +_2 8 +_2 6$.

(d) Show that you will not lose in Nim if after each move, you leave the Nim sum of the sizes of the heaps at 0.

(e) Show that the unique winning move from the position in Figure 4.3 is to remove 5 counters from Heap 2.

2. **The orthogonal group.** Consider a 2×2 orthogonal matrix

$$A = \begin{bmatrix} a & b \\ c & d \end{bmatrix}.$$

(a) Show that the orthogonality condition $A^T A = I$ reduces to the three equations

$$a^2 + c^2 = 1 \tag{4.27}$$
$$b^2 + d^2 = 1 \tag{4.28}$$
$$ab + cd = 0. \tag{4.29}$$

(b) Show that the solutions (a, b, c, d) to the simultaneous equations (4.27) through (4.29) are of two types, $(\cos\theta, -\sin\theta, \sin\theta, \cos\theta)$ as in the picture on the left, or $(\cos\theta, \sin\theta, \sin\theta, -\cos\theta)$ as in the picture on the right:

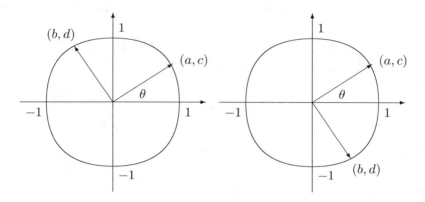

(c) Show that the set

$$SO_2(\mathbb{R}) = \left\{ \begin{bmatrix} \cos\theta & -\sin\theta \\ \sin\theta & \cos\theta \end{bmatrix} \;\middle|\; 0 \leq \theta < 2\pi \right\}$$

forms a subgroup of the orthogonal group. (The group $SO_2(\mathbb{R})$ is called the *special orthogonal group* of dimension 2.)

(d) Show that the orthogonal group is the union of $SO_2(\mathbb{R})$ and the coset

$$SO_2(\mathbb{R}) \begin{bmatrix} 1 & 0 \\ 0 & -1 \end{bmatrix}.$$

3. **Dihedral groups.** Consider an integer $n > 2$.

(a) Show that the set

$$C_n = \left\{ \begin{bmatrix} \cos(2r\pi/n) & -\sin(2r\pi/n) \\ \sin(2r\pi/n) & \cos(2r\pi/n) \end{bmatrix} \;\middle|\; 0 \leq r < n \right\}$$

forms a commutative subgroup of the special orthogonal group.

(b) Show that the union

$$D_n = C_n \cup C_n \begin{bmatrix} 1 & 0 \\ 0 & -1 \end{bmatrix}$$

forms a noncommutative subgroup of the orthogonal group. (The group D_n is called the *dihedral group* of degree n.)

(c) Consider the set

$$P_n = \left\{ \begin{bmatrix} \cos(2s\pi/n) \\ \sin(2s\pi/n) \end{bmatrix} \;\middle|\; 0 \leq s < n \right\}$$

of column vectors from $\mathbb{R}_2^1$. Using a column vector $\begin{bmatrix} x_1 \\ x_2 \end{bmatrix}$ to specify the point (x_1, x_2) of the plane $\mathbb{R}^2$, show that the set P_n specifies the points of a regular n-gon in the plane.

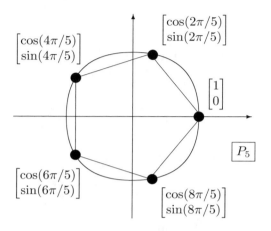

(d) For a matrix A in the dihedral group D_n, and for a column vector **p** in P_n (a vertex of the n-gon), show that the matrix product

$$L_A(\mathbf{p}) = A\mathbf{p}$$

is again a vertex of the n-gon. Conclude that the dihedral group D_n forms the full group of symmetries of the regular n-gon P_n.

4.11   Notes

Section 4.3

N.H. Abel was a Norwegian mathematician who lived from 1802 to 1829.

Section 4.4

R. Descartes was a French mathematician and philosopher who lived from 1596 to 1650. J.S. Hadamard was a French mathematician who lived from 1865 to 1963.

Section 4.7

J.L. Lagrange was a French mathematician who lived from 1736 to 1813.

Chapter 5

HOMOMORPHISMS

A study of sets inevitably leads to a study of functions between sets. Similarly, a study of algebraic structures such as semigroups, monoids, or groups entails a study of the functions that preserve the algebraic structure. These functions are known as *homomorphisms* (literally "same shape").

5.1   Homomorphisms

Consider the exponential function $\exp : \mathbb{R} \to \mathbb{R}; x \mapsto e^x$. By the law of exponents,

$$\exp(x + y) = e^{x+y} = e^x \cdot e^y = \exp(x) \cdot \exp(y). \qquad (5.1)$$

Now the domain of the exponential function is the semigroup $(\mathbb{R}, +)$ of real numbers under addition. The codomain of the exponential function is the semigroup $(\mathbb{R}, \cdot)$ of real numbers under multiplication. The equation (5.1) says that we may either add two real numbers x and y in the domain, and then map across to $\exp(x + y)$ in the codomain, or else map x and y across individually to $\exp(x)$, $\exp(y)$ in the codomain, and then multiply these two numbers in the codomain. Either way, we get the same answer.

DEFINITION 5.1 (Homomorphisms and isomorphisms of semigroups, monoids, and groups.)

(a) *Let $\theta : (X, \circ) \to (Y, *)$ be a function from a semigroup $(X, \circ)$ to a semigroup $(Y, *)$. Then θ is said to be a* semigroup homomorphism *if*

$$\theta(x_1 \circ x_2) = \theta(x_1) * \theta(x_2)$$

for all x_1, x_2 in X.

(b) *Let $\theta : (X, \circ, e) \to (Y, *, f)$ be a function from a monoid $(X, \circ, e)$ to a monoid $(Y, *, f)$. Then θ is said to be a* monoid homomorphism *if it is a semigroup homomorphism $\theta : (X, \circ,) \to (Y, *)$ with $\theta(e) = f$.*

(c) *Let $\theta : (X, \circ, e) \to (Y, *, f)$ be a function from a group $(X, \circ, e)$ to a group $(Y, *, f)$. Then θ is said to be a* group homomorphism *if it is a* *monoid homomorphism* $\theta : (X, \circ, e) \to (Y, *, f)$ *with* $\theta(x^{-1}) = (\theta(x))^{-1}$ *for all x in X.*

(d) *Bijective semigroup, monoid, and group homomorphisms are described respectively as* semigroup, monoid, *and* group isomorphisms.

The relationship of isomorphism between semigroups, monoids, or groups X and Y is often denoted by

$$X \cong Y$$

— compare (2.25). The context should make clear what kind of isomorphism is presented: of sets, semigroups, monoids, or groups.

Example 5.2 (The exponential function.)

The law of exponents (5.1) shows that $\exp : (\mathbb{R}, +) \to (\mathbb{R}, \cdot)$ is a semigroup homomorphism from the semigroup of real numbers under addition to the semigroup of real numbers under multiplication. Furthermore, the equation

$$\exp(0) = 1$$

shows that $\exp : (\mathbb{R}, +, 0) \to (\mathbb{R}, \cdot, 1)$ is a monoid homomorphism from the monoid of real numbers under addition to the monoid of real numbers under multiplication. ⬜

Example 5.3 (Inclusion of a subgroup.)

Let H be a subgroup of a group G. Then the inclusion function

$$j : H \hookrightarrow G; h \mapsto h$$

is a group homomorphism. ⬜

Example 5.4 (Projections.)

Given sets X and Y, define the respective *projections*

$$\pi_1 : X \times Y \to X; (x, y) \mapsto x$$

and

$$\pi_2 : X \times Y \to Y; (x, y) \mapsto y$$

to the first and second factors. If X and Y are semigroups, monoids, or groups, then the projections are homomorphisms of semigroups, monoids, and groups respectively. ⬜

PROPOSITION 5.5 (Semigroup homomorphisms between groups.)

Let $\theta : (X, \circ) \to (Y, *)$ be a semigroup homomorphism between two groups $(X, \circ, e)$ and $(Y, *, f)$. Then θ is a group homomorphism.

PROOF Since θ is a semigroup homomorphism, the equation

$$\theta(e) * \theta(e) = \theta(e \circ e) = \theta(e)$$

holds in Y. However, since f is the identity element of Y, and $\theta(e)$ is an element of Y, the identity property (4.5) in Y gives

$$\theta(e) * f = \theta(e).$$

It follows that $\theta(e) * \theta(e) = \theta(e) * f$, so $\theta(e) = f$ by Corollary 4.54, and θ is a monoid homomorphism.

Now for each element x of X, we have

$$\theta(x) * \theta(x^{-1}) = \theta(x \circ x^{-1}) = \theta(e) = f.$$

But $\theta(x) * (\theta(x))^{-1} = f$, so Corollary 4.54 again gives $\theta(x^{-1}) = (\theta(x))^{-1}$, making θ a group homomorphism. ⬚

In contrast with Proposition 5.5, a semigroup homomorphism between monoids need not be a monoid homomorphism (compare Exercise 1).

THEOREM 5.6 (Monoid homomorphisms and groups of units.)
Let $\theta : (M, \circ, e) \to (N, *, f)$ be a monoid homomorphism. Then θ restricts to a group homomorphism $\theta^* : M^* \to N^*$ between the corresponding groups of units.

PROOF Suppose that u lies in M^*, with $u \circ v = e = v \circ u$ for some v in M. Then

$$\theta(u) * \theta(v) = \theta(u \circ v) = \theta(e) = f = \theta(v) * \theta(u),$$

so that $\theta(u)$ lies in N^*. The restriction

$$\theta^* : M^* \to N^*; u \mapsto \theta(u)$$

is a semigroup homomorphism between the respective groups of units. By Proposition 5.5, it is then a group homomorphism. ⬚

Example 5.7 (Determinants.)
The determinant function

$$\det : \mathbb{R}^2_2 \to \mathbb{R}; \begin{bmatrix} a & b \\ c & d \end{bmatrix} \mapsto ad - bc \tag{5.2}$$

is a monoid homomorphism from the monoid of 2×2 real matrices under multiplication to the monoid of real numbers under multiplication (compare Exercise 7). It restricts to a group homomorphism from $GL(2, \mathbb{R})$ to the group of nonzero real numbers under multiplication. ⬚

A function $f : X \to Y$ between sets is fully described by its *graph*, the subset

$$\{(x, f(x)) \mid x \text{ in } X\} \tag{5.3}$$

of $X \times Y$. Homomorphisms may then be recognized by their graphs.

PROPOSITION 5.8 (The graph of a homomorphism.)

Let $(X, \circ)$ and $(Y, *)$ be semigroups. Then a function $f : X \to Y$ is a semigroup homomorphism if and only if the graph (5.3) is a subsemigroup of the direct product semigroup $X \times Y$.

PROOF If f is a semigroup homomorphism, and x_1, x_2 are elements of X, then

$$(x_1, f(x_1))(x_2, f(x_2)) = (x_1 \circ x_2, f(x_1) * f(x_2)) = (x_1 \circ x_2, f(x_1 \circ x_2)),$$

so the graph is closed under multiplication. Conversely, suppose the graph is closed under multiplication. Then for elements x_1 and x_2 of X, the graph contains both $(x_1 \circ x_2, f(x_1 \circ x_2))$ and

$$(x_1, f(x_1))(x_2, f(x_2)) = (x_1 \circ x_2, f(x_1) * f(x_2)).$$

However, since f is a function, there is a unique element $(x, f(x))$ of the graph for each element x of X. By this uniqueness for the element $x_1 \circ x_2$ of X, we have $f(x_1 \circ x_2) = f(x_1) * f(x_2)$, so that f is a semigroup homomorphism. $\square$

COROLLARY 5.9

Let $(X, \circ, e)$ and $(Y, *, f)$ be monoids. Then a function $f : X \to Y$ is a monoid homomorphism if and only if the graph (5.3) is a submonoid of the direct product monoid $X \times Y$.

5.2   Normal subgroups

Let $f : X \to Y$ be a function. The image $f(X) = \{f(x) \mid x \text{ in } X\}$ is a subset of the codomain Y. If f is a homomorphism of semigroups, monoids, or groups, the image will carry the corresponding algebra structure.

PROPOSITION 5.10 (Images of homomorphisms.)

Let $f : (X, \cdot) \to (Y, \cdot)$ be a semigroup homomorphism.

(a) *The image $f(X)$ is a subsemigroup of Y.*

(b) *If $f : (X, \cdot, e_X) \to (Y, \cdot, e_Y)$ is a monoid homomorphism, then $f(X)$ is a submonoid of Y.*

(c) *If $f : (X, \cdot, e_X) \to (Y, \cdot, e_Y)$ is a group homomorphism, then $f(X)$ is a subgroup of Y.*

PROOF (a): For elements x and x' of X, we have $f(x)f(x') = f(xx')$, showing that $f(X)$ is closed under multiplication.

(b): In this case $e_Y = f(e_X)$ lies in $f(X)$.

(c): For an element x of X, we have $f(x)^{-1} = f(x^{-1})$, so that $f(X)$ is closed under inversion. $\square$

Now consider a group homomorphism $f : X \to Y$ from a group $(X, \cdot, e_X)$ to a group $(Y, \cdot, e_Y)$. As a function $f : X \to Y$ from the domain set X to the codomain set Y, the homomorphism $f : X \to Y$ specifies a kernel relation **ker** f on X, with

$$x \ \mathbf{ker} \ f \ x' \qquad \text{if and only if} \qquad f(x) = f(x')$$

(3.3). The equivalence class $[e_X]_{\mathbf{ker} \, f}$ of the identity element e_X of X is the inverse image

$$f^{-1}\{f(e_X)\}$$

— compare (3.5) and Exercise 3 of Chapter 3. Since $f : X \to Y$ is a group homomorphism, this equivalence class may be expressed in the form

$$[e_X]_{\mathbf{ker} \, f} = f^{-1}\{e_Y\} \tag{5.4}$$

as the inverse image of the identity element e_Y of the codomain group Y.

PROPOSITION 5.11 (Kernel class of the identity.)
Let $f : (X, \cdot, e_X) \to (Y, \cdot, e_Y)$ be a group homomorphism.

(a) *The equivalence class (5.4) forms a subgroup N of X.*

(b) *For all x in X and n in N,*

$$xnx^{-1} \qquad \text{lies in} \qquad N. \tag{5.5}$$

PROOF (a): First note that N is nonempty, since it contains the element e_X. Then for elements n and n' of N, the homomorphic properties of f give

$$f(n'n^{-1}) = f(n')f(n^{-1}) = f(n')f(n)^{-1} = e_Y e_Y^{-1} = e_Y,$$

so that N is a subgroup of X by Proposition 4.43.

(b): The homomorphic properties of f give

$$f(xnx^{-1}) = f(x)f(n)f(x^{-1}) = f(x)e_Y f(x)^{-1} = f(x)f(x)^{-1} = e_Y \, ,$$

so that xnx^{-1} lies in N as required. ⬚

DEFINITION 5.12 (Normal subgroups, group kernels.)   *Let X be a group.*

(a) *A subgroup N of X satisfying the additional closure property* (5.5) *is called a* normal subgroup *of X.*

(b) *For a group homomorphism $f : X \to Y$ with domain X, the normal subgroup $f^{-1}\{e_Y\}$ of X is called the* (group) kernel Ker f *of f.*

Note the distinction between the kernel relation **ker** f (lower case "k") and the kernel subgroup Ker f (upper case "K").

Example 5.13 (The Klein 4-group.)
The Klein 4-group V_4 is a normal subgroup of the symmetric group S_4. Indeed, the nonidentity elements of V_4 consist of all 3 possible products $\alpha \circ \alpha'$ with disjoint 2-cycles α and α'. Then for any permutation β in S_4, we have

$$\beta \circ (\alpha \circ \alpha') \circ \beta^{-1} = (\beta \circ \alpha \circ \beta^{-1}) \circ (\beta \circ \alpha' \circ \beta^{-1}). \tag{5.6}$$

By Exercise 38 in Chapter 2, the two factors on the right-hand side of (5.6) are again disjoint 2-cycles. Thus V_4 satisfies the additional closure property (5.5). On the other hand, $\{(0), (0\ 1)\}$ is a subgroup of S_4 which is not normal (compare Exercise 14). ⬚

The easy proof of the following result is left as Exercise 15.

PROPOSITION 5.14 (Normal subgroups of abelian groups.)
In an abelian group G, every subgroup is normal.

Consider a group homomorphism $f : (G, \cdot, e_X) \to (Y, \cdot, e_Y)$. According to Definition 5.12(b), the equivalence class $[e_X]_{\mathbf{ker}\,f}$ of the identity element e_X of X under the kernel relation **ker** f is the group kernel Ker f. More generally, each equivalence class under the kernel relation **ker** f is a coset of the group kernel Ker f.

PROPOSITION 5.15 (Kernel classes are cosets.)
Let $f : X \to Y$ be a group homomorphism, with kernel relation **ker** f *and group kernel $N = \text{Ker}\,f$. Let x be an element of X. Then the equivalence class $[x]_{\mathbf{ker}\,f}$ under the kernel relation* **ker** f *is the coset Nx.*

PROOF As usual, in proving the equality of the two sets $[x]_{\mathbf{ker}\, f}$ and Nx, we will show that each is contained in the other.

First, consider an element y of the equivalence class $[x]_{\mathbf{ker}\, f}$, so $f(x) = f(y)$. Then by the homomorphic property of f,

$$f(yx^{-1}) = f(y)f(x)^{-1} = e_Y \,,$$

so yx^{-1} is some member n of $N = f^{-1}\{e_Y\}$. As $yx^{-1} = n$, we obtain y as the member nx of the coset Nx.

Conversely, consider a member nx of the coset Nx, with n in N. Then

$$f(nx) = f(n)f(x) = e_Y f(x) = f(x) \,,$$

whence $nx\ \mathbf{ker}\, f\ x$, and nx lies in $[x]_{\mathbf{ker}\, f}$ by the symmetry of $\mathbf{ker}\, f$. ☐

Proposition 5.15 allows us to recognize the set $X_{\mathbf{ker}\, f}$ of equivalence classes as the set

$$X/N = \{Nx \mid x\ \text{in}\ \ X\}$$

of right cosets of the normal subgroup $N = \text{Ker}\, f$, and to recognize the surjection

$$s : X \to X_{\mathbf{ker}\, f}; x \mapsto [x]_{\mathbf{ker}\, f}$$

(3.12) from the First Isomorphism Theorem as the map

$$s : X \to X/N; x \mapsto Nx \,.$$

In the next section, it will be shown that each normal subgroup N of a group X yields a group structure on its set X/N of cosets.

5.3   Quotients

For subsets A and B of a group $(X, \cdot, e_X)$, consider the multiplication

$$A \cdot B = \{ab \mid a\ \text{in}\ \ A,\ b\ \text{in}\ \ B\} \tag{5.7}$$

— compare (4.23).

PROPOSITION 5.16 (Recognizing subgroups.)
Let X be a group.

(a) *The multiplication (5.7) is associative.*

(b) *A nonempty subset H of X is a subgroup if and only if $H \cdot H = H$ and $H^{-1} = H$.*

PROOF   (a): Let A, B, and C be subsets of X. Then

$$A \cdot (B \cdot C) = \{a(bc) \mid a \text{ in } A, b \text{ in } B, c \text{ in } C\}$$
$$= \{(ab)c \mid a \text{ in } A, b \text{ in } B, c \text{ in } C\} = (A \cdot B) \cdot C.$$

(b): If nonempty H satisfies the two equalities, then x and y in H imply xy^{-1}
lies in $H \cdot H^{-1} = H \cdot H = H$, so H is a subgroup by Proposition 4.43.

Now suppose that H is a subgroup. Then $H \cdot H \subseteq H$ by the closure under
multiplication. Conversely, each element h of H can be written as eh in $H \cdot H$.
Also $H^{-1} \subseteq H$ since H is closed under inversion. On the other hand, each
element h of H can be written as the element $(h^{-1})^{-1}$ of H^{-1}.   □

PROPOSITION 5.17 (Cosets of normal subgroups.)
Let N be a normal subgroup of a group X. Then the set

$$X/N = \{Nx \mid x \text{ in } X\}$$

of right cosets is a group $(X/N, \cdot, N)$ under the multiplication (5.7), with

$$(Nx)^{-1} = Nx^{-1} \tag{5.8}$$

for x in X.

PROOF   First note that $N = xNx^{-1}$ for any element x of X. Certainly

$$xNx^{-1} \subseteq N$$

by the closure property (5.5) of N. Conversely, if n lies in N, then so does
$n' = xnx^{-1}$. Thus $n = x^{-1}(xnx^{-1})x = x^{-1}n'x = x^{-1}n'(x^{-1})^{-1}$ lies in
$x^{-1}Nx$ by the closure property (5.5).

Now we have

$$Nx \cdot Ny = N \cdot xNx^{-1} \cdot xy = N \cdot Nxy = Nxy, \tag{5.9}$$

so X/N is closed under the associative multiplication of cosets. For any x in
X, we have $N \cdot Nx = (N \cdot N)x = Nx$ and

$$Nx \cdot N = N \cdot xNx^{-1} \cdot x = N \cdot Nx = Nx,$$

so that N is an identity element. Finally

$$Nx \cdot Nx^{-1} = N \cdot xNx^{-1} = N \cdot N = N$$

and

$$Nx^{-1} \cdot Nx = N \cdot (x^{-1})N(x^{-1})^{-1} = N \cdot N = N,$$

so that Nx^{-1} is the inverse of the coset Nx.   □

By Proposition 5.11, group kernels are normal subgroups. The converse is now seen to be true: normal subgroups are group kernels.

COROLLARY 5.18
Let N be a normal subgroup of a group X. Then there is a homomorphism

$$X \to X/N; x \mapsto Nx$$

with group kernel N.

PROOF   By (5.9), the map $x \mapsto Nx$ is a group homomorphism. Its group kernel is $\{x \mid Nx = N\} = N$. $\Box$

DEFINITION 5.19 (Quotient groups.)   *Let N be a normal subgroup of a group X. Then the group*

$$(X/N, \cdot, N)$$

of Proposition 5.17 is called the quotient of X by the normal subgroup N.

Example 5.20 (Modular arithmetic.)
Let d be a positive integer. In the group $(\mathbb{Z}, +, 0)$ of integers under addition, the subgroup $d\mathbb{Z}$ of multiples of d is normal. The quotient group $\mathbb{Z}/d\mathbb{Z}$ is the set $\mathbb{Z}_{\mathbf{mod}\, d}$, with the addition

$$(d\mathbb{Z} + a) + (d\mathbb{Z} + b) = d\mathbb{Z} + (a + b)$$

given in (3.21). Inverses are given by the negation

$$-(d\mathbb{Z} + a) = d\mathbb{Z} - a,$$

while the identity element is the subgroup $d\mathbb{Z}$.

In fact, the set $\mathbb{Z}/d\mathbb{Z}$ carries more structure, the multiplication

$$(d\mathbb{Z} + a) \cdot (d\mathbb{Z} + b) = d\mathbb{Z} + (a \cdot b)$$

given in (3.22). Under this multiplication, the set $\mathbb{Z}/d\mathbb{Z}$ becomes a monoid, with identity element $d\mathbb{Z} + 1$. Furthermore, the set $\mathbb{Z}/_d$ inherits the group and monoid structure from $\mathbb{Z}/d\mathbb{Z} = \mathbb{Z}_{\mathbf{mod}\, d}$ via the isomorphism (3.18). (See Exercises 5 and 23.) $\Box$

5.4 The First Isomorphism Theorem for Groups

The results of the preceding sections may be summarized to show how a group homomorphism factorizes under a strengthened version of the First Isomorphism Theorem for Sets (Theorem 3.11, page 57).

THEOREM 5.21 (First Isomorphism Theorem for Groups.)
Let $f : (X, \cdot, e_X) \to (Y, \cdot, e_Y)$ be a group homomorphism.

(a) *The group kernel $N = f^{-1}\{e_Y\}$ is a normal subgroup of the domain group X.*

(b) *The image $f(X)$ is a subgroup of the codomain group Y.*

(c) *In the factorization*

$$f = j \circ b \circ s$$

given by the First Isomorphism Theorem for Sets, the surjection s may be taken as the surjective homomorphism

$$s : X \to X/N; x \mapsto Nx$$

of Corollary 5.18, the bijection b is the well-defined group isomorphism

$$b : X/N \to f(X); Nx \mapsto f(x)$$

from the quotient X/N to the image $f(X)$, and the injection j is the injective group homomorphism

$$j : f(X) \hookrightarrow Y; f(x) \mapsto f(x)$$

of Example 5.3.

If the domain of the group homomorphism in the First Isomorphism Theorem is finite, then the bijection b may be used to count the size of the image.

COROLLARY 5.22
Let $f : X \to Y$ be a group homomorphism with group kernel N and finite domain X. Then the size $|f(X)|$ of the image of f is the index

$$|X/N| = |X|/|N|$$

of the subgroup N of X.

Example 5.23 (The special linear group.)
Consider the group homomorphism

$$\det : \mathrm{GL}(2, \mathbb{R}) \to \mathbb{R}^*; \begin{bmatrix} a & b \\ c & d \end{bmatrix} \mapsto ad - bc$$

(compare Example 5.7 and Exercise 7). The group kernel is the set $\mathrm{SL}(2, \mathbb{R})$ of 2×2 real matrices of determinant 1. This group $\mathrm{SL}(2, \mathbb{R})$ is called the (*real*) *special linear group* of dimension 2. The First Isomorphism Theorem for Groups exhibits the isomorphism

$$\mathrm{GL}(2, \mathbb{R})/\mathrm{SL}(2, \mathbb{R}) \cong \mathbb{R}^*$$

from the quotient group to the group of nonzero real numbers under multiplication. ▯

An important application of the First Isomorphism Theorem for Groups is the classical Chinese Remainder Theorem.

THEOREM 5.24 (Chinese Remainder Theorem.)
Let a and b be coprime positive integers. Then there are isomorphisms

$$\mathbb{Z}/ab\mathbb{Z} \cong \mathbb{Z}/a\mathbb{Z} \times \mathbb{Z}/b\mathbb{Z}$$

of sets, groups under addition, and monoids under multiplication.

PROOF Consider the map

$$p : \mathbb{Z}/ab\mathbb{Z} \to \mathbb{Z}/a\mathbb{Z} \times \mathbb{Z}/b\mathbb{Z} ;$$
$$ab\mathbb{Z} + x \mapsto (a\mathbb{Z} + x, b\mathbb{Z} + x) .$$

It is clearly well-defined, since

$$ab \mid (x - x') \text{ implies } a \mid (x - x') \text{ and } b \mid (x - x') .$$

It is certainly a (semi)group and monoid homomorphism. For an element $ab\mathbb{Z} + x$ of the group kernel $\mathrm{Ker}\, p$, the representative integer x is a multiple of both a and b. Since a and b are coprime, their lowest common multiple is ab (compare Exercise 43 in Chapter 1). Thus $ab \mid x$, and the group kernel $\mathrm{Ker}\, p$ is trivial. It follows that the group homomorphism is injective, since the classes of the kernel relation $\mathrm{ker}\, p$ are the cosets of the subgroup $\mathrm{Ker}\, p$. Since the domain and codomain have the same finite number of elements, Corollary 5.22 shows that the map p is surjective. ▯

5.5   The Law of Exponents

Let x be an element of a monoid $(M, \cdot, e)$. Then for natural numbers n, the *powers* x^n are defined recursively by

$$x^0 = e \qquad \text{and} \qquad x^{n+1} = x^n \cdot x \tag{5.10}$$

(compare Section 2.4 for powers in a semigroup or monoid of functions).

- In the monoid $(\mathbb{R}, \cdot, 1)$ of real numbers under multiplication, the notation (5.10) agrees with the usual power notation for real numbers x.

- In the monoid $(\mathbb{N}, +, 0)$ of natural numbers under addition, the power notation x^n for a natural number x translates to the multiple notation nx.

Generally, for a (commutative) monoid $(M, +, 0)$ written using additive notation, the recursive definition (5.10) of powers translates to a recursive definition

$$0x = 0 \qquad \text{and} \qquad (n+1)x = nx + x$$

of *multiples*.

For an element x of a monoid $(M, \cdot, e)$, and natural numbers m, n, the *Law of Exponents*

$$x^{m+n} = x^m \cdot x^n \tag{5.11}$$

may be proved by induction on n (Exercise 28). The Law of Exponents underlies the following theorem, which shows the special role played by the monoid $(\mathbb{N}, +, 0)$ of natural numbers under addition, and the number 1 as an element of that monoid.

THEOREM 5.25 (Universality of natural numbers.)
Let x be an element of a monoid $(M, \cdot, e)$. Then there is a unique monoid homomorphism

$$f : (\mathbb{N}, +, 0) \to (M, \cdot, e); n \mapsto x^n \tag{5.12}$$

with $f(1) = x$.

PROOF   The map f of (5.12) is a monoid homomorphism, since

$$f(0) = x^0 = e$$

by definition, and

$$f(m+n) = x^{m+n} = x^m \cdot x^n = f(m) \cdot f(n)$$

for natural numbers m and n by the Law of Exponents (5.11). Now suppose that $\varphi : (\mathbb{N}, +, 0) \to (M, \cdot, e)$ is a monoid homomorphism with $\varphi(1) = x$. For natural numbers n, the equation $\varphi(n) = f(n)$ follows by induction on n. Certainly $\varphi(0) = e = f(0)$, since φ is a monoid homomorphism. Then if $\varphi(n) = f(n)$, we have

$$\varphi(n+1) = \varphi(n) \cdot \varphi(1) = \varphi(n) \cdot x = f(n) \cdot x = x^n \cdot x = x^{n+1} = f(n+1) \,,$$

the first equation holding since φ is a semigroup homomorphism. ⬜

Now let x be an element of a group $(G, \cdot, e)$. For each positive integer n, define

$$x^{-n} = (x^{-1})^n \,.$$

The Law of Exponents (5.11) for monoids may then be extended to a *Law of Exponents*

$$x^{m+n} = x^m \cdot x^n \tag{5.13}$$

for groups, holding for all integers m and n (Exercise 30). The analogue of Theorem 5.25 holds (Exercise 31). It points out the special role played by the group $(\mathbb{Z}, +, 0)$ of integers under addition, and the element 1 of that group.

THEOREM 5.26 (Universality of integers.)
Suppose that x is an element of a group $(G, \cdot, e)$. Then there is a unique group homomorphism

$$\exp_x : (\mathbb{Z}, +, 0) \to (G, \cdot, e); n \mapsto x^n \tag{5.14}$$

with $\exp_x(1) = x$.

Example 5.27 (Exponentiation.)
Consider the element e of the group $(\mathbb{R}^*, \cdot, 1)$ of nonzero real numbers under multiplication. Then

$$\exp_e(n) = e^n$$

for each integer n. For the element 2 of $\mathbb{R}^*$, we have $\exp_2(n) = 2^n$ for each integer n. Thus the uniquely specified group homomorphism (5.14) may be considered as an "exponentiation to base x" in the group G. ⬜

In Theorem 5.26, the group kernel $\mathrm{Ker}(\exp_x)$ of the homomorphism $\exp_x$ is a subgroup of the group $(\mathbb{Z}, +, 0)$ of integers. By Theorem 4.46 (page 81), the group $\mathrm{Ker}(\exp_x)$ is the set of multiples of a natural number d_x.

DEFINITION 5.28 (Cyclic group generated, order of element.)
Let x be an element of a group $(G, \cdot, e)$.

(a) *The image $\langle x \rangle$ of the group homomorphism $\exp_x$ in (5.14) is called the (cyclic) subgroup of G generated by x.*

(b) *If $d_x = 0$, the element x is said to be of* infinite order.

(c) *If d_x is a positive integer, the element x is said to be of* finite order d_x.

Note that part (b) of the First Isomorphism Theorem for Groups, applied to the group homomorphism $\exp_x : \mathbb{Z} \to G$, confirms that the image

$$\langle x \rangle = \{ \ldots,\ x^{-2},\ x^{-1},\ x^0 = e,\ x^1 = x,\ x^2,\ x^3,\ \ldots \}$$

of $\exp_x$, the set of all powers of the element x, really is a subgroup of G.

Two cases arise:

• **If x has infinite order**, the group kernel of $\exp_x$ is trivial. Thus $\exp_x$ is injective, and the powers

$$\ldots,\ x^{-2},\ x^{-1},\ x^0 = e,\ x^1 = x,\ x^2,\ x^3,\ \ldots$$

of x are all distinct. Part (c) of the First Isomorphism Theorem for Groups, applied to the group homomorphism $\exp_x : \mathbb{Z} \to G$, then yields the group isomorphism

$$b : \mathbb{Z} \to \langle x \rangle; n \mapsto x^n$$

between the infinite cyclic group $\langle x \rangle$ and the group of integers $(\mathbb{Z}, +, 0)$ under addition. In general, any group C_∞ isomorphic to the group of integers is described as an *infinite cyclic group*.

• **If x has finite order d_x**, the bijection b in the First Isomorphism Theorem for Groups shows that x has precisely d_x distinct powers

$$x^0 = e,\ x^1 = x,\ x^2,\ x^3,\ \ldots,\ x^{d_x - 1}.$$

Since the classes of the kernel relation $\mathbf{ker}(\exp_x)$ are the cosets of the subgroup $d_x \mathbb{Z}$, two powers x^m and x^n of x are equal if and only if the difference $m - n$ is a multiple of the order d_x. In this case, we may also consider the indices n in powers x^n of x as integers modulo d_x. In other words, when x has finite order d, the bijection b in the First Isomorphism Theorem for Groups yields the group isomorphism $b : \mathbb{Z}/d \to \langle x \rangle; n \mapsto x^n$ between the cyclic group $\langle x \rangle$ of size d and the group of integers $(\mathbb{Z}/d, +, 0)$ modulo d under addition.

In general, any group C_d isomorphic to the group of integers modulo d under addition is described as a *cyclic group* of finite order d. This is consistent with the nomenclature of Example 2.30 (page 38) — compare Exercise 34.

REMARK 5.29 Let x be an element of a group G. Whether x has finite or infinite order, this order is just the size (or cardinality)

$$|\langle x \rangle| \tag{5.15}$$

of the cyclic group $\langle x \rangle$ generated by x. It is convenient to use (5.15) as a standard notation for the order of a group element x. ⬚

5.6 Cayley's Theorem

Abstract groups were introduced as a generalization of the concept of a group of permutations. It will now be shown that the generalization has not strayed too far:

> Every group is isomorphic to a group of permutations.

To put groups in context, we begin by examining semigroups.

Let $(S, \cdot)$ be a semigroup. Then for each element s of S, we define the *left multiplication* by s to be the map

$$\lambda_s : S \to S; x \mapsto s \cdot x\,. \tag{5.16}$$

Example 5.30 (Real shifts.)

Let $(\mathbb{R}.+)$ be the semigroup of real numbers under addition. Then for each real number r, the left multiplication λ_r by r is the shift σ_r of Example 2.5 (page 30). ▯

Example 5.31 (Cycles.)

Let n be a positive integer. Let $(\mathbb{Z}/_n, +)$ be the semigroup of integers modulo n under addition. Then λ_1 is the cycle

$$\begin{pmatrix} 0 & 1 & 2 & \ldots & (n-1) \end{pmatrix}$$

from the cyclic group C_n (Example 2.30, page 38). ▯

PROPOSITION 5.32

Let $(S, \cdot)$ be a semigroup. Consider the semigroup $(S^S, \circ)$ of all functions from the set S to itself, with the operation of function composition. Then the map

$$\Lambda : (S, \cdot) \to (S^S, \circ); s \mapsto \lambda_s$$

is a semigroup homomorphism.

PROOF For elements s, t, and x of S, the associative law yields

$$(\lambda_s \circ \lambda_t)(x) = \lambda_s(\lambda_t(x)) = \lambda_s(t \cdot x) = s \cdot (t \cdot x) = (s \cdot t) \cdot x = \lambda_{s \cdot t}(x)\,.$$

Thus the composite map $\lambda_s \circ \lambda_t$ is equal to the single map $\lambda_{s \cdot t}$. Rewriting in terms of Λ, we obtain $\Lambda(s) \circ \Lambda(t) = \Lambda(s \cdot t)$, showing that Λ is a semigroup homomorphism. ▯

Example 5.33

Let X be a set. Define an operation $\cdot$ on X by

$$x \cdot y = y$$

for all x, y in X. This operation is associative, since

$$x \cdot (y \cdot z) = x = x \cdot y = (x \cdot y) \cdot z$$

for x, y, z in X. In the semigroup $(X, \cdot)$, we have $\lambda_x = \mathrm{id}_X$ for all x in X. The map Λ of Proposition 5.32 becomes the constant map

$$\Lambda : x \mapsto \mathrm{id}_X$$

in this case.   □

For groups, the collapse observed in Example 5.33 cannot happen. In fact, we obtain the desired isomorphism of each abstract group with a group of permutations.

THEOREM 5.34 (Cayley's Theorem.)
Let $(G, \cdot, e)$ be a group.

(a) *The semigroup homomorphism*

$$\Lambda : (G, \cdot) \to (G^G, \circ); x \mapsto \lambda_x$$

  is injective.

(b) *The image of Λ is a group of permutations on the set G.*

(c) *The abstract group G is isomorphic to the group $\Lambda(G)$ of permutations of the set G.*

PROOF   (a): Suppose $\lambda_x = \lambda_y$ for elements x and y of G. Then

$$x = x \cdot e = \lambda_x(e) = \lambda_y(e) = y \cdot e = y \,.$$

(b): For each element x of G, the map λ_x is invertible, with two-sided inverse $\lambda_{x^{-1}}$. Indeed, for each element g of G, we have

$$\lambda_x \circ \lambda_{x^{-1}}(g) = x \cdot x^{-1} \cdot g = g = \mathrm{id}_X(g) \,,$$

so that $\lambda_x \circ \lambda_{x^{-1}} = \mathrm{id}_X$. Considering x^{-1} in place of x yields $\lambda_{x^{-1}} \circ \lambda_x = \mathrm{id}_X$.

(c): The map $(G, \cdot, e) \to (\Lambda(G), \circ, \mathrm{id}_X); x \mapsto \lambda_x$ is a bijective semigroup homomorphism between groups. As such, it is a group isomorphism.   □

Example 5.35 (Position vectors and translation vectors.)
Consider the group $\mathbb{R}^2$ of 2-dimensional real vectors (Example 4.32, page 75). An element (x_1, x_2) of $\mathbb{R}^2$ represents a *position vector*. Its image $\lambda_{(x_1,x_2)}$ in $\Lambda(\mathbb{R}^2)$ under the isomorphism of Theorem 5.34(c) becomes the corresponding *translation vector*

$$\lambda_{(x_1,x_2)} : \mathbb{R}^2 \to \mathbb{R}^2; (a_1, a_2) \mapsto (x_1 + a_1, x_2 + a_2)$$

(Figure 5.1). Thus according to Cayley's Theorem, the group of position vectors under addition is isomorphic to the group of translation vectors under addition. ∎

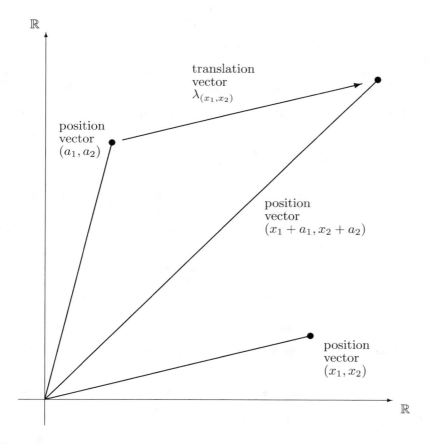

FIGURE 5.1: Position vectors and translation vectors.

5.7 Exercises

1. Define $\theta : \mathcal{P}(\underline{1}) \to \mathcal{P}(\underline{2}); X \mapsto X$ (compare Example 4.11 and Exercise 6 in Chapter 4).

   (a) Show that $\theta : (\mathcal{P}(\underline{1}), \cap) \to (\mathcal{P}(\underline{2}), \cap)$ is a semigroup homomorphism.

   (b) Show that θ is not a monoid homomorphism.

2. Show that the group $(\mathbb{R}^4, +, (0,0,0,0))$ of 4-dimensional real vectors is isomorphic to the group

$$\left(\mathbb{R}_2^2, +, \begin{bmatrix} 0 & 0 \\ 0 & 0 \end{bmatrix}\right)$$

   of real 2×2 matrices under addition.

3. Show that the group $\Sigma_{\mathbb{R}}$ of shifts on $\mathbb{R}$ (compare Example 2.28, page 37) is isomorphic to the group $(\mathbb{R}, +, 0)$ of real numbers under addition.

4. Let $f : G \to H$ and $g : H \to K$ be semigroup homomorphisms. Show that the composite $g \circ f : G \to K$ is also a semigroup homomorphism.

5. Let $\theta : X \to Y$ be a bijection from the set X to the set Y.

   (a) If X carries a semigroup structure $(X, \circ)$, show that there is a unique way of defining an associative multiplication $*$ on Y to yield a semigroup isomorphism $\theta : (X, \circ) \to (Y, *)$.

   (b) Suppose further that $(X, \circ, e)$ is a monoid. Show that $(Y, *, \theta(e))$ is a monoid.

   (c) Suppose further that $(X, \circ, e)$ is a group. Show that $(Y, *, \theta(e))$ is a group.

   In (a), (b), and (c), we say that the set Y *inherits* the semigroup, monoid, or group structure *from X via θ*.

6. Show that the group $((\mathbb{Z}/_2)^2, +, 00)$ of length 2 bit strings (compare Example 4.35, page 78) is isomorphic to the Klein four-group V_4.

7. For real numbers $a, b, c, d, a', b', c', d'$, verify the identity

$$(ad - bc)(a'd' - b'c') = (aa' + bc')(cb' + dd') - (ab' + bd')(ca' + dc').$$

8. Define the *trace* function

$$\mathrm{tr} : \mathbb{R}_2^2 \to \mathbb{R}; \begin{bmatrix} a & b \\ c & d \end{bmatrix} \mapsto a + d \tag{5.17}$$

from the group $\mathbb{R}_2^2$ of 2×2 real matrices under componentwise addition (4.17) to the group of real numbers under addition. Show that the trace function is a group homomorphism.

9. Prove Corollary 5.9.

10. Let X and Y be groups. Show that a function $f : X \to Y$ is a group homomorphism if and only if the graph of f is a subgroup of the direct product group $X \times Y$.

11. Let n be a positive integer. Show that there is a group homomorphism

$$\rho : C_n \to \mathrm{GL}(2, \mathbb{R})$$

with

$$\rho\Big((0\ 1\ 2\ \ldots\ (n-2)\ (n-1))^r\Big) = \begin{bmatrix} \cos(2r\pi/n) & -\sin(2r\pi/n) \\ \sin(2r\pi/n) & \cos(2r\pi/n) \end{bmatrix}$$

for $1 \le r \le n$. (A homomorphism ρ from a group G to a group of matrices is known as a *matrix representation* of the group G.)

12. Show that there is an injective monoid homomorphism from the monoid A of affine functions to the monoid $\mathbb{R}_2^2$ of real 2×2 matrices under multiplication, given by mapping the function

$$f : \mathbb{R} \to \mathbb{R}; x \mapsto m \cdot x + c$$

to the matrix

$$\begin{bmatrix} m & c \\ 0 & 1 \end{bmatrix}.$$

13. Show that the injective monoid homomorphism of Exercise 12 restricts to a matrix representation ρ of the group $\mathrm{Aff}(\mathbb{R})$ of affine functions (compare Exercise 45 in Chapter 2).

14. (a) Show that $\{(0), (0\ 1)\}$ is a subgroup of the symmetric group S_3.

 (b) Show that $\{(0), (0\ 1)\}$ is not a normal subgroup of S_3.

15. Prove Proposition 5.14.

16. Let N be a subgroup of a group G. Let x and y be elements of G. Show that the cosets Nx and Ny are equal if and only if xy^{-1} is an element of N.

17. Show that a subgroup N of a group G is normal if and only if each right coset Nx of N with an element x of G is equal to the left coset xN.

18. Show that a subgroup N of a group G is normal if and only if each right coset Nx of N with an element x of G is equal to the left coset yN with some element y of G.

Introduction to Abstract Algebra

19. Give an example of a group G, a subgroup H, and an element x of G such that the right coset Hx and left coset xH are distinct subsets of G.

20. Let M and N be normal subgroups of a group G. Show that MN is also a normal subgroup of G.

21. For a normal subgroup N of a group G, the equation (5.8) defines the inverse $(Nx)^{-1}$ of the coset Nx in the quotient group G/N. Show that the equation is also consistent with a different interpretation of the left-hand side, namely as the set

$$(Nx)^{-1} = \{(nx)^{-1} \mid n \text{ in } N\}$$

of inverses (in G) of elements of the coset Nx.

22. Let $f : (X, \cdot, e_X) \to (Y, \cdot, e_Y)$ be a group homomorphism. Show that f is injective if and only if $\operatorname{Ker} f = \{e_X\}$. (Compare Exercise 4 in Chapter 3.)

23. Verify the claims of Example 5.20.

24. (a) Show that the intersection

$$\mathrm{SO}_2(\mathbb{R}) = \mathrm{SL}(2, \mathbb{R}) \cap \mathrm{O}_2(\mathbb{R})$$

is a normal subgroup of the group $\mathrm{O}_2(\mathbb{R})$ of orthogonal matrices. (Compare Study Project 2 in Chapter 4.)

(b) Show that there is an isomorphism

$$\mathrm{O}_2(\mathbb{R})/\mathrm{SO}_2(\mathbb{R}) \cong \{\pm 1\}$$

from the quotient group to the group of real numbers $\{\pm 1\}$ under multiplication.

[Hint: Use the First Isomorphism Theorem for Groups to obtain both (a) and (b) directly.]

25. Let M and N be normal subgroups of a group G.

(a) Show that the map

$$\theta : G \to G/M \times G/N; x \mapsto (Mx, Nx)$$

is a group homomorphism.

(b) Show that the group kernel of θ is the intersection $M \cap N$ of the normal subgroups M and N.

(c) Conclude that $M \cap N$ is a normal subgroup of G.

26. Let G be a group. Consider the subset

$$\widehat{G} = \{(g,g) \mid g \text{ in } G\}$$

of $G \times G$, the so-called *diagonal*.

   (a) Show that $\widehat{G}$ is a subgroup of $G \times G$. (Hint: If you have done Exercise 10, you may apply it by noting that the diagonal is the graph of the group homomorphism id_G.)

   (b) Show that G is commutative if and only if $\widehat{G}$ is a normal subgroup of $G \times G$.

27. Find a solution x to the simultaneous congruences

$$x \equiv 2 \quad \text{mod } 7\,,$$
$$x \equiv 7 \quad \text{mod } 10\,.$$

28. Prove the Law of Exponents (5.11) for an element x of a monoid $(M, \cdot, e)$.

29. Show that in additive notation, the Law of Exponents takes the form

$$(m+n)x = mx + nx$$

of a *right distributive law*.

30. Prove the Law of Exponents (5.13) for an element x of a group $(G, \cdot, e)$. (Hint: Beyond the monoid version, the additional cases to verify are when one or both of m, n is negative.)

31. Prove Theorem 5.26.

32. Let d be a positive integer. Show that the order of $d\mathbb{Z}+1$ in $\mathbb{Z}/d\mathbb{Z}$ is d.

33. Suppose that m and n are coprime positive integers. Show that there is a group isomorphism

$$C_{mn} \cong C_m \times C_n\,.$$

34. Consider a cycle

$$\alpha = (x_1 \ x_2 \ x_3 \ \ldots \ x_a)$$

of length a in the symmetric group S_n. Show that α has order a.

35. Let r be a positive integer. Show that the order of the matrix

$$\begin{bmatrix} \cos(2\pi/r) & -\sin(2\pi/r) \\ \sin(2\pi/r) & \cos(2\pi/r) \end{bmatrix}$$

in $GL(2, \mathbb{R})$ is r.

36. Let x be an element of a finite group G of size n. Applying Lagrange's Theorem to the subgroup $\langle x \rangle$, show that the order of x is a divisor of n.

37. Let x be an element of finite order in a group G. Show that the order of each power of x is a divisor of the order of x.

38. Let x be an element of a group G. Show that the order of x^{-1} is the order of x.

39. Let x be an element of odd finite order in a group G. Show that x^2 has the same order as x.

40. In the affine group $\mathrm{Aff}(\mathbb{R})$, determine precisely which elements have finite order. (Hint: You may wish to use the matrix representation ρ from Exercise 13.)

41. Let $f : G \to H$ be a group homomorphism with finite domain. For each element x of G, show that the order of $f(x)$ is finite, and a divisor of the order of x.

42. In a group G, consider elements x and y of coprime finite orders a and b. Suppose that $xy = yx$. Show that there is a group isomorphism
$$\mathbb{Z}/_a \times \mathbb{Z}/_b \to \langle xy \rangle; (r, s) \mapsto x^r y^s \,.$$

43. In the symmetric group S_n, consider disjoint cycles α and β of respective lengths a and b, with a and b coprime. Show that $\alpha \circ \beta$ has order ab.

44. Show that every monoid is isomorphic to a monoid of functions.

45. Let X and Y be isomorphic sets.

    (a) Show that the monoids X^X and Y^Y are isomorphic. (Compare Example 2.2, page 29.)

    (b) Show that the symmetric groups $X!$ and $Y!$ are isomorphic.

5.8 Study projects

1. **Error-correcting codes.** Storing information for future retrieval, and transmitting information from one location to another, are two basic tasks of information technology. These tasks are complicated by the occurrence of errors. Information storage media are subject to damage and deterioration, while communication channels are subject to noise and interference. Error-correcting codes are designed to compensate

for the effects of errors that are not too serious, enabling the original information to be recovered. (They are not directly related to secret codes as discussed in Section 2.9.)

The simplest piece of information is a single yes/no dichotomy, a binary digit or bit taking the value 1 or 0 (page 59). It could represent, say, the verdict of a jury in a criminal trial: "Guilty" or "Not guilty." If an error occurs, it will change the bit from 1 to 0 or vice versa. The information in the bit will be lost completely. To protect the information from errors changing a single bit, redundancy is added. The single information bit 1 is encoded as the bit string 111, while 0 is encoded as 000. The full set of encoded messages is the *code*

$$C = \{000, 111\}. \tag{5.18}$$

Now if an error changes a bit, say changes 111 to 011, the original information bit 1 may be recovered as the commonest bit in the string 011 ("majority vote"). The price paid for this robustness is a tripling in the space needed to record the information (3 bits instead of 1), or the time taken to transmit it.

Groups play a key role in the design of coding schemes. Consider the group $((\mathbb{Z}/2)^3, +, 000)$ of bit strings of length 3 (Example 4.35, page 78). First, confirm that the code C is a subgroup of this group. Now changing the first bit from the right in a bit string of length 3 means adding the error 001, thus

$$111 + 001 = 110.$$

There are three possible single-bit errors:

$$e_1 = 001, \qquad e_2 = 010, \qquad e_3 = 100,$$

labeled by the error location, counting from the right. For completeness, the zero "error" is

$$e_0 = 000.$$

Adding this "error" means no change to the bit string. The full set of errors is

$$E = \{e_0 = 000, \ e_1 = 001, \ e_2 = 010, \ e_3 = 100\}.$$

Although E is a subset of the set $(\mathbb{Z}/2)^3$ of all bit strings of length 3, it is not a subgroup of $((\mathbb{Z}/2)^3, +, 000)$. (Why not?)

However, there is a bijection

$$\theta : (\mathbb{Z}/2)^2 \to E;$$

$$00 \mapsto e_{00} = 000, \ 01 \mapsto e_{01} = 001, \ 10 \mapsto e_{10} = 010, \ 11 \mapsto e_{11} = 100$$

taking a bit string of length 2, the binary representation of a number $0 \le i < 2^2$, to the corresponding error e_i. According to Exercise 5(c), this means that the set E inherits a group structure $(E, *, 000)$ making θ a group isomorphism. Show that the multiplication table of the error set E under $*$ is as given in Figure 5.2.

$*$	000	001	010	100
000	000	001	010	100
001	001	000	100	010
010	010	100	000	001
100	100	010	001	000

FIGURE 5.2: Length 3 error set E under $*$.

The code (5.18) is recovered from the set of errors as:

$$000 = e_{00} + e_{01} + e_{00} * e_{01} = 000 + 001 + 000 * 001 = 000$$
$$111 = e_{01} + e_{10} + e_{01} * e_{10} = 001 + 010 + 001 * 010 = 111 \,.$$

Further, show that:

(a) A group homomorphism

$$s : \left((\mathbb{Z}/_2)^3, +, 000\right) \to (E, *, 000) \tag{5.19}$$

is uniquely defined by $e_i \mapsto e_i$ for $0 < i < 4$. [The homomorphism (5.19) is known as the *syndrome*.]

(b) The code C is the group kernel of the syndrome. For example,

$$s(111) = s(e_1 + e_2 + e_3) = e_1 * e_2 * e_3 = 000 \,.$$

(c) When a bit string x is received, the single error that is assumed to have occurred is $s(x)$. For example, if $x = 101$ is received, the most probable single error is $s(101) = s(e_1 + e_3) = e_1 * e_3 = e_2 = 010$. The word $x = 101$ is decoded as $x + s(x) = 101 + 010 = 111$.

2. **Hamming codes.** For a positive integer r, define $l = 2^r - 1$. The integer r is known as the *redundancy*, and the integer l is known as the *channel length*. The set $(\mathbb{Z}/_2)^l$ of bit strings of length l is known as the *channel*. The errors are defined to be e_0, the length l string of zeros, and e_i for $1 \le i \le l$ as the length l bit string with zeros everywhere except for

a 1 in the i-th position counting from the right. (This notation extends the case $r = 2$ discussed in Study Project 1.) Usually, we consider the indices $0 \leq i < 2^r$ as being expressed in binary notation, as bit strings of length r.

(a) Show that the *error set*

$$E = \{e_i \mid 0 \leq i < 2^r\}$$

is not a subgroup of $((\mathbb{Z}/2)^l, +, 0)$, unless $r = 1$.

(b) Show that E inherits a group structure $(E, *, e_0)$ from the bijection

$$\theta : (\mathbb{Z}/2)^r \to E; i \mapsto e_i \,.$$

(c) Show that $e_i * e_j = e_k$, where k is the Nim sum of i and j — compare Study Project 1 in Chapter 4.

(d) Show that a group homomorphism

$$s : ((\mathbb{Z}/2)^l, +, e_0) \to (E, *, e_0) \tag{5.20}$$

is uniquely defined by $e_i \mapsto e_i$ for $0 < i \leq l$. The homomorphism (5.20) is known as the *syndrome*.

(e) Show that the group kernel $C = \operatorname{Ker} s$ has 2^{l-r} elements. The subgroup C of the channel $((\mathbb{Z}/2)^l, +, e_0)$ is known as the *Hamming code* of *dimension* $l - r$.

(f) Write out the 16 elements of the Hamming code of dimension 4 in the channel of length 7. (Note that these may be used to encode hexadecimal digits.)

(g) Show that the elements of the Hamming code may be obtained equally well, either as elements of the group kernel of the syndrome, or from expressions such as $e_i + e_j + e_i * e_j$ for $1 \leq i \neq j < 2^3$.

(h) If the word 1101100 is received in the channel of length 7, which is the most likely element of the Hamming code to have been transmitted?

(i) If the word 1101101 is received in the channel of length 7, which is the most likely element of the Hamming code to have been transmitted?

3. **Dyadic integers.** When we do mathematics, we carry a picture of the set $\mathbb{Z}$ of integers much like (1.1). Although we normally work with relatively small integers, represented by their decimal expansions, we do have the capacity to contemplate arbitrarily large numbers, with a positive or negative sign. In a computer, the representation of integers is different. The representation corresponds to a binary expansion, and

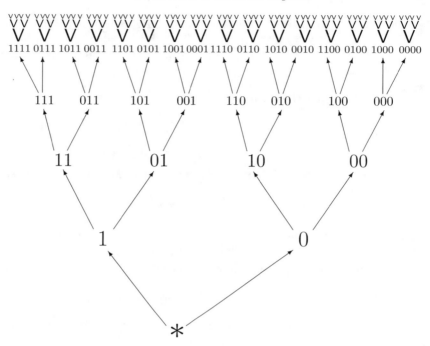

FIGURE 5.3: Paths in the binary tree.

there is a limit on the size of the integers that can be handled directly, say $2^{64} - 1$. Computer representations of integers are modeled by the set $\mathbb{Z}_2$ of *dyadic* or 2-*adic integers*. To emphasize the distinction, members of the set $\mathbb{Z}$ are described as *rational integers*.

Replacing (1.1), the typical picture is displayed in Figure 5.3. The starting point is the asterisk at the bottom, the *root* of the binary tree. At the next level up are the two binary digits 0 and 1, forming the set $\mathbb{Z}/2$ of integers modulo 2. Above them is the set $\mathbb{Z}/4$ of integers modulo 4, with the elements written in their binary expansions:

$$\mathbb{Z}/_{2^2} = \{00, 01, 10, 11\}.$$

One level higher is

$$\mathbb{Z}/_{2^3} = \{000, 001, \ldots, 111\},$$

and so on. Each node in the tree is a bit string b (considering $*$ as the empty string or string of length zero). Two arrows emerge from the node b, one going to $0b$ and one to $1b$.

Now consider a particular integer, say 6. In the set $\mathbb{Z}/_2$ of integers modulo 2, the integer 6 belongs to the class of even integers, represented by the binary digit 0. In $\mathbb{Z}/_{2^2}$, the integer 6 is represented by $6 \bmod 2^2 = 2$, but written in binary notation as 10. In $\mathbb{Z}/_{2^3}$, the integer 6 is represented by $6 \bmod 2^3 = 110$. In $\mathbb{Z}/_{2^4}$, the integer 6 is represented by $6 \bmod 2^4 = 0110$. From then on, 6 is represented by $0\ldots0110$. Altogether, the integer 6 is represented by the path

$$* \to 0 \to 10 \to 110 \to 0110 \to 00110 \to \cdots \to 0\ldots0110 \to \ldots$$

in the binary tree. A general rational integer a is represented by the path

$$* \to a \bmod 2 \to a \bmod 2^2 \to \cdots \to a \bmod 2^r \to a \bmod 2^{r+1} \to \ldots$$

in the binary tree. The set $\mathbb{Z}_2$ of dyadic integers consists of the full set of all paths from the root in the binary tree. In a particular computer implementation, the paths have to stop after a certain point, say after 64 steps.

(a) Find the path representing the integer 13.

(b) Find the path representing the integer -1.

(c) Find the path representing the integer -5.

(d) Show that the path representing a natural number n eventually has each step of the form

$$\overbrace{00\ldots00}^{k \text{ zeroes}} b \to \overbrace{000\ldots00}^{k+1 \text{ zeroes}} b$$

for some finite bit string b that is a binary expansion of n.

(e) Show that the path representing a negative number n eventually has each step of the form

$$\overbrace{11\ldots11}^{k \text{ ones}} b \to \overbrace{111\ldots11}^{k+1 \text{ ones}} b$$

for some finite bit string b. What is the relation of the bit string b to n?

(f) Show that the path

$$* \to 1 \to 01 \to 101 \to 0101 \to 10101 \to 010101 \to 1010101 \to \ldots$$

represents a dyadic integer which is not a rational integer.

(g) Show that the map

$$(\mathbb{Z}/_{2^{r+1}}, +, 0) \to (\mathbb{Z}/_{2^r}, +, 0); a \bmod 2^{r+1} \mapsto a \bmod 2^r$$

is a well-defined group homomorphism for each positive integer r.

(h) Show that the map

$$\left(\mathbb{Z}/_{2^{r+1}}, \cdot, 1\right) \to \left(\mathbb{Z}/_{2^r}, \cdot, 1\right); a \bmod 2^{r+1} \mapsto a \bmod 2^r$$

is a well-defined monoid homomorphism for each integer $r > 0$.

(i) Given two paths

$$* \to a_1 \to a_2 \to \cdots \to a_r \to a_{r+1} \to \dots$$

and

$$* \to b_1 \to b_2 \to \cdots \to b_r \to b_{r+1} \to \dots$$

(with a_k, b_k in $\mathbb{Z}/_{2^k}$ for $k > 0$), show that there is a well-defined sum

$$* \to a_1 + b_1 \to a_2 + b_2 \to \cdots \to a_r + b_r \to a_{r+1} + b_{r+1} \to \dots \quad (5.21)$$

and a well-defined product

$$* \to a_1 \cdot b_1 \to a_2 \cdot b_2 \to \cdots \to a_r \cdot b_r \to a_{r+1} \cdot b_{r+1} \to \dots \quad (5.22)$$

of paths.

(j) Show that the set $\mathbb{Z}_2$ of dyadic integers forms an additive group under the sum (5.21).

(k) Show that the set $\mathbb{Z}_2$ of dyadic integers forms a monoid under the multiplication (5.22).

(l) Show that the map $\mathbb{Z} \to \mathbb{Z}_2$ taking a rational integer a to the path

$$* \to a \bmod 2 \to a \bmod 2^2 \to \cdots \to a \bmod 2^r \to a \bmod 2^{r+1} \to \dots$$

is a group homomorphism for sums and a monoid homomorphism for products.

4. **The Euler ϕ-function.** The *Euler ϕ-function* is defined by setting $\phi(d)$, for a positive integer d, to be the number

$$|(\mathbb{Z}/_d)^*|$$

of elements in the group of units of the monoid $(\mathbb{Z}/_d, \cdot, 1)$ of integers modulo d under multiplication.

(a) Let d be a positive integer. For a nonzero integer a, show that $a \bmod d$ is a unit of $(\mathbb{Z}/_d, \cdot, 1)$ if and only if $\gcd(a, d) = 1$.

(b) For a prime number p, show that $\phi(p) = p - 1$.

(c) For a power p^e of a prime number p, with $e > 0$, show that there are p^{e-1} elements of the set

$$\mathbb{Z}/_{p^e} = \{0, 1, 2, \ldots, p^e - 1\}$$

which are multiples of p. Conclude that

$$\phi(p^e) = p^e - p^{e-1} = p^e \left(1 - \frac{1}{p}\right).$$

(d) Use Theorem 5.6 and the Chinese Remainder Theorem to show that

$$(\mathbb{Z}/_{mn})^* \cong (\mathbb{Z}/_m \times \mathbb{Z}/_n)^*$$

for coprime positive integers m and n.

(e) Use Theorem 4.33 (page 76) to conclude that

$$\phi(mn) = \phi(m)\phi(n) \tag{5.23}$$

for coprime positive integers m and n. The property (5.23) is sometimes described as the *multiplicativity* of the Euler ϕ-function.

(f) Given a positive integer n, consider its factorization

$$n = p_1^{e_1} p_2^{e_2} \ldots p_r^{e_r}$$

into a product of powers of distinct prime numbers, as given by the Fundamental Theorem of Arithmetic. Show that the value of the Euler ϕ-function is

$$\phi(n) = n \left(1 - \frac{1}{p_1}\right)\left(1 - \frac{1}{p_2}\right)\ldots\left(1 - \frac{1}{p_r}\right). \tag{5.24}$$

(g) For $d > 1$, let $1 = d_1, d_2, d_3, \ldots, d_{s-1}, d_s = n$ be a full list of the positive divisors of d. (Compare Figure 1.3 for $d = 72$.) Show that

$$\phi(d_1) + \phi(d_2) + \cdots + \phi(d_s) = d.$$

5. Consider the set

$$X = \left\{ \begin{bmatrix} n & m \\ 0 & n \end{bmatrix} \;\middle|\; n, m \text{ in } \mathbb{Z}, \; n \neq 0 \right\}$$

of upper triangular matrices, and the division function

$$d : X \to \mathbb{R}; \quad \begin{bmatrix} n & m \\ 0 & n \end{bmatrix} \mapsto n^{-1}m \tag{5.25}$$

(compare Section 3.3).

(a) Show that $(X, \cdot, I_2)$ forms a monoid under the usual matrix multiplication.

(b) Show that $(X, \circ)$ forms a monoid under Hadamard multiplication (4.18) of matrices. In particular, specify the identity element of this monoid.

(c) Show that the division function

$$d : (X, \circ) \to (\mathbb{R}, \cdot)$$

of (5.25) is a monoid homomorphism to the monoid of real numbers under multiplication.

(d) Show that the division function

$$d : (X, \cdot, I_2) \to (\mathbb{R}, +, 0)$$

of (5.25) is a monoid homomorphism to the monoid of real numbers under addition. Is it a group homomorphism?

(e) Apply the First Isomorphism Theorem for Sets (page 57) to the division function d of (5.25), obtaining the factorization

$$
\begin{array}{ccc}
X & \xrightarrow{\;d\;} & \mathbb{R} \\
{\scriptstyle s}\big\downarrow & & \big\uparrow{\scriptstyle j} \\
X_{\ker d} & \xrightarrow[\;b\;]{} & \mathbb{Q}
\end{array}
$$

in which the function b is invertible. Use the inverse isomorphism (of sets)

$$b^{-1} : \mathbb{Q} \to X_{\ker d} \, ,$$

together with the technique of Exercise 5, to create an additive group structure $(X_{\ker d}, +)$ and a multiplicative monoid structure $(X_{\ker d}, \cdot)$ on the set $X_{\ker d}$ of equivalence classes. Show that the surjection s yields monoid homomorphisms

$$s : (X, \cdot) \to (X_{\ker d}, +)$$

and

$$s : (X, \circ) \to (X_{\ker d}, \cdot) \, .$$

In particular, note the indirect confirmation that the well-defined multiplication (3.9) and addition (3.10) are associative operations.

5.9 Notes

Section 5.3

Quotient groups are sometimes described as "factor groups," particularly in the older literature. Unfortunately, this usage leads to confusion with the designation of X and Y as the factors in the direct product $X \times Y$.

Section 5.4

The Chinese Remainder Theorem is usually attributed to the third century mathematician Sun Zi (not to be confused with the author of *The Art of War*).

Section 5.5

In contrast with the convention of (5.15), some authors use the notation $|x|$ for the order of a group element x. For an element Ng of the quotient G/N of a group G by a normal subgroup N, this notation leads to confusion between the order $|\langle Ng \rangle|$ of the quotient group element Ng and the size $|Ng|$ of the coset Ng.

Section 5.6

A. Cayley was an English mathematician who lived from 1821 to 1895.

Section 5.8

R. Hamming was an American mathematician who lived from 1915 to 1998. L. Euler was a Swiss mathematician, later moving to Russia, who lived from 1707 to 1783.

Chapter 6

RINGS

Many of the sets encountered so far — such as the integers $\mathbb{Z}$, the reals $\mathbb{R}$, or the 2×2 real matrices $\mathbb{R}_2^2$ — have carried an additive group structure and a multiplicative semigroup or monoid structure. These two structures often combine to form a richer structure, known as a ring.

6.1 Rings

In Definition 6.1 below, a ring is defined as a set with two operations, an addition $x + y$ and a multiplication $x \cdot y$ (or often just xy) of elements x and y of R. In compound expressions involving both additions and multiplications, the multiplications are to be carried out first (following the convention used when working with integers and real numbers). For example, the right-hand side of (6.1) below is computed as $(x \cdot r) + (y \cdot r)$, and not as $x(r + y)r$. We say that the multiplication *binds more strongly* than the addition.

DEFINITION 6.1 (Distributive laws, unital and nonunital rings.)
Suppose that a set R carries a (commutative) additive group structure $(R, +, 0)$ and a multiplicative semigroup structure $(R, \cdot)$.

(a) *The combined structure $(R, +, \cdot)$ is said to satisfy the* right distributive law *if*

$$(x + y) \cdot r = x \cdot r + y \cdot r \tag{6.1}$$

for all x, y, r in R.

(b) *The structure $(R, +, \cdot)$ is said to satisfy the* left distributive law *if*

$$r \cdot (x + y) = r \cdot x + r \cdot y \tag{6.2}$$

for all x, y, r in R.

(c) *The structure $(R, +, \cdot)$ is said to be a* (nonunital) ring *if it satisfies both the right and left distributive laws.*

(d) *A ring $(R, +, \cdot)$ is said to be a* (unital) ring *if it forms a monoid $(R, \cdot, 1)$ under multiplication.*

(e) *A ring $(R, +, \cdot)$ is said to be* commutative *if the semigroup $(R, \cdot)$ is commutative.*

REMARK 6.2 Note that the group structure $(R, +, 0)$ of a ring $(R, +, \cdot)$ is always commutative. The issue of commutativity in a ring — Definition 6.1(e) — only arises in connection with the semigroup structure $(R, \cdot)$. For a commutative ring $(R, +, \cdot)$, the left and right distributive laws coincide. In a general ring $(R, +, \cdot)$, to say that two elements x and y *commute* means that $x \cdot y = y \cdot x$ (compare Definition 4.2, page 68). ▯

The identity element 0 of the additive group $(R, +, 0)$ of a ring $(R, +, \cdot)$ is known as the *zero* of the ring R. If R is unital, then the identity element 1 of the monoid $(R, \cdot, 1)$ is known as the *identity* or the *one* of the ring R. Unital rings are sometimes described as *rings with a one*, while nonunital rings are called *rings without a one*. In a unital ring R, the invertible elements — the units of the monoid $(R, \cdot, 1)$ — are called the *units* of the ring R. The group of units of a unital ring R is written as R^*, consistently with the notation of Proposition 4.20 (page 72).

REMARK 6.3 According to Definition 6.1(c), all rings are nonunital, regardless of whether they do or do not possess an identity element. When a ring R which actually has an identity element is described as "nonunital," the identity element is being disregarded. ▯

Example 6.4 (Integers.)
The integers form a unital commutative ring $(\mathbb{Z}, +, \cdot)$ under the usual addition and multiplication. Note that the right distributive law

$$(m + n)r = mr + nr$$

reduces to the Law of Exponents in the additive group $(\mathbb{Z}, +, 0)$ — compare Exercise 29 in Chapter 5. ▯

Example 6.5 (Reals.)
The set $\mathbb{R}$ of real numbers forms a commutative, unital ring $(\mathbb{R}, +, \cdot)$ under the usual addition and multiplication. ▯

Example 6.6 (Zero rings.)
Let $(A, +, 0)$ be an abelian group (written additively). Define a new, trivial

multiplication on the set A by

$$x \cdot_0 y = 0$$

for all x, y in A. Then $(A, +, \cdot_0)$ is a nonunital, commutative ring, known as the *zero ring* on the abelian group $(A, +, 0)$. Note that the distributive laws are satisfied trivially, since each side of the equations (6.1) and (6.2) in A reduces to 0. ▯

Example 6.7 (The trivial ring.)
The zero ring on the trivial abelian group $\{0\}$ is unital, with 0 as the identity element. It is known as the *trivial ring*. ▯

The following example does not exhibit a ring. It shows that having an additive group structure and a semigroup structure connected by one of the distributive laws is not enough to guarantee that the other distributive law will hold (and thus to yield a ring).

Example 6.8
Let $(A, +, 0)$ be a nontrivial additive group, say with nonzero element a. Define the semigroup operation

$$x \cdot y = y$$

on A, as in Example 5.33 (page 110). Note that the left distributive law (6.1) holds trivially, since each side reduces to

$$x + y.$$

On the other hand, the right distributive law reduces to

$$r = r + r,$$

which does not hold for $r = a$. ▯

Example 6.9 (Integers modulo d.)
Let d be a positive integer. Then the set $\mathbb{Z}/d\mathbb{Z}$ or $\mathbb{Z}/_d$ of integers modulo d forms a commutative unital ring $(\mathbb{Z}/_d, +, \cdot)$ under modular addition and multiplication. Using (3.21) and (3.22), the distributive law for $\mathbb{Z}/_d$ follows from the distributive law for the integers (Exercise 2). Note that for $d = 1$, the ring $\mathbb{Z}/_d$ is the unital zero ring of Example 6.7. ▯

Example 6.10 (Matrix rings.)
For a nonunital ring R, let R_2^2 denote the set of 2×2 matrices

$$\begin{bmatrix} r_{11} & r_{12} \\ r_{21} & r_{22} \end{bmatrix}$$

with entries r_{ij} from R. Then R_2^2 forms an additive (commutative) group
under the componentwise addition (4.17), and a noncommutative semigroup
under the usual multiplication (2.7) of matrices. The distributive laws also
hold (Exercise 3). If the ring R is unital, then so is the corresponding matrix
ring R_2^2. Its identity element is the matrix

$$I_2 = \begin{bmatrix} 1 & 0 \\ 0 & 1 \end{bmatrix}$$

in which the entries are the zero and identity of the unital ring R.   ▯

Example 6.11 (Direct products.)
Let $(R, +, \cdot)$ and $(S, +, \cdot)$ be nonunital rings. The product group $(R \times S, +)$
and semigroup $(R \times S, \cdot)$ combine to form a ring $(R \times S, +, \cdot)$, the (*direct*)
product of the rings R and S. Note that the distributive laws in the product
follow componentwise from the distributive laws in the factors R and S (Exercise 5). If R and S are unital, then $R \times S$ is unital, with componentwise
identity element $(1, 1)$.   ▯

Example 6.12 (Function rings.)
Let X be a set, and let $(S, +, \cdot)$ be a ring. According to Definition 4.34 (page
77) and Exercise 23 in Chapter 4, the set S^X of all functions $f : X \to S$ from
X to S carries a componentwise additive group structure $(S^X, +, z)$ — with
the constant zero function

$$z : X \to S; x \mapsto 0$$

— and a componentwise semigroup structure $(S^X, \cdot)$. Now for functions
$f, g, h : X \to S$, the right distributive law in S implies

$$[(f + g) \cdot h](x) = [f(x) + g(x)] \cdot h(x)$$
$$= f(x) \cdot h(x) + g(x) \cdot h(x) = [f \cdot h + g \cdot h](x)$$

for each element x of X. Thus the right distributive law

$$(f + g) \cdot h = f \cdot h + g \cdot h$$

holds in $(S^X, +, \cdot)$. By a similar argument, the left distributive law also holds
(Exercise 6). The set S^X becomes a ring, the X-th *power* of the ring S, or
the *ring of* S-valued *functions* on the set X. If S is unital, then so is the
power S^X. Its identity element is the function $u : X \to S; x \mapsto 1$ which takes
a constant value of the identity in S at each element x of X. For example,
the set $\mathbb{R}^{\mathbb{R}}$ of all real functions $f : \mathbb{R} \to \mathbb{R}$ forms a unital ring.   ▯

6.2 Distributivity

What is the significance of the distributive laws in a ring $(R, +, \cdot)$? For an element r of R, consider the *left multiplication*

$$R \to R; x \mapsto r \cdot x \tag{6.3}$$

by r. The left distributive law (6.2) states that the function (6.3) is a semigroup homomorphism from $(R, +)$ to itself. By Proposition 5.5, it follows that the left multiplication by r is a group homomorphism from the additive group $(R, +, 0)$ of R to itself. In other words, the multiplication preserves the zero:

$$r \cdot 0 = 0 \tag{6.4}$$

and the negation:

$$r \cdot (-s) = -(rs)$$

for s in R. Furthermore,

$$r \cdot (x - y) = r \cdot x - r \cdot y \tag{6.5}$$

for x and y in R (Exercise 7). In similar fashion, the *right multiplication*

$$R \to R; x \mapsto x \cdot s \tag{6.6}$$

by an element s of R is also a group homomorphism from $(R, +, 0)$ to itself. Thus

$$(-r) \cdot (-s) = r \cdot s \tag{6.7}$$

holds for any r and s in R (Exercise 8). Another useful property is the equation

$$r \cdot 0 = 0 = 0 \cdot r \tag{6.8}$$

for any element r of R. Indeed,

$$0 + r \cdot 0 = r \cdot 0 = r \cdot (0 + 0) = r \cdot 0 + r \cdot 0,$$

the first two equations holding by the group axioms, and the third by left distributivity. Cancellation in the group $(R, +, 0)$ then yields $0 = r \cdot 0$. The other equation in (6.8) is proved similarly (Exercise 9).

In a ring $(R, +, \cdot)$, it is useful to have a so-called *sigma notation* for repeated sums. Let m be an integer. Suppose that x_i is an element of R, for integers $i = m, m+1, m+2, \ldots$. By induction on n, define

$$\sum_{i=m}^{l} x_i = 0$$

for any integer $l < m$, and

$$\sum_{i=m}^{n+1} x_i = x_{n+1} + \sum_{i=m}^{n} x_i \, .$$

Thus

$$\sum_{i=1}^{5} x_i = x_1 + x_2 + x_3 + x_4 + x_5 \, ,$$

for example. Note that the index i in the sigma notation is a *bound* or *dummy variable*, which may be replaced by any other symbol, e.g.,

$$\sum_{i=1}^{n} x_i = \sum_{j=1}^{n} x_j = \sum_{k=0}^{n-1} x_{k+1} \, .$$

Using the sigma notation, we formulate an extension of the distributive laws. (The proof by induction is assigned as Exercise 11.)

PROPOSITION 6.13 (Generalized distributive law.)
Let x_i and y_i be elements of a ring R, for $i = 1, 2, \dots$. Then

$$\left(\sum_{i=1}^{m} x_i \right) \cdot \left(\sum_{j=1}^{n} y_j \right) = \sum_{i=1}^{m} \sum_{j=1}^{n} x_i y_j \tag{6.9}$$

for natural numbers m and n.

COROLLARY 6.14
Let x and y be elements of a ring $(R, +, \cdot)$. Then for integers m and n,

$$(mx) \cdot (ny) = (mn)xy \, . \tag{6.10}$$

PROOF   The proof divides naturally into four cases:

- For $m, n > 0$, set $x_i = x$ and $y_j = y$ in Proposition 6.13.

- For $m < 0$, $n > 0$, set $x_i = -x$ and $y_j = y$ in Proposition 6.13.

- For $m > 0$, $n < 0$, set $x_i = x$ and $y_j = -y$ in Proposition 6.13.

- For $m, n < 0$, set $x_i = -x$ and $y_j = -y$ in Proposition 6.13.

Note that (6.10) is trivial if $m = 0$ or $n = 0$. $\square$

6.3 Subrings

The concept of a subring in a ring combines the concepts of subgroups, subsemigroups, and submonoids.

DEFINITION 6.15 (Unital and nonunital subrings.)

(a) *A subset S of a nonunital ring $(R, +, \cdot)$ is said to be a* (nonunital) *subring of R if S is a subgroup of $(R, +, 0)$ and a subsemigroup of $(R, \cdot)$.*

(a) *A subset S of a unital ring $(R, +, \cdot)$ is said to be a* (unital) *subring of R if S is a subgroup of $(R, +, 0)$ and a submonoid of $(R, \cdot, 1)$.*

It is often left implicit as to whether a subring is asserted to be nonunital or unital. For example, in any ring R, the subset R itself forms a subring, the *improper* subring. It will be always be a nonunital subring, and will be a unital subring if R itself is unital. On the other hand, the trivial ring $\{0\}$ (compare Example 6.6) is a nonunital subring of each ring R. Although the trivial ring is unital, it is not a unital subring of any nontrivial unital ring (in which $0 \neq 1$).

Example 6.16 (Subrings of the integers.)
By Theorem 4.46 (page 81), each subgroup of the additive group $(\mathbb{Z}, +, 0)$ of integers is the set $d\mathbb{Z}$ of multiples of some natural number d. Since the divisibility relation is transitive, each subgroup of $(\mathbb{Z}, +, 0)$ is also a subsemigroup of $(\mathbb{Z}, \cdot)$, and hence a nonunital subring of the unital ring of integers. In fact, since $1\mathbb{Z} = \mathbb{Z}$, the only unital subring of $\mathbb{Z}$ is the improper subring. ▯

To check that a subset S of a ring R is a (nonunital) subring of R, Proposition 4.43 and Remark 4.44 (page 80) show that three properties of S have to be verified:

- S is nonempty;
- x and y in S imply $x - y$ in S;
- x and y in S imply $x \cdot y$ in S.

Example 6.17 (The ring $R[i]$, complex numbers, Gaussian integers.)
Let R be a unital ring, and let $R[i]$ be the set of 2×2 matrices of the form

$$\begin{bmatrix} x & -y \\ y & x \end{bmatrix}$$

for x and y in R. Then $R[i]$ is a unital subring of the ring R_2^2 of all 2×2 matrices over R. Certainly the identity matrix I_2 lies in $R[i]$, and $R[i]$ is closed under the (componentwise) subtraction of matrices. Now

$$\begin{bmatrix} x & -y \\ y & x \end{bmatrix} \cdot \begin{bmatrix} u & -v \\ v & u \end{bmatrix} = \begin{bmatrix} xu - yv & -xv - yu \\ yu + xv & -yv + xu \end{bmatrix} = \begin{bmatrix} (xu - yv) & -(yu + xv) \\ (yu + xv) & (xu - yv) \end{bmatrix}, \quad (6.11)$$

so that $R[i]$ is also closed under multiplication. If R is commutative, it follows from (6.11) that $R[i]$ is also commutative. Here are two important special cases:

- The ring $\mathbb{R}[i]$ is the ring $\mathbb{C}$ of *complex numbers* (compare Study Project 1);

- The subring $\mathbb{Z}[i]$ of $\mathbb{R}[i]$ is known as the ring of *Gaussian integers*.

Also, see Example 7.18 (page 163).   ∏

Let R be a unital ring, with the identity element (exceptionally) denoted by u. The cyclic subgroup $\langle u \rangle$ of $(R, +, 0)$ generated by u, the set of all integral multiples nu of u, forms a unital subring of $(R, +, \cdot)$. Indeed for integers m and n, Corollary 6.14 yields

$$(mu) \cdot (nu) = (mn)(uu) = (mn)u \,.$$

The subring $\langle u \rangle$ is known as the *prime subring* of R. If u has infinite order, the ring R is said to have *characteristic* 0. If u has finite order d, the ring R is said to have *characteristic d*.

Let y be an element of a unital ring $(R, +, \cdot)$. Continuing to write u for the identity element of R, and considering a natural number m, Corollary 6.14 yields

$$(mu)x = m(ux) = mx \qquad (6.12)$$

on setting $n = 1$ and $x = u$. Similarly, we get

$$x(mu) = m(xu) = mx \,. \qquad (6.13)$$

Thus:

PROPOSITION 6.18 (Characteristic and additive order.)
Let $(R, +, \cdot)$ be a unital ring.

(a) *Each element of the prime subring of R commutes with each element of R.*

(b) *If R has finite characteristic d, then in the abelian group $(R, +, 0)$, each element x has finite order dividing d.*

In a unital ring R with identity u and characteristic d, (6.12) and (6.13) imply that we may consistently identify the multiple nu with the element $n + d\mathbb{Z}$ of the integers modulo d for each natural number n. Thus the prime subring of R may be considered as the set $\{n \mid n \text{ in } \mathbb{Z}\}$, with $m = n$ if and only if d divides $m - n$.

Example 6.19

Consider the ring $\mathbb{Z}/2\mathbb{Z} \times \mathbb{Z}/2\mathbb{Z}$, the direct product of two copies of the ring of integers modulo 2. This ring is unital, with identity element $(1,1)$. The prime subring is formed by the diagonal subgroup (compare Exercise 26 in Chapter 5). The characteristic of $\mathbb{Z}/2\mathbb{Z} \times \mathbb{Z}/2\mathbb{Z}$ is 2. $\quad\square$

6.4 Ring homomorphisms

Let $(R, +, \cdot)$ and $(S, +, \cdot)$ be (nonunital) rings.

DEFINITION 6.20 (Ring homomorphisms, isomorphisms.)

(a) *A function $f : R \to S$ is said to be a (nonunital) ring homomorphism if it forms both a group homomorphism $f : (R, +, 0) \to (S, +, 0)$ and a semigroup homomorphism $f : (R, \cdot) \to (S, \cdot)$.*

(b) *If R and S are unital rings, then a ring homomorphism $f : R \to S$ is described as* unital *if $f(1) = 1$, so that $f : (R, \cdot, 1) \to (S, \cdot, 1)$ is a monoid homomorphism.*

(c) *A bijective ring homomorphism is called a (ring) isomorphism. The notation $R \cong S$ means that rings R and S are isomorphic.*

The image $f(R)$ of a ring homomorphism $f : R \to S$ is a subring of the codomain ring S (nonunital or unital, according to whether f is nonunital or unital).

Example 6.21 (Inclusion of subrings.)

Let S be a (nonunital) subring of a ring R. Then the inclusion map

$$j : S \hookrightarrow R; x \mapsto x$$

from S to R is a (nonunital) ring homomorphism. If R is unital, and S is a unital subring, then $j : S \hookrightarrow R$ is a unital ring homomorphism. $\quad\square$

Example 6.22 (Abelian group homomorphisms.)
Let $(A,+,0)$ and $(B,+,0)$ be abelian groups. Let $f : (A,+,0) \to (B,+,0)$
be a group homomorphism. Then there is a (nonunital) ring homomorphism

$$f : (A,+,\cdot_0) \to (B,+,\cdot_0); x \mapsto f(x)$$

of the corresponding zero rings.   ▯

Example 6.23 (Projections and insertions.)
Let R and S be rings. Then the *projections*

$$\pi_1 : R \times S \to R; (x,y) \mapsto x$$

and

$$\pi_2 : R \times S \to S; (x,y) \mapsto y$$

are ring homomorphisms (compare Example 5.4). If R and S are unital, the
projections are unital ring homomorphisms. The *insertions*

$$\iota_1 : R \to R \times S; x \mapsto (x,0)$$

and

$$\iota_2 : S \to R \times S; y \mapsto (0,y)$$

are also ring homomorphisms. (The notation is quite standard: ι is the Greek
letter "iota.") However, even if R and S are unital, the insertion ι_1 is only
unital if S is trivial, since the identity of $R \times S$ is $(1,1)$.   ▯

Example 6.24 (Determinants.)
Although the determinant function (5.2) is a monoid homomorphism for the
usual matrix multiplication, it does not form a ring homomorphism, since

$$\det(A + B) \neq \det A + \det B$$

for general matrices A and B.   ▯

Example 6.25 (Scalar multiples of the identity matrix.)
Let R be a unital ring, and let $R[i]$ be the corresponding ring of matrices
from Example 6.17. Then there is an injective unital ring homomorphism

$$e : R \to R[i]; x \mapsto \begin{bmatrix} x & 0 \\ 0 & x \end{bmatrix}. \tag{6.14}$$

Note that for each element x of R, the image $e(x)$ is the scalar multiple xI_2
of the identity matrix I_2 by the scalar x. It is convenient to identify the ring
R with its isomorphic image $e(R)$ under (6.14), so that R becomes a subring
of the ring $R[i]$.   ▯

We conclude this section with a further examination of the relationship between unital and nonunital rings. It has already been noted that a unital ring becomes nonunital, simply by disregarding the special role of the identity element (Remark 6.3). Now suppose that R is a nonunital ring. Write $R + \mathbb{Z}$ for the direct product set $\{(r, m) \mid r \text{ in } R,\ m \text{ in } \mathbb{Z}\}$. with componentwise additive group structure. Define a multiplication on $R + \mathbb{Z}$ by

$$(r, m) \cdot (s, n) = (rs + ms + nr, mn). \tag{6.15}$$

This multiplication is associative, and distributive over the componentwise addition on each side (Exercise 25). Now $(0, 1) \cdot (r, m) = (r, m) = (r, m) \cdot (0, 1)$ for r in R and m in $\mathbb{Z}$, so $R + \mathbb{Z}$ becomes a unital ring $(R + \mathbb{Z}, +, \cdot)$ with identity element $(0, 1)$. Consider the map $e_R : R \to R + \mathbb{Z}; r \mapsto (r, 0)$. This map is clearly an injective group homomorphism, and from (6.15) it is also seen to be a semigroup homomorphism. Thus f is an injective ring homomorphism. A nonunital ring R is often identified with its isomorphic image $e_R(R)$ under e_R. One may then say that each nonunital ring R embeds into a unital ring $R + \mathbb{Z}$.

6.5 Ideals

Consider a ring homomorphism $f : R \to S$. The group kernel $\operatorname{Ker} f$ or

$$J = \{x \text{ in } X \mid f(x) = 0\}$$

has the so-called *absorptive property*:

$$j \text{ in } J \quad \text{and} \quad x \text{ in } R \quad \text{imply} \quad jx,\ xj \text{ in } J. \tag{6.16}$$

Indeed, for j in J and x in X,

$$f(jx) = f(j)f(x) = 0 \cdot f(x) = 0 = f(x) \cdot 0 = f(x)f(j) = f(xj).$$

DEFINITION 6.26 (Ideal of a ring.) *A subset J of a ring $(R, +, \cdot)$ is said to be an* ideal *of R, written $J \triangleleft R$, if:*

- *It forms a subgroup of $(R, +, 0)$;*
- *It satisfies the absorptive property (6.16).*

Thus we have the ring-theoretic analogue of Proposition 5.11.

PROPOSITION 6.27 (Kernels of homomorphisms are ideals.)
The group kernel $\operatorname{Ker} f$ of a ring homomorphism $f : R \to S$ forms an ideal in the domain ring R.

In the context of Proposition 6.27, the group kernel Ker f is described as the *ring kernel* of the ring homomorphism $f : R \to S$.

Example 6.28 (Ideals of zero rings.)
Let $(A, +, 0)$ be an (additive) abelian group. If J is a subgroup of A, then J trivially satisfies the absorptive property (6.16) under the multiplication $\cdot_0$ of the zero ring $(A, +, \cdot_0)$. Thus J forms an ideal of the zero ring. In fact, J is the ring kernel of the ring homomorphism $(A, +, \cdot_0) \to (A/J, +, \cdot_0)$ furnished by the group homomorphism

$$A \to A/J; x \mapsto x + J$$

(compare Example 6.22). ▯

Since an ideal J of a ring R is by definition a subgroup of $(R, +, 0)$, the absorptive property (6.16) specializes (considering x from J rather than from anywhere in R) to show that ideals are certainly subrings. By (6.8), it is seen that the trivial subring of a ring R forms an ideal of R. The improper subring R of R also forms an ideal, the *improper* ideal. If R is a unital ring, the absorptive property (6.16) shows that an ideal containing the identity element is improper.

DEFINITION 6.29 (Simple rings.)   *A ring R is said to be* simple *if it has no proper, nontrivial ideals.*

Example 6.30 (Rings of prime order are simple.)
Let R be a ring with a prime number of elements. By Proposition 4.51, the additive group $(R, +, 0)$ of the ring R has no proper, nontrivial subgroups. Thus the ring R has no proper, nontrivial ideals: It is simple. For instance, the ring of integers $\mathbb{Z}/p\mathbb{Z}$ (compare Example 6.9) modulo a prime number p forms a simple ring. Similarly, the zero ring $(A, +, \cdot_0)$ determined by an abelian group with a prime number of elements is also simple. ▯

In contrast to simple rings, the ring of integers has many ideals.

Example 6.31 (Ideals in the ring of integers.)
Each subring of the ring $(\mathbb{Z}, +, \cdot)$ is just the set $d\mathbb{Z}$ of multiples of a certain natural number d (Example 6.16). Since divisibility is a transitive relation, it follows that each subring of $\mathbb{Z}$ is actually an ideal of $\mathbb{Z}$. Conversely, since each ideal is a subring, each ideal is of the form $d\mathbb{Z}$ for some natural number d. ▯

The divisibility relation on the set of natural numbers translates nicely to the subset relationship on the set of ideals of $\mathbb{Z}$.

PROPOSITION 6.32 (Ideals and divisibility.)
Let c and d be natural numbers. Then d divides c if and only if the ideal $c\mathbb{Z}$ is a subset of the ideal $d\mathbb{Z}$.

PROOF If c is a multiple of d, then c lies in $d\mathbb{Z}$, so the ideal $c\mathbb{Z}$ is a subset of the ideal $d\mathbb{Z}$. Conversely, if $c\mathbb{Z}$ is a subset of $d\mathbb{Z}$, then $c \cdot 1 = c$ belongs to the set $d\mathbb{Z}$ of multiples of d. $\square$

The concept of an ideal arose in an attempt to extend divisibility properties of integers to more general rings. It transpired that the sets of multiples were more amenable than the actual elements themselves, so these sets were considered as "ideal numbers."

6.6 Quotient rings

If J is an ideal in a ring $(R, +, \cdot)$, then we may form the quotient group R/J, the set of cosets $x + J$ of the (normal) subgroup J in the abelian group $(R, +, 0)$. Now since J is an ideal in R, there is a well-defined semigroup multiplication on R/J given by

$$(x + J)(y + J) = xy + J$$

for x, y in R. Suppose that the respective cosets $x + J$ and $y + J$ are also represented by ring elements x' and y', so that $x' = x + j$ and $y' = y + k$ for elements j and k of J. Then

$$x'y' + J = (x + j)(y + k) + J = xy + (xk + jy + jk + J) = xy + J,$$

since $xk + jy + jk$ is an element of the ideal J. The distributivity of the multiplication over the addition in R/J follows from the distributivity in R (Exercise 34). Thus the quotient group R/J actually forms a *quotient ring*. The group homomorphism

$$s : R \to R/J; x \mapsto x + J \tag{6.17}$$

becomes a ring homomorphism. If R is unital, then so is the quotient R/J, with identity element $1 + J$. In this case the function (6.17) becomes a unital ring homomorphism.

Example 6.33 (Modular arithmetic.)
For a positive integer d, the ring $\mathbb{Z}/d\mathbb{Z}$ of integers modulo d (Example 6.9) is the quotient of the ring $\mathbb{Z}$ of integers by the ideal $d\mathbb{Z}$ consisting of all multiples of d. $\square$

Just as for groups (Section 5.4), a ring homomorphism factorizes under a strengthened version of the First Isomorphism Theorem.

THEOREM 6.34 (First Isomorphism Theorem for Rings.)
Let $f : (R, +, \cdot) \to (S, +, \cdot)$ be a ring homomorphism.

(a) *The group kernel $K = f^{-1}\{0\}$ is a ideal in the domain ring R.*

(b) *The image $f(R)$ is a subring of the codomain ring S.*

(c) *In the factorization*

$$f = j \circ b \circ s$$

given by the First Isomorphism Theorem for Sets, the surjection s may be taken as the surjective homomorphism

$$s : X \to X/K; x \mapsto x + K$$

of (6.17), the bijection b is the well-defined ring isomorphism

$$b : R/K \to f(R); x + K \mapsto f(x)$$

from the quotient R/K to the image $f(R)$, and the injection j is the injective ring homomorphism

$$j : f(R) \hookrightarrow S; f(x) \mapsto f(x)$$

of Example 6.21.

Example 6.35 (Kernels and images of projections and insertions.)
Let R and S be rings. In Example 6.23, the ring kernel of the surjective projection $\pi_2 : R \times S \to S$ is the image $\iota_1(R)$ of the ring R under the insertion ι_1. Thus the First Isomorphism Theorem exhibits an isomorphism

$$(R \times S)/\iota_1(R) \cong S .$$

Similarly, $(R \times S)/\iota_2(S) \cong R$ (Example 35).   ⬚

6.7   Polynomial rings

Let R be a ring. An *indeterminate* X (over R) is a symbol that is not related to any element of R. A *polynomial* over R in an indeterminate X is an expression of the form

$$p(X) = p_n X^n + p_{n-1} X^{n-1} + \cdots + p_2 X^2 + p_1 X + p_0 . \qquad (6.18)$$

Here n is a natural number, while p_n, p_{n-1}, $\ldots$, p_2, p_1, p_0 are elements of R known as the *coefficients* of the polynomial $p(X)$. Specifically, p_i (for $0 \leq i \leq n$) is called the *coefficient of X^i* in $p(X)$. The individual summands $p_i X^i$ in (6.18) are called the *terms* of the polynomial. (The final summand p_0 may be written as $p_0 X^0$.) If R is unital, a term $1X^i$ may be written just as X^i. If a coefficient p_i happens to be zero, then the term $p_i X^i$ need not be written explicitly. For example, $1X^2 + 1$ and $X^2 + 2X + 1$ denote the same polynomial if R is the ring of integers modulo 2. Continuing this convention, two polynomials, say $p(X)$ as in (6.18) and

$$q(X) = q_m X^m + \cdots + q_1 X + q_0 , \qquad (6.19)$$

are defined to be equal if

$$p_r = q_r , \ p_{r-1} = q_{r-1} , \ \ldots, \ p_2 = q_2 , \ p_1 = q_1 , \ p_0 = q_0$$

for $r = \min(m, n)$ and

$$p_i = 0, \quad q_i = 0$$

for $i > r$. (This is the process of *equating coefficients*.) For example, the polynomials

$$25X^2 + X - 3$$

and

$$0X^3 + 25X^2 + X - 3$$

are equal if R is the ring $\mathbb{Z}$ of integers. The set of all polynomials in X over R is written as $R[X]$. It includes the ring R itself, as the set of *constant polynomials* p_0. (Note that the single zero coefficient cannot be omitted from the constant polynomial 0.)

The set $R[X]$ of polynomials in X over R inherits a componentwise additive abelian group structure $(R[X], +, 0)$ from R. Thus if $n = m$, the sum of $p(X)$ from (6.18) and $q(X)$ from (6.19) is

$$(p + q)(X) = (p_n + q_n)X^n + (p_{n-1} + q_{n-1})X^{n-1} + \ldots$$
$$\cdots + (p_2 + q_2)X^2 + (p_1 + q_1)X + (p_0 + q_0) .$$

The negative $-p(X)$ of $p(X)$ is the polynomial

$$(-p)(X) = -p_n X^n - p_{n-1} X^{n-1} - \cdots - p_2 X^2 - p_1 X - p_0$$

obtained from $p(X)$ by negating all the coefficients. Using the sigma notation introduced in Section 6.2, the polynomial $p(X)$ of (6.18) may be written as

$$p(X) = \sum_{i=0}^{n} p_i X^i$$

(admitting $i = 0$ as a possible index).

A semigroup structure $(R[X], \cdot)$ will now be defined, in such a way that the multiplication $\cdot$ is distributive over the addition in the group $(R[X], +, 0)$. To specify such a multiplication, it suffices to set

$$X^i \cdot X^j = X^{i+j}$$

and

$$r \cdot X^i = rX^i = X^i \cdot r$$

for elements r of R and for natural numbers i, j. Thus for the polynomials $p(X)$ from (6.18) and $q(X)$ from (6.19), distributivity gives

$$\left(\sum_{i=0}^{n} p_i X^i\right) \cdot \left(\sum_{i=0}^{m} q_i X^i\right) = \sum_{k=0}^{n+m} \left(\sum_{j=0}^{k} p_j q_{k-j}\right) X^k \,. \qquad (6.20)$$

Note that coefficients such as p_{n+m} appearing in (6.20) (on the right-hand side for $k = n + m$ and $j = k$), but not in (6.18) or (6.19), are taken to be zero. In effect, (6.20) is modeled on the usual product of real polynomials in $\mathbb{R}[X]$, systematically collecting all the terms involving a specific power X^r of the indeterminate X. For example, (6.20) gives

$$(5X + 1)(7X^2 + 3X + 2)$$
$$= (1 \cdot 2) + (1 \cdot 3 + 5 \cdot 2)X + (1 \cdot 7 + 5 \cdot 3)X^2 + (5 \cdot 7)X^3$$
$$= 35X^3 + 22X^2 + 13X + 2 \,.$$

The multiplication may be displayed schematically as follows.

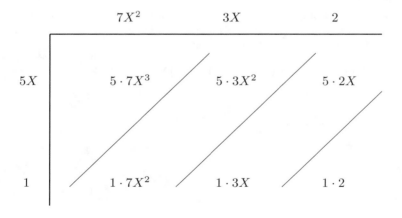

Equal powers of the indeterminate X are collected along the diagonal stripes, corresponding to the right-hand side of (6.20).

THEOREM 6.36

Let R be a ring, and let X be an indeterminate.

(a) *The set $R[X]$ of polynomials over R in an indeterminate X forms a ring with componentwise additive group structure, and with the multiplication (6.20).*

(b) *The ring $(R, +, \cdot)$ is a subring of $(R[X], +, \cdot)$, namely the subring of constant polynomials.*

(c) *If R is commutative, then the ring $(R[X], +, \cdot)$ is commutative.*

(d) *If R is unital, then the ring $(R[X], +, \cdot)$ is unital, with identity element given by the constant polynomial 1.*

PROOF   Most of Theorem 6.36 is straightforward. We will just see how the associativity of the multiplication (6.20) is verified. To do this, it is helpful to rewrite (6.20) in the form

$$\left(\sum_{i=0}^{n} p_i X^i \right) \cdot \left(\sum_{j=0}^{m} q_j X^j \right) = \sum_{h=0}^{n+m} \left(\sum_{i+j=h} p_i q_j \right) X^h \qquad (6.21)$$

with the convention that sums like those in the coefficient of X^k on the right-hand side of (6.21) are understood to be taken over all natural numbers i and j satisfying the specified condition $i + j = k$. Consider the polynomials (6.18), (6.19), and

$$r(X) = \sum_{k=0}^{l} r_k X^k .$$

Using the new convention, we have

$$\left[\left(\sum_{i=0}^{n} p_i X^i \right) \cdot \left(\sum_{j=0}^{m} q_j X^j \right) \right] \cdot \left(\sum_{k=0}^{l} r_k X^k \right)$$

$$= \left[\sum_{h=0}^{n+m} \left(\sum_{i+j=h} p_i q_j \right) X^h \right] \cdot \left(\sum_{k=0}^{l} r_k X^k \right)$$

$$= \sum_{g=0}^{n+m+l} \left[\sum_{h+k=g} \left(\sum_{i+j=h} p_i q_j \right) r_k \right] X^g$$

$$= \sum_{g=0}^{n+m+l} \left(\sum_{i+j+k=g} p_i q_j r_k \right) X^g$$

$$= \sum_{g=0}^{n+m+l} \left[\sum_{i+h=g} p_i \left(\sum_{j+k=h} q_j r_k \right) \right] X^g$$

$$= \left(\sum_{i=0}^{n} p_i X^k \right) \cdot \left[\sum_{h=0}^{m+l} \left(\sum_{j+k=h} q_j r_k \right) X^h \right]$$

$$= \left(\sum_{i=0}^{n} p_i X^k \right) \cdot \left[\left(\sum_{j=0}^{m} q_j X^j \right) \cdot \left(\sum_{k=0}^{l} r_k X^k \right) \right]$$

as verification of the associative law in $R[X]$.   □

DEFINITION 6.37 (The ring of polynomials.)   *The ring $(R[X], +, \cdot)$ of Theorem 6.36 is called the* polynomial ring *or* ring of polynomials *over R in the indeterminate X.*

Example 6.38 (Binomial coefficients.)
Let n be a natural number. The *binomial coefficients* are defined as the coefficients $\binom{n}{r}$ of the polynomial

$$(1+X)^n = \sum_{r=0}^{n} \binom{n}{r} X^r \tag{6.22}$$

in $\mathbb{Z}[X]$. Note that

$$\binom{n}{0} = 1 = \binom{n}{n}$$

for all natural numbers n. Further,

$$\binom{n}{r} = 0$$

for $r > n$, by the convention allowing additional zero terms in the polynomial (6.22). Since

$$(1+X)^{n+1} = (1+X)^n \cdot (1+X) \tag{6.23}$$

by (5.10) in the monoid $(\mathbb{Z}[X], \cdot, 1)$, we obtain the *recurrence relation*

$$\binom{n+1}{r+1} = \binom{n}{r+1} + \binom{n}{r} \tag{6.24}$$

for natural numbers r from the definition (6.20) of the multiplication in the polynomial ring $\mathbb{Z}[X]$, equating coefficients of X^{r+1} on each side of (6.23). □

6.8 Substitution

In Section 6.7, polynomials were described as formal combinations of powers of an indeterminate X. On the other hand, polynomials appear in calculus as special kinds of functions. Functions are obtained from polynomials as follows.

THEOREM 6.39 (**Substitution Principle.**)
Suppose that

$$\theta : R \to S$$

is a homomorphism between commutative rings R and S. Then for each fixed element c of the ring S, there is a unique homomorphism

$$\theta_c : R[X] \to S$$

with $\theta_c(X) = c$. If $\theta : R \to S$ is a homomorphism of unital rings, then so is $\theta_c : R[X] \to S$.

PROOF   Consider a polynomial

$$p(X) = p_n X^n + p_{n-1} X^{n-1} + \cdots + p_2 X^2 + p_1 X + p_0$$

in $R[X]$. If $\theta_c : R[X] \to S$ is any ring homomorphism with $\theta_c(X) = c$, then

$$\theta_c\big(p(X)\big) = \theta(p_n)c^n + \theta(p_{n-1})c^{n-1} + \cdots + \theta(p_2)c^2 + \theta(p_1)c + \theta(p_0), \quad (6.25)$$

so that θ_c is unique. Indeed, since $\theta : R \to S$ is a ring homomorphism, it is straightforward to check that the map

$$\theta_c : R[X] \to S$$

defined by (6.25) is a ring homomorphism (Exercise 37).   ▯

DEFINITION 6.40 (**Evaluation, roots.**)   *Let S be a subring of a ring R, and let*

$$j : S \hookrightarrow R; x \mapsto x$$

be the inclusion map from S to R (compare Example 6.21). Let $p(X)$ be a polynomial in $S[X]$.

(a) *For each element c of R, the image $j_c\big(p(X)\big)$ in R is written as $p(c)$, and described as the* value *of the polynomial $p(X)$ at the element c.*

(b) *The function*

$$R \to R; c \mapsto p(c)$$

is called evaluation *of the polynomial $p(X)$, or the* polynomial function *determined by the polynomial $p(X)$.*

(c) *The element c of R is said to be a* root *or* zero *of the polynomial $p(X)$ if $p(c) = 0$.*

Example 6.41
Consider the polynomial $X^2 - 2$ in $\mathbb{Z}[X]$. Then the real number $\sqrt{2}$ is a root of $X^2 - 2$ in $\mathbb{R}$. ☐

Example 6.42
Consider the polynomial $X^2 - 1$ in $\mathbb{Z}/8[X]$. Then 1, 3, 5, and 7 are four distinct roots of the quadratic polynomial $X^2 - 1$ in $\mathbb{Z}/8$. ☐

Example 6.43
Consider the polynomial $p(X) = X^2 + X$ in $\mathbb{Z}/2[X]$. Then 0 and 1 are roots of $X^2 + X$ in $\mathbb{Z}/2$. In other words, although the polynomial $p(X)$ is nonzero, it determines the zero function

$$\mathbb{Z}/2 \to \mathbb{Z}/2; 0 \mapsto 0, \; 1 \mapsto 0$$

as its polynomial function. ☐

The final result shows that any particular choice of an indeterminate is irrelevant.

PROPOSITION 6.44 (Indeterminacy of indeterminates.)
Let R be a commutative ring. Let X and Y be indeterminates over R. Then the polynomial rings $R[X]$ and $R[Y]$ are isomorphic.

PROOF   Let $j : R \hookrightarrow R[X]$ be the inclusion of R as a subring of $R[X]$ — compare Theorem 6.36(b). Let $k : R \hookrightarrow R[Y]$ be the corresponding inclusion in $R[Y]$. Then the ring homomorphisms

$$j_Y : R[X] \to R[Y]; p(X) \mapsto p(Y)$$

and

$$k_X : R[Y] \to R[X]; p(Y) \mapsto p(X)$$

are mutually inverse. ☐

6.9 Exercises

1. Give an example of a set A with an additive group structure $(A, +, 0)$ and a semigroup structure $(A, \cdot)$ such that the right distributive law is satisfied, but not the left. Justify your claims.

2. Let d be a positive integer. Use (3.21) and (3.22), along with the right distributive law for $(\mathbb{Z}, +, \cdot)$, to give a formal proof of the left distributive law in $(\mathbb{Z}/d\mathbb{Z}, +, \cdot)$.

3. Verify the claims of Example 6.10.

4. Let R be a zero ring (as in Example 6.6). Show that the ring R_2^2 of 2×2 matrices over R is commutative.

5. Verify the claims of Example 6.11.

6. Let X be a set, and let S be a ring. Give a careful proof that the left distributive law holds in $(S^X, +, \cdot)$.

7. Prove that (6.5) holds in each ring R.

8. Prove that (6.7) holds in each ring R.

9. Prove that the right-hand equation in (6.8) holds in each ring R.

10. Write the left-hand side of (1.8) using sigma notation.

11. Prove Proposition 6.13.

12. Show that the right and left distributive laws are both special cases of the generalized distributive law.

13. Suppose that

$$xx = x$$

for each element x of a ring R.

   (a) Show that

$$x + x = 0$$

   for all x in R.

   (b) Show that R is commutative.

14. Let S be a subring of a ring R. Show that S_2^2 is a subring of R_2^2.

15. Let X be a subset of a ring R. Show that

$$C_R(X) = \{r \text{ in } R \mid rx = xr \text{ for all } x \text{ in } X\}$$

is a subring of R. [The subring $C_R(X)$ is known as the *commutant* of X in R.]

16. Let S be the set of all functions $f : \mathbb{R} \to \mathbb{R}$ with $f(0) = 0$. Show that S forms a nonunital subring of the function ring $\mathbb{R}^{\mathbb{R}}$ of all real-valued functions. Does it form a unital subring? Does it form a unital ring?

17. Consider the subset $S = \{0, 2, 4, 6, 8\}$ of $\mathbb{Z}/10\mathbb{Z}$.

    (a) Show that $S = \{0, 2, 4, 6, 8\}$ forms a subring of $\mathbb{Z}/10\mathbb{Z}$.

    (b) Show that S forms a unital ring.

    (c) Show that S is not a unital subring of $\mathbb{Z}/10\mathbb{Z}$.

18. Specify the group of units of the monoid of Gaussian integers under multiplication.

19. Let d be a positive integer. Consider the ring $(\mathbb{Z}/d)_2^2$ of 2×2 matrices over the ring $\mathbb{Z}/d$ of integers modulo d.

    (a) Show that the prime subring of $(\mathbb{Z}/d)_2^2$ is the set

$$\left\{ \begin{bmatrix} x & 0 \\ 0 & x \end{bmatrix} \,\middle|\, x \text{ in } \mathbb{Z}/d \right\}$$

    of diagonal matrices with equal diagonal entries.

    (b) Show that the characteristic of $(\mathbb{Z}/d)_2^2$ is d.

20. What is the prime subring of the ring $\mathbb{R}_2^2$ of 2×2 real matrices?

21. Suppose $d > 1$.

    (a) Show that the ring $\mathbb{Z}/d\mathbb{Z} \times \mathbb{Z}/d\mathbb{Z}$ is unital, with identity element $(1, 1)$.

    (b) Show that the characteristic of $\mathbb{Z}/d\mathbb{Z} \times \mathbb{Z}/d\mathbb{Z}$ is d.

22. What is the characteristic of the ring $\mathbb{Z}/4\mathbb{Z} \times \mathbb{Z}/2\mathbb{Z}$?

23. What is the characteristic of the ring $\mathbb{Z}/3\mathbb{Z} \times \mathbb{Z}/2\mathbb{Z}$?

24. Suppose that $x^3 = x$ for each element x of a unital ring R. Show that R has a finite characteristic d that is a divisor of 6.

25. Verify that the multiplication (6.15) is associative, and that it distributes on each side over the componentwise addition on $R + \mathbb{Z}$.

26. Let R be a ring. Identify the prime subring of the unital ring $R + \mathbb{Z}$.

27. Let r be an element of a unital ring R. Show that

$$(r, 0) \cdot (1, -1) = (0, 0)$$

in $R + \mathbb{Z}$.

28. Let R be a unital ring having finite characteristic d. Determine the characteristic of the unital ring $R + \mathbb{Z}$.

29. Let R be a ring. Show that there is a surjective ring homomorphism

$$p : R + \mathbb{Z} \to \mathbb{Z}; (r, m) \mapsto m.$$

What is the ring kernel $\operatorname{Ker} p$?

30. Let x be an element of a commutative, unital ring R. Show that xR is an ideal of R.

31. Let x and y be elements of a commutative, unital ring R. Show that $xR + yR$ is an ideal of R.

32. Let a and b be positive integers. Show that $a\mathbb{Z} + b\mathbb{Z} = \gcd(a, b)\mathbb{Z}$.

33. Let x be an element of a unital ring R. Show that

$$xR + Rx + RxR$$

is an ideal of R. Note that RxR means $\{\sum_{j=0}^{n} s_j x t_j \mid n \text{ in } \mathbb{N}; \ s_j, t_j \text{ in } R\}$.

34. Let J be an ideal of a ring R. Show that the right and left distributive laws hold in R/J.

35. Let R and S be rings. Show that $(R \times S)/\iota_2(S) \cong R$.

36. Let n, m, and r be natural numbers. Use the definition (6.20) of the product in the polynomial ring $\mathbb{Z}[X]$ to prove the identity

$$\binom{n+m}{r} = \sum_{j=0}^{r} \binom{n}{j} \binom{m}{r-j}$$

for binomial coefficients.

37. Complete the proof of Theorem 6.39: Show that the map

$$\theta_c : R[X] \to S$$

defined by (6.25) is a homomorphism of the additive group and semi-group (or monoid) structures on $R[X]$ and S.

38. Determine all the roots of the polynomial $X^2 + 1$ in the ring $\mathbb{Z}/_5$ of integers modulo 5.

39. Determine all the roots of the polynomial $X^4 - X$ in the ring $\mathbb{Z}/_4$ of integers modulo 4.

40. Prove the identity

$$\sum_{r=0}^{n} \binom{n}{r} = 2^n$$

for natural numbers n. [Hint: Evaluate (6.22) at the integer 1.]

41. Prove the identity

$$\sum_{r=0}^{n} (-1)^r \binom{n}{r} = 0$$

for positive integers n.

42. Prove the identity

$$\sum_{r=0}^{n/2} \binom{n}{2r} = \sum_{r=1}^{n/2} \binom{n}{2r-1}$$

for even natural numbers n.

43. For natural numbers $r \leq n$, show that

$$\binom{n}{r} = \frac{n!}{(n-r)!r!} . \qquad (6.26)$$

In particular, conclude that (6.26) gives the number of r-element subsets of an n-element set.

44. For a prime number p and an integer r with $0 < r < p$, show that the binomial coefficient

$$\binom{p}{r}$$

is divisible by p.

45. Let R be a commutative, unital ring of prime characteristic p. Show that the *Frobenius map*

$$\varphi : R \to R; x \mapsto x^p \qquad (6.27)$$

is a ring homomorphism.

46. Show that for each element r of a commutative, unital ring R, there is a unique unital ring homomorphism $\phi : \mathbb{Z}[X] \to R$ with $\phi(X) = r$.

6.10 Study projects

1. **Complex numbers.** Consider the commutative ring $\mathbb{R}[i]$ of all matrices

$$z = \begin{bmatrix} x & -y \\ y & x \end{bmatrix} \tag{6.28}$$

with real entries x and y (compare Example 6.17, page 133). These matrices are known as *complex numbers*, and the set $\mathbb{R}[i]$ is written as $\mathbb{C}$. Recall that the ring $\mathbb{R}$ of real numbers appears as a subring of $\mathbb{C}$, namely as the ring of scalar multiples xI_2 of the identity matrix I_2 by real scalars x (Example 6.25, page 136). There is an additive group isomorphism

$$\mathbb{C} \to \mathbb{R}^2; \begin{bmatrix} x & -y \\ y & x \end{bmatrix} \mapsto (x, y) \tag{6.29}$$

from the set of complex numbers z to the real plane $\mathbb{R}^2$. Each complex number (6.28) is often identified with its image (x, y) in the plane.

Define the *modulus* of the complex number z as the square root

$$|z| = \left\{ \det \begin{bmatrix} x & -y \\ y & x \end{bmatrix} \right\}^{\frac{1}{2}} = \sqrt{x^2 + y^2}$$

of the (nonnegative) determinant. In the plane representation (6.29), the modulus of z is the distance of the point (x, y) from the origin $(0, 0)$. Write the transpose of the matrix z as the *complex conjugate*

$$\overline{z} = \begin{bmatrix} x & -y \\ y & x \end{bmatrix}^T = \begin{bmatrix} x & y \\ -y & x \end{bmatrix}.$$

In the plane representation (6.29), complex conjugation corresponds to reflection $(x, y) \mapsto (x, -y)$ in the x-axis.

(a) Show that
$$|z|^2 = z \cdot \overline{z}.$$

(b) Show that
$$z = 0 \quad \text{if and only if} \quad |z| = 0.$$

(c) Show that the set
$$S^1 = \{z \mid 1 = |z|\} \tag{6.30}$$

of complex numbers of unit modulus is the special orthogonal group $SO_2(\mathbb{R})$ — compare Study Project 2 in Chapter 4. Note that in the plane representation (6.29), the set (6.30) is the *unit circle*, the circle of points at radius 1 from the origin.

(d) Given a second element

$$w = \begin{bmatrix} u & -v \\ v & u \end{bmatrix}$$

of $\mathbb{C}$, show that

$$|z \cdot w| = |z| \cdot |w|.$$

[Hint: Use Example 5.7 (page 97) and Exercise 7 of Chapter 5.]

(e) Rewrite the multiplication formula (6.11) for the product $z \cdot w$ of two complex numbers z and w using the plane representations (x, y) and (u, v).

(Note: Although this format is often used for the definition of the product of complex numbers, it gives no hint as to why the product should be associative. On the other hand, the definition of the complex numbers as a subring of the ring of real 2×2 matrices removes the need for any extra verification. It also enables the modulus and complex conjugation to be reduced to the standard matrix concepts of determinant and transpose.)

(f) For the complex number

$$i = \begin{bmatrix} 0 & -1 \\ 1 & 0 \end{bmatrix},$$

show that $i^2 = -1$ (recalling that the identity 1 of $\mathbb{C}$ is the matrix I_2).

(g) Show that each complex number can be written in the form

$$\begin{bmatrix} x & -y \\ y & x \end{bmatrix} = z = x + iy. \tag{6.31}$$

(h) Show that each nonzero complex number z is invertible, and that its inverse is given by

$$z^{-1} = |z|^{-2} \cdot \overline{z}. \tag{6.32}$$

(i) For a nonzero complex number z, show that the complex number $|z|^{-1}z$ is an element

$$\begin{bmatrix} \cos\theta & -\sin\theta \\ \sin\theta & \cos\theta \end{bmatrix}$$

of the unit circle S^1.

(j) Show that each nonzero complex number z can be written in the *polar form*

$$z = r(\cos\theta + i\sin\theta)$$

with real $r = |z| > 0$ and $0 \leq \theta < 2\pi$.

2. **Check digits.** The rings $\mathbb{Z}/d$ of integers modulo a positive number d are often used to mitigate the effect of errors in the recording and reporting of long sequences of digits. For example, consider a 13-digit EAN barcode, as illustrated in Figure 6.1.

FIGURE 6.1: A (simulated) 13-digit EAN barcode.

Here, it is actually the sequence

$$a_{12} \;\|\; a_{11} \, a_{10} \;\ldots\; a_6 \;\|\; a_5 \;\ldots\; a_2 \, a_1 \qquad \| \qquad (6.33)$$

of the left-most 12 decimal digits which carries the information (type of product, manufacturer, etc.). The final digit a_0 is known as the *check digit*. It is chosen so that the equation

$$1 \cdot a_{12} + 3 \cdot a_{11} + 1 \cdot a_{10} + \cdots + 1 \cdot a_2 + 3 \cdot a_1 + 1 \cdot a_0 = 0 \qquad (6.34)$$

is satisfied in the ring $\mathbb{Z}/10$ of integers modulo 10.

(a) Confirm whether the condition (6.34) is satisfied by the example illustrated in Figure 6.1.

(b) Confirm whether the condition (6.34) is satisfied by the 13-digit EAN barcode on this book.

(c) Show that for each 12-digit sequence (6.33) of decimal digits, there is a unique check digit a_0 that may be added to ensure the satisfaction of (6.34).

(d) Suppose that a full 13-digit sequence

$$a_{12} \;\|\; a_{11} \, a_{10} \;\ldots\; a_6 \;\|\; a_5 \;\ldots\; a_2 \, a_1 \, a_0 \;\|$$

satisfying (6.34) was placed on a product, but that a certain digit a_k (with $12 \geq k \geq 0$) becomes illegible. Show that the value of a_k may be recovered from the 12 remaining legible digits. (Hint: In the ring $\mathbb{Z}/10$, the element 3 is a unit.)

(e) Explain why the error-correction capability described in (d) above is weaker than that given by the error-correcting codes introduced in Section 5.8.

(f) Suppose that a full 13-digit sequence

$$a_{12} \parallel a_{11} \ a_{10} \ \ldots \ a_6 \parallel a_5 \ \ldots \ a_2 \ a_1 \ a_0 \parallel$$

satisfying (6.34) is being reported manually, but a keyboard error transposes two (distinct) adjacent digits $a_{k+1} \ a_k$ (with $12 > k \geq 0$), so that they appear in the order $a_k \ a_{k+1}$ instead. Show that a check of the validity of (6.34) for the erroneous sequence

$$a_{12} \ \ldots \ a_{k+2} \ \overleftrightarrow{a_k \ a_{k+1}} \ a_{k-1} \ \ldots \ a_0$$

will indicate that an error occurred, unless $|a_{k+1} - a_k| = 5$.

(g) The *International Standard Book Number* (ISBN) is a 10-digit code

$$a_{10} \ a_9 \ \ldots \ a_3 \ a_2 \text{ - } a_1 \tag{6.35}$$

in which the final digit a_1 is a check digit. The 9 information digits $a_{10}, \ldots, a_2$ are actually decimal, but they are interpreted as if they were undecimal (base 11). Working in the ring of integers $\mathbb{Z}/11$ modulo 11, the ISBN (6.35) is required to satify the condition

$$\sum_{k=1}^{10} k \cdot a_k = 0. \tag{6.36}$$

Show that for each sequence

$$a_{10} \ a_9 \ \ldots \ a_3 \ a_2$$

of information digits, there is a unique element a_1 of $\mathbb{Z}/11$ that may be appended to ensure that (6.36) is satisfied. (The Roman numeral X is used as a single undecimal digit for 10.)

(h) Confirm that ISBN 981-02-4942-X satisfies the condition (6.36).

(i) Confirm that the ISBN of this book satisfies the condition (6.36).

(j) Compute the full ISBN when the sequence of information digits is 83-89656-20.

(k) Show that if a single digit in a valid ISBN is illegible, it can be recovered uniquely from the remaining 9 legible digits.

(l) Investigate the ability of ISBN encoding to recognize the occurrence of a transposition error, as discussed in (e) above for the EAN code.

3. **Board games.** Certain board games are played with two dice. The board is marked with squares. When it is their turn, players move their markers forward between 2 and 12 steps, according to the total of the two numbers between 1 and 6 shown on each die.

0	1	2	3	4	5	6	7	8	9	10	11	12

FIGURE 6.2: Part of a board game.

During the course of the game, players frequently end up with their markers on the square labeled 0 in Figure 6.2. To help win the game, it is useful to be able to predict which of the squares 2 through 12 these players will reach after their next move.

Assuming that a die is fair, each of the six possible numbers 1, 2, ..., 6 on its faces is equally likely. Thus the chance of throwing any given number is $\frac{1}{6}$. However, when two dice are thrown, the possible totals 2, 3, ..., 12 are not equally likely. The chances of throwing a given total may be computed within the ring $\mathbb{R}[X]$ of real polynomials.

(a) Let
$$p_1(X) = p_2(X) = \frac{1}{6}X^6 + \frac{1}{6}X^5 + \frac{1}{6}X^4 + \frac{1}{6}X^3 + \frac{1}{6}X^2 + \frac{1}{6}X,$$
so the chance of throwing r with die i (for $i = 1, 2$) is given by the coefficient of X^r in the polynomial $p_i(X)$.

(b) Show that the chance of throwing a total of t with the two dice is given by the coefficient of X^t in the product polynomial
$$p_1(X) \cdot p_2(X).$$

(c) Compute the product $p_1(X) \cdot p_2(X)$.

(d) If a player starts with their marker on square 0 in Figure 6.2, on which square are they most likely to land at their next turn?

(e) If a player starts with their marker on square 0 in Figure 6.2, on which squares are they least likely to land at their next turn?

(f) By what factor are they more likely to land on the most likely square, as opposed to one of the least likely squares?

4. **Finite support.** Let X be a set, and let $(R, +, \cdot)$ be a ring. Consider a function
$$f : X \to R.$$

The set
$$\operatorname{supp} f = \{x \text{ in } X \mid f(x) \neq 0\}$$

of elements of X at which the function f takes a nonzero value is called the *support* of f. The function $f : X \to R$ is said to have *finite support* if the set $\operatorname{supp} f$ is finite. Let $f : X \to R$ and $g : X \to R$ be functions with finite support.

(a) Show that the zero constant function has empty support.

(b) Show that $\operatorname{supp}(f - g)$ is a subset of the union of $\operatorname{supp} f$ and $\operatorname{supp} g$.

(c) Show that $\operatorname{supp}(f \cdot g)$ is a subset of the intersection of $\operatorname{supp} f$ and $\operatorname{supp} g$.

(d) Conclude that the set R_0^X of functions $f : X \to R$ with finite support forms a nonunital subring of the ring of R-valued functions on X.

(e) If X is infinite and R is unital, show that the set R_0^X of functions $f : X \to R$ with finite support does not form a unital subring of the ring of R-valued functions on X.

6.11   Notes

Section 6.1

Many authors do not distinguish explicitly between unital and nonunital rings, leaving the determination to the context.

Section 6.3

C.F. Gauss was a German mathematician and physicist who lived from 1777 to 1855.

Section 6.9

F.G. Frobenius was a German mathematician who lived from 1849 to 1917.

Chapter 7

FIELDS

Amongst all rings, the ring $\mathbb{Z}$ of integers and the ring $\mathbb{R}$ of real numbers have special properties that make them easy to manipulate. For example, nonzero integers may be canceled from equations, and nonzero real numbers have multiplicative inverses. The special properties are formalized in the concepts of "integral domain" and "field."

7.1 Integral domains

In a general ring R, let $R^\times$ denote the set

$$\{x \text{ in } R \mid x \neq 0\}$$

of nonzero elements of R. The concepts of "integral domain" and "field" address the algebraic properties of the set $R^\times$.

DEFINITION 7.1 (Integral domains.) *A ring $(R, +, \cdot)$ is said to be an* integral domain *(abbreviated as:* ID) *if it is both commutative and unital, and if the set $R^\times$ of nonzero elements of R forms a monoid $(R^\times, \cdot, 1)$ under the multiplication of the ring.*

Example 7.2 (The ring of integers.)
The ring $\mathbb{Z}$ of integers is the prototypical integral domain. ⬛

Example 7.3 (The ring of real numbers.)
The ring $\mathbb{R}$ of real numbers forms an integral domain. ⬛

Example 7.4 (A product ring.)
Consider the nonzero elements $(1, 0)$ and $(0, 1)$ in the direct product $\mathbb{Z}/_2 \times \mathbb{Z}/_2$. Then
$$(1, 0) \cdot (0, 1) = (1 \cdot 0, 0 \cdot 1) = (0, 0),$$
so the commutative, unital ring $\mathbb{Z}/_2 \times \mathbb{Z}/_2$ is not an integral domain. ⬛

Example 7.5 (Integers modulo 3.)
The multiplication table for the set $\mathbb{Z}/_3^\times$ of nonzero elements of the ring $\mathbb{Z}/_3$ of integers modulo 3 is as follows:

$\cdot$	1	2
1	1	2
2	2	1

Thus $\mathbb{Z}/_3$ forms an integral domain.   ▯

Example 7.6 (Integers modulo 4.)
The multiplication table for the set $\mathbb{Z}/_4^\times$ of nonzero elements of the ring $\mathbb{Z}/_4$ of integers modulo 4 is as follows:

$\cdot$	1	2	3
1	1	2	3
2	2	0	2
3	3	2	1

Note that $2 \cdot 2 = 0$: The set $\mathbb{Z}/_4^\times$ of nonzero elements of $\mathbb{Z}/_4$ is not closed under multiplication. Thus $\mathbb{Z}/_4$ does not form an integral domain.   ▯

The commutative, unital rings of Examples 7.4 and 7.6 fail to be integral domains because of the behavior of special elements: 2 in $\mathbb{Z}/_4$, or $(1,0)$ and $(0,1)$ in $\mathbb{Z}/_2 \oplus \mathbb{Z}/_2$. In each case, 0 is a multiple of these nonzero elements.

DEFINITION 7.7 (Zero divisors.)   *In a ring $(R, +, \cdot)$, a given element r is said to be a* zero *divisor if:*

(a) *r is nonzero;*

(b) *There is a nonzero element s of R with*

$$r \cdot s = 0 \quad or \quad s \cdot r = 0. \tag{7.1}$$

Integral domains are then characterized amongst all commutative, unital, and nontrivial rings by the absence of zero divisors.

PROPOSITION 7.8 (Integral domains and zero divisors.)
Let R be a commutative, unital, nontrivial ring. The following conditions are equivalent:

(a) *R is an integral domain;*

(b) *R has no zero divisors.*

PROOF (a) $\Rightarrow$ (b): Suppose that R is an integral domain. The existence of nonzero elements r and s with (7.1) would violate the closure of the monoid $(R^\times, \cdot, 1)$ under multiplication.

(b) $\Rightarrow$ (a): Suppose that R has no zero divisors. Since R is unital and nontrivial, the set $R^\times$ contains the identity element 1. The absence of zero divisors then guarantees that $R^\times$ forms a submonoid of $(R, \cdot, 1)$. $\quad\square$

REMARK 7.9 In Proposition 7.8, the hypothesis of nontriviality is necessary. The trivial ring R is commutative, unital, and devoid of zero divisors, but the empty set $R^\times$ does not form a monoid. $\quad\square$

In Proposition 4.53 (page 85), it was observed that elements of a group may be canceled from equations. In integral domains, nonzero elements may also be canceled, even though the multiplicative structure does not form a group.

PROPOSITION 7.10 (Cancellation in integral domains.)
Let R be an integral domain, with elements a, b_1, b_2. Suppose that a is nonzero. If
$$a \cdot b_1 = a \cdot b_2 \,, \tag{7.2}$$
then $b_1 = b_2$.

PROOF If (7.2) holds, then $a \cdot b_1 - a \cdot b_2 = 0$. The left distributive law now yields
$$a \cdot (b_1 - b_2) = 0 \,. \tag{7.3}$$
Since a is nonzero, and R has no zero divisors, the factor $b_1 - b_2$ in (7.3) must be zero. Thus $b_1 = b_2$. $\quad\square$

Recall that integral domains are commutative, by definition. Thus given the cancellation from the left in Proposition 7.10, there is no need for a separate discussion of cancellation from the right, as in Proposition 4.53(b) for groups (page 85).

7.2   Degrees

For a ring R, consider a polynomial

$$p(X) = p_n X^n + p_{n-1} X^{n-1} + \cdots + p_2 X^2 + p_1 X + p_0 \qquad (7.4)$$

in the polynomial ring $R[X]$. Suppose that the coefficient p_n is nonzero. It is then called the *leading coefficient* of $p(X)$, and the integer n is called the *degree* $\deg p$ or $\deg p(X)$ of the polynomial $p(X)$. In particular, nonzero constant polynomials p_0 — elements of the ring R — have degree 0. The zero constant polynomial 0 is deemed to have leading coefficient 0, and degree $-\infty$. For any natural number n, we have

$$-\infty + n = -\infty = n + (-\infty) \qquad (7.5)$$

and $-\infty < n$, so $\max(-\infty, n) = n$. Also $-\infty + (-\infty) = -\infty$.

The behavior of the degree of polynomials under the algebraic operations on $R[X]$ is described as follows.

PROPOSITION 7.11 (Inequalities for degrees.)
Let R be a ring. Consider polynomials $p(X)$ and $q(X)$ over R.

(a) $\deg \big(p(X) + q(X)\big) \leq \max \big(\deg p(X), \deg q(X)\big)$;

(b) $\deg \big(p(X) \cdot q(X)\big) \leq \deg p(X) + \deg q(X)$.

PROOF   If at least one of $p(X)$, $q(X)$ is zero, say $p(X) = 0$, then (a) reduces to the equality $\deg q(X) = \max \big(- \infty, \deg q(X)\big) = \deg q(X)$, while (b) reduces to the equality $-\infty = -\infty + \deg q(X)$. Otherwise, suppose that

$$p(X) = p_n X^n + p_{n-1} X^{n-1} + \ldots$$

and

$$q(X) = q_m X^m + q_{m-1} X^{m-1} + \ldots$$

with $p_n \neq 0$ and $q_m \neq 0$. Without loss of generality, suppose $n \geq m$ — if not, interchange the roles of $p(X)$ and $q(X)$.

(a): Since $p(X) + q(X) = (p_n + q_n)X^n + (p_{n-1} + q_{n-1})X^{n-1} + \ldots$ if $n = m$, and $p(X) + q(X) = p_n X^n + \ldots$ if $n > m$, we have

$$\deg \big(p(X) + q(X)\big) \leq n = \max \big(\deg p(X), \deg q(X)\big),$$

proving that (a) holds.

(b): Since

$$p(X) \cdot q(X) = \left(p_n X^n + p_{n-1} X^{n-1} + \dots \right) \cdot \left(q_m X^m + q_{m-1} X^{m-1} + \dots \right)$$
$$= p_n q_m X^{n+m} + (p_{n-1} q_m + p_n q_{m-1}) X^{n+m-1} + \dots , \qquad (7.6)$$

we have

$$\deg \left(p(X) \cdot q(X) \right) \le n + m = \deg p(X) + \deg q(X)$$

proving that (b) holds. ⬜

Example 7.12 (Strict inequality for degree of sum.)
Consider the polynomials $p(X) = X^2 + 1$ and $q(X) = -X^2 + 2X$ in $\mathbb{R}[X]$. Both have degree 2, but $\deg \left(p(X) + q(X) \right) = \deg(2X + 1) = 1$. Thus the inequality in Proposition 7.11(a) may be strict. ⬜

Example 7.13 (Strict inequality for degree of product.)
Consider the polynomials $p(X) = 2X^2$ and $q(X) = 2X^2 + 1$ in $\mathbb{Z}/4[X]$. Both of the polynomials $p(X)$ and $q(X)$ have degree 2, but $\deg \left(p(X) \cdot q(X) \right) = \deg(2X^2) = 2 < \deg p(X) + \deg q(X)$. Thus again in Proposition 7.11(b), the inequality may be strict. ⬜

For polynomials over integral domains, the inequality in Proposition 7.11(b) becomes equality.

COROLLARY 7.14 (Equality for degree of product over an ID.)
Let $p(X)$ and $q(X)$ be elements of the ring $R[X]$ of polynomials over an integral domain R. Then

$$\deg \left(p(X) \cdot q(X) \right) = \deg p(X) + \deg q(X) . \qquad (7.7)$$

In particular, the ring of polynomials over an integral domain is itself an integral domain.

PROOF If $p(X)$ or $q(X)$ is zero, it was already seen that Proposition 7.11(b) becomes equality. Otherwise, both leading coefficients p_n of $p(X)$ and q_m of $q(X)$ are nonzero. Since R is an integral domain, it follows that their product $p_n q_m$ is nonzero. By (7.6), this nonzero product is the leading coefficient of $p(X) \cdot q(X)$, so

$$\deg \left(p(X) \cdot q(X) \right) = n + m = \deg p(X) + \deg q(X)$$

as required for (7.7). Finally, if $p(X)$ and $q(X)$ are nonzero, their degrees are natural numbers. Then by (7.7), the degree $\deg \left(p(X) \cdot q(X) \right)$ is also a natural number, so that $p(X) \cdot q(X) \ne 0$. ⬜

7.3   Fields

In the commutative, unital ring $\mathbb{R}$ of real numbers, the set $\mathbb{R}^\times$ of nonzero elements forms a group $(\mathbb{R}^\times, \cdot, 1)$ under multiplication. In other words, each nonzero real number has a multiplicative inverse: The group of units $\mathbb{R}^*$ of the full monoid $(\mathbb{R}, \cdot, 1)$ is the set $\mathbb{R}^\times$ of nonzero reals.

DEFINITION 7.15 (Fields, field homomorphisms.)   *A ring* $(R, +, \cdot)$ *is said to be a* field *if it is both commutative and unital, and if the set* $R^\times$ *of nonzero elements of* R *forms a group* $(R, \cdot, 1)$ *under the multiplication of the ring. A* field homomorphism *is a unital ring homomorphism between fields.*

Note the distinction between integral domains and fields. Let $(R, +, \cdot)$ be a commutative, unital ring. For R to be an integral domain, the set $R^\times$ of nonzero elements has to form a monoid under multiplication. For R to be a field, $R^\times$ has to satisfy the stronger requirement of being a group. The ring $(\mathbb{Z}, +, \cdot)$ of integers is an integral domain, but it is not a field, since 2 does not have a multiplicative inverse in the set $\mathbb{Z}^\times$ of nonzero integers. On the other hand, 2 does have a multiplicative inverse 2^{-1} in the field $\mathbb{R}$ of real numbers.

> Do not confuse additive inverses with multiplicative inverses. Each element r of a ring R has an additive inverse $-r$ in the additive group $(R, +, 0)$ of the ring $(R, +, \cdot)$, but unless r is a nonzero element of a field R, there is no guarantee of a multiplicative inverse r^{-1} for r.

For finite commutative, unital rings, there is no distinction between integral domains and fields.

PROPOSITION 7.16 (Finite integral domains.)
A finite integral domain is a field.

PROOF   Let R be a finite integral domain. By definition, the set $R^\times$ of nonzero elements of R forms a monoid $(R^\times, \cdot, 1)$ under multiplication. It remains to be shown that each nonzero element r of R is invertible.

Consider the set $r^{\mathbb{N}} = \{1, r^1, r^2, r^3, \dots, r^m, \dots, r^n, \dots\}$ of powers of r. Since $r^{\mathbb{N}}$ is a subset of the finite set $R^\times$ of nonzero elements of R, there are natural numbers $m < n$ such that $r^m = r^n$. This equation may be written in the form

$$r^m \cdot 1 = r^m \cdot \left(r \cdot r^{n-m-1}\right).$$

Cancellation within the integral domain R (Proposition 7.10) then yields the equation $1 = r \cdot r^{n-m-1}$, so that r is invertible.   ▯

Example 7.17 (The fields $\mathbb{Z}/_p$.)

Let p be a prime number. The nontrivial commutative, unital ring $\mathbb{Z}/_p$ of integers modulo p is an integral domain. Indeed, if $(a + p\mathbb{Z}) \cdot (b + p\mathbb{Z}) = p\mathbb{Z}$ for integers a and b, the prime number p divides the product ab. It follows that p divides a or p divides b, so that at least one of the factors $a + p\mathbb{Z}$ or $b + p\mathbb{Z}$ is zero. By Proposition 7.16, it follows that the finite integral domain $(\mathbb{Z}/_p, +, \cdot)$ is a field. [For an alternative argument showing that $\mathbb{Z}/_p$ is a field, see Exercise 10(c) below, and Example 6.30 on page 138.] ⬜

Example 7.18 (The field $\mathbb{Z}/_3[i]$.)

Consider the commutative, unital ring $\mathbb{Z}/_3[i]$ of 2×2 matrices

$$\begin{bmatrix} x & -y \\ y & x \end{bmatrix} \tag{7.8}$$

with entries from the field $\mathbb{Z}/_3$ (compare Example 6.17, page 133). The ring has 9 elements, since there are 3 independent choices from $\mathbb{Z}/_3$ for each of the entries x and y of a matrix (7.8).

The determinant of the matrix (7.8) is $x^2 + y^2$. Now in $\mathbb{Z}/_3$, we have $0^2 = 0$ and $1^2 = 2^2 = 1$, so x^2 and y^2 lie in the set $\{0, 1\}$. Furthermore, the only solution of $x^2 + y^2 = 0$ is $x = y = 0$. Thus nonzero matrices (7.8) have nonzero determinants. Since the determinant of a product of these matrices is the product of their individual determinants, it follows that $\mathbb{Z}/_3[i]$ is a finite integral domain, and hence is a field. ⬜

Example 7.19 (Complex numbers.)

The ring $\mathbb{C}$ of complex numbers forms a field — compare (6.32) and Study Project 1 in Chapter 6. ⬜

Example 7.20 (Bit strings of length 2.)

In Example 7.6, it was seen that the ring $\mathbb{Z}/_4$ of integers modulo 4 did not even form an integral domain, let alone a field. Nevertheless, there are fields with 4 elements. Consider the set $(\mathbb{Z}/_2)^2$ of bit strings of length 2 (Example 4.35, page 78). This set forms a group with componentwise addition, an isomorphic copy of the Klein 4-group V_4. With multiplication $*$ as given in the table of Figure 7.1, the set of length 2 bit strings forms a field $((\mathbb{Z}/_2)^2, +, *)$. Certainly the nonzero elements form a 3-element cyclic group under the multiplication. It remains to verify the distributivity, to check that the left multiplication (6.3) by each element r is an additive group homomorphism. This is trivial for $r = 00$ and $r = 01$. For $r = 10$, the left multiplication is an isomorphism from the additive group to itself, with left multiplication by 11 as the inverse isomorphism (Exercise 12). ⬜

*	00	01	10	11
00	00	00	00	00
01	00	01	10	11
10	00	10	11	01
11	00	11	01	10

FIGURE 7.1: Multiplication table of a 4-element field.

7.4 Polynomials over fields

If F is a field, the ring of polynomials $F[X]$ over F in the indeterminate X has some special properties. Since fields are integral domains, Corollary 7.14 shows that $F[X]$ is an integral domain. In fact, $F[X]$ behaves even more like the ring of integers: It admits a division algorithm, much like the Division Algorithm for $\mathbb{Z}$ (Section 1.4).

THEOREM 7.21 (Division Algorithm for polynomials over fields.)
Let F be a field. Let

$$d(X) = d_m X^m + \cdots + d_1 X + d_0$$

be a nonzero polynomial (the divisor*) in $f[X]$, with leading coefficient $d_m \neq 0$. Then for each polynomial*

$$a(X) = a_n X^n + \cdots + a_1 X + a_0$$

(a dividend*) in $F[X]$, there is a unique polymomial $q(X)$ (the* quotient*) in $F[X]$, and a unique polynomial $r(X)$ (the* remainder*) in $f[X]$, such that*

$$\deg r(X) < \deg d(X) \tag{7.9}$$

and

$$a(X) = d(X)q(X) + r(X). \tag{7.10}$$

PROOF   The proof proceeds by induction on the degree n of the dividend.

Induction Basis: If $\deg a(X) < \deg d(X)$, then (7.9) and (7.10) specify $q(X) = 0$ and $r(X) = a(X)$ uniquely.

Induction Step: Suppose that the Division Algorithm has been established for all dividends of degree less than the degree n of the given dividend $a(X)$. In particular, suppose that

$$n = \deg a(X) \geq \deg d(X) = m$$

(since otherwise, the Induction Basis applies). Consider the polynomial

$$\alpha(X) = a(X) - a_n b_m^{-1} X^{n-m} d(X).$$

The coefficient of X^n in $\alpha(X)$ is $a_n - a_n b_m^{-1} b_m = 0$, so $\deg \alpha(X) < \deg a(X)$. By the induction hypothesis, there are polynomials $\gamma(X)$ and $r(X)$ with $\deg r(X) < \deg d(X)$ and

$$\alpha(X) = d(X)\gamma(X) + r(X).$$

Then

$$a(X) = d(X)\big(a_n b_m^{-1} X^{n-m} + \gamma(X)\big) + r(X),$$

yielding (7.10) as required, with $q(X) = a_n b_m^{-1} X^{n-m} + \gamma(X)$.

Uniqueness: Suppose

$$a(X) = d(X)q_1(X) + r_1(X) = d(X)q_2(X) + r_2(X)$$

with $\deg r_1(X)$, $\deg r_2(X) < \deg d(X)$. Then

$$r_1(X) - r_2(X) = d(X)\big(q_2(X) - q_1(X)\big),$$

as a multiple of $d(X)$ with degree less than $\deg d(X)$, must be zero. Thus $r_1(X) = r_2(X)$ and $q_1(X) = q_2(X)$. □

The induction used to prove Theorem 7.21 yields a recursive procedure, *Long Division*, for dividing one polynomial by another in $F[X]$. As an example, consider the division of $3X^3 - X^2 + 4$ by $2X^2 + 1$ in $\mathbb{Z}/5[X]$. The calculation is displayed as follows:

$$
\begin{array}{r}
4X + 2 \\
2X^2 + 1\,\overline{)\ 3X^3 - X^2 \qquad + 4} \\
3X^3 \qquad + 4X \\
\hline
4X^2 + X + 4 \\
4X^2 \qquad + 2 \\
\hline
X + 2
\end{array}
$$

At the first step, the divisor $2X^2 + 1$ is multiplied by $4X$ to obtain $3X^3 + 4X$. This multiple is subtracted from the dividend $3X^3 - X^2 + 4$ to yield $4X^2 + X + 4$. Next, the divisor $2X^2 + 1$ is multiplied by 2 to obtain $4X^2 + 2$. This multiple is subtracted from the intermediate dividend $4X^2 + X + 4$ to yield $X + 2$. Since the degree of $X + 2$ is less than the degree of the divisor, the polynomial $X + 2$ is the remainder. Thus the equation (7.10) takes the form

$$3X^3 - X^2 + 4 = (2X^2 + 1)(4X + 2) + (X + 2)$$

for this example.

For linear divisors of the form $(X - c)$, the results of division are particularly transparent.

PROPOSITION 7.22 (Evaluating the remainder.)
Let $f(X)$ be a polynomial over a field F. Then for an element c of F, the Division Algorithm yields

$$f(X) = (X - c)q(X) + f(c) \qquad (7.11)$$

for some quotient $q(X)$.

PROOF   The remainder

$$f(X) - (X - c)q(X) \qquad (7.12)$$

after division by the polynomial $(X - c)$ must have degree less than 1, so it is some constant k in F. The constant is obtained as $k = f(c) - (c-c)q(c) = f(c)$ on evaluating (7.12) at the field element c. $\square$

COROLLARY 7.23 (Roots and linear divisors.)
In $F[X]$, a polynomial $f(X)$ has a field element c as a root if and only if the linear polynomial $(X - c)$ divides $f(X)$.

Example 6.42 (page 146) indicated that the polynomial $X^2 - 1$ of degree 2 in $\mathbb{Z}/_8[X]$ has (at least) 4 roots in the ring $\mathbb{Z}/_8$. The final corollary of the Division Algorithm shows that such a surfeit of roots cannot happen with nonzero polynomials over fields.

PROPOSITION 7.24 (When degrees bound the number of roots.)
Let $f(X)$ be a nonzero polynomial over a field F. Then the number of roots of $f(X)$ in F does not exceed $\deg f(X)$.

PROOF   The proof proceeds by induction on the degree. The induction basis comprises the nonzero constant polynomials, which have degree 1, and no roots. For the induction step, consider a polynomial $f(X)$ of degree $n > 0$, and suppose that no polynomial of degree $n - 1$ has more than $n - 1$ roots. Suppose that $f(X)$ has a root, say c, in F. By Corollary 7.23,

$$f(X) = (X - c)q(X) \qquad (7.13)$$

for some polynomial $q(X)$ of degree $n - 1$. If b is a root of $f(X)$ distinct from c, then evaluation of (7.13) at b gives $0 = f(b) = (b-c)q(b)$. Since $(b-c) \neq 0$, it follows that $q(b) = 0$. By the induction hypothesis, there are at most $n - 1$ possible such roots b distinct from c. Thus $f(X)$ has at most n roots. $\square$

7.5   Principal ideal domains

Section 7.4 shows that, like the ring of integers, the ring of polynomials $F[X]$ over a field F admits a Division Algorithm. It is helpful to formalize the similarities between the rings $\mathbb{Z}$ and $F[X]$.

DEFINITION 7.25 (PID.)   *Let R be a commutative, unital ring.*

(a) *An ideal J of R is described as* principal *if it is the set dR of multiples of a single element d of R.*

(b) *The ring R is said to be a* principal ideal domain (*abbreviated as:* PID) *if it is an integral domain in which each ideal is principal.*

Example 7.26 (Integers form a PID.)
As each ideal of $\mathbb{Z}$ has the form $d\mathbb{Z}$ for some natural number d (compare Example 6.31, page 138), the integral domain $\mathbb{Z}$ is a principal ideal domain. □

THEOREM 7.27 (A polynomial ring over a field is a PID.)
The ring $F[X]$ of polynomials over a field F is a principal ideal domain.

PROOF   Since the field F is an integral domain, Corollary 7.14 shows that $F[X]$ is also an integral domain. The trivial ideal $0F[X]$ is principal. Consider a nontrivial ideal J of $F[X]$, containing a nonzero element $n(X)$. Define the subset
$$S = \{\deg f(x) \mid 0 \neq f(X) \text{ in } J\} \tag{7.14}$$
of the set $\mathbb{N}$ of natural numbers. Since S contains $\deg n(X)$, it is nonempty. The Well-Ordering Principle shows that S contains a least element, say the degree $\deg d(X)$ of a nonzero element $d(X)$ of J. It will be shown that J is the principal ideal $d(X)F[X]$.

Since $d(X)$ is an element of J, and J has the absorptive property, the set $d(X)F[X]$ of multiples of $d(X)$ is a subset of J. Conversely, consider an element $a(X)$ of J. It will be shown that $a(X)$ is a multiple of $d(X)$. The Division Algorithm yields
$$a(X) = d(X)q(X) + r(X) \tag{7.15}$$
with $\deg r(X) < \deg d(X)$. But $r(X)$, as the difference $a(X) - d(X)q(X)$ of the elements $a(X)$ and $d(X)q(X)$ of J, is itself an element of J. Since the degree of $d(X)$ is not greater than the degree of any other nonzero element of J, it follows that $r(X)$ is zero. Thus (7.15) actually expresses $a(X)$ as a multiple of $d(X)$. □

A polynomial over a unital ring R is said to be *monic* if its leading coefficient is 1. Suppose that J is a nontrivial ideal of the ring $F[X]$ of polynomials over a field F. In the proof of Theorem 7.27, let M denote the set of all nonzero polynomials $d(X)$ in J with $\deg d(X)$ as the least element of the set S in (7.14). Then amongst the polynomials in the set M, just one is monic (Exercise 17).

DEFINITION 7.28 (Minimal polynomial of a nontrivial ideal.)   *A monic polynomial $d(X)$ is called the* minimal polynomial *of the ideal $d(X)F[X]$.*

Example 7.29 (Complex numbers.)
Consider the evaluation homomorphism

$$\mathbb{R}[X] \to \mathbb{C}; p(X) \mapsto p(i)$$

to the field of complex numbers from the ring of polynomials over the reals. The polynomial $X^2 + 1$ lies in the kernel ideal, and has minimal degree there. (Indeed, Proposition 7.22 shows that the quotient of $\mathbb{R}[X]$ by the ideal of multiples of a linear polynomial would be isomorphic to $\mathbb{R}$.) Thus the kernel ideal is the set $(X^2 + 1)\mathbb{R}$ of multiples of the minimal polynomial $X^2 + 1$. The First Isomorphism Theorem yields

$$\mathbb{R}[X]/(X^2 + 1)\mathbb{R}[X] \to \mathbb{C}; \qquad (7.16)$$
$$p(X) + (X^2 + 1)\mathbb{R}[X] \mapsto p(i)$$

as an isomorphism of the field of complex numbers with the quotient of the ring of real polynomials by the ideal of multiples of $X^2 + 1$. $\Box$

Example 7.30 (The bit-string field.)
Consider the evaluation homomorphism

$$\mathbb{Z}/2[X] \to (\mathbb{Z}/2)^2; p(X) \mapsto p(10)$$

to the bit-string field of Example 7.20 from the ring of polynomials over the field of integers modulo 2. The minimal polynomial of the kernel ideal is $X^2 + X + 1$, since $10^2 + 10 + 1 = 11 + 10 + 01 = 0$. Thus

$$\mathbb{Z}/2[X]/(X^2 + X + 1)\mathbb{Z}/2[X] \to (\mathbb{Z}/2)^2; \qquad (7.17)$$
$$p(X) + (X^2 + X + 1)\mathbb{Z}/2[X] \mapsto p(10)$$

is an isomorphism of fields. $\Box$

In order to work with quotient rings $F[X]/J$ of rings of polynomials over a field F, it is useful to establish a representation of their elements.

PROPOSITION 7.31 (Representing elements of $F[X]/J$.)
*Let J be a nonzero ideal in the ring of polynomials over a field F. Let $d(X)$
be the minimal polynomial of J, with $\deg d(X) = n$. Then the map*

$$F^n \to F[X]/J; \tag{7.18}$$

$$(r_{n-1}, r_{n-2}, \dots, r_1, r_0) \mapsto r_{n-1}X^{n-1} + r_{n-2}X^{n-2} + \cdots + r_1X + r_0 + J$$

is an isomorphism of additive groups.

PROOF The map (7.18) is certainly a group homomorphism, since both
F^n and $F[X]$ have their additive group structures defined componentwise.
Now suppose that the n-tuple $(r_{n-1}, r_{n-2}, \dots, r_1, r_0)$ lies in the kernel of
(7.18). Then the polynomial

$$r(X) = r_{n-1}X^{n-1} + r_{n-2}X^{n-2} + \cdots + r_1X + r_0 \tag{7.19}$$

is a multiple of $d(X)$. Since $\deg r(X) < n = \deg d(X)$, it follows that the
polynomial $r(X)$ is zero, so the corresponding n-tuple

$$(r_{n-1}, r_{n-2}, \dots, r_1, r_0)$$

is also zero. Thus the group homomorphism (7.18) is injective, since its group
kernel is zero. Finally, consider an arbitrary coset $a(X) + J$ in the quotient
$F[X]/J$. Using the Division Algorithm to write

$$a(X) = d(X)q(X) + r(X)$$

with $\deg r(X) < n$, say $r(X)$ written as in (7.19), the coset

$$a(X) + J = r(X) + J$$

appears as the image of the n-tuple $(r_{n-1}, r_{n-2}, \dots, r_1, r_0)$ under (7.18). Thus
the map (7.18) is surjective. ☐

Example 7.32 (Complex numbers.)
With $F = \mathbb{R}$ and $d(X) = X^2 + 1$, the composite of the group isomorphism
(7.18) with the field isomorphism (7.16) is the map

$$\mathbb{R}^2 \to \mathbb{C}; (r_1, r_0) \mapsto r_0 + r_1 i$$

— compare (6.29). ☐

Example 7.33 (The bit-string field.)
Consider $F = \mathbb{Z}/2$ and $d(X) = X^2 + X + 1$. Then the composite of the group
isomorphism (7.18) with the field isomorphism (7.17) reduces to the identity
map on the set $(\mathbb{Z}/2)^2$ of pairs or length 2 bit strings. ☐

7.6   Irreducible polynomials

An integer $p > 1$ was defined to be irreducible if the only expressions of p as a product $p = ab$ of positive integers a and b are when $a = 1$ or $b = 1$ (1.30). Example 7.17 then showed that the quotient $\mathbb{Z}/p\mathbb{Z}$ of the principal ideal domain of integers by the ideal of multiples of an irreducible element p was actually a field. (Recall that the terms "prime" and "irreducible" are synonymous for positive integers, by Proposition 1.11.) This section pursues the analogy between the principal ideal domains $\mathbb{Z}$ and $F[X]$ for a field F. The definition of irreducibility is extended to polynomials, and irreducible polynomials are then used to construct new fields. Polynomials which are not irreducible (according to Definition 7.34 below) are described as *reducible*. Reducible polynomials are nonconstant polynomials that admit a nontrivial factorization.

DEFINITION 7.34 (Irreducible polynomials.)   *Let F be a field. A nonconstant polynomial $p(X)$ in $F[X]$ is said to be* irreducible *(over F) if*

$$p(X) = a(X)b(X) \quad implies \quad \left(\deg a(X) = 0 \quad or \quad \deg b(X) = 0 \right) \quad (7.20)$$

for polynomials $a(X)$ and $b(X)$ in $F[X]$.

Example 7.35 (The polynomial $X^2 + 1$ over $\mathbb{R}$ and $\mathbb{C}$.)
The polynomial $X^2 + 1$ is irreducible over the field $\mathbb{R}$ of real numbers. Indeed, if $X^2 + 1$ admitted a nontrivial factorization, the factors would be linear, and then $X^2 + 1$ would have real roots, according to Corollary 7.23. However, the square of each real number is nonnegative, so there are no real numbers r with $r^2 + 1 = 0$.

On the other hand, the polynomial $X^2 + 1$ is reducible over the field of complex numbers, admitting the nontrivial factorization

$$X^2 + 1 = X^2 - i^2 = (X + i)(X - i)$$

in the polynomial ring $\mathbb{C}[X]$.   ▯

In general, it can be very tricky to decide whether a given polynomial is irreducible or not. Fortunately, there is a fairly simple criterion for quadratic and cubic polynomials.

PROPOSITION 7.36 (Irreducibility of quadratics and cubics.)
Let F be a field, and let $p(X)$ be a polynomial in $F[X]$ with degree 2 or 3. Then $p(X)$ is irreducible if and only if it has no roots in F.

PROOF If $p(X)$ is reducible and admits a nontrivial factorization, then one of the factors is linear, say

$$p(X) = (a_1 X + a_0) \cdot b(X)$$

with a_0 and (nonzero) a_1 in F, and $b(X)$ in $F[X]$. Then

$$p\big(-a_1^{-1} a_0 \big) = \big(a_1 \cdot (-a_1^{-1} a_0) + a_0 \big) \cdot b\big(a_1^{-1} a_0 \big) = 0\,,$$

so $p(X)$ has the root $-a_1^{-1} a_0$ in F.

Conversely, suppose $p(X)$ has a root c in F. Then Corollary 7.23 shows that $p(X)$ has a linear factor, and therefore is reducible. ☐

Example 7.37 (An irreducible quadratic.)
The polynomial $p(X) = X^3 + X + 1$ is an irreducible element of $\mathbb{Z}/2[X]$, since $p(0) = 0 + 0 + 1 = 1 \neq 0$ and $p(1) = 1 + 1 + 1 = 1 \neq 0$. ☐

Example 7.38 (Failure of the test for quartics.)
Proposition 7.36 does not work for polynomials $p(X)$ of degree larger than 3. For instance, the reducible polynomial

$$(X^2 + 1)(X^2 + 1)$$

in $\mathbb{R}[X]$ has no real roots. ☐

The main use of irreducible polynomials is to bootstrap the construction of new fields from a given starter field.

THEOREM 7.39 (The Bootstrap Theorem.)
Let F be a field, and let $p(X)$ be an irreducible polynomial in $F[X]$. Let J be the ideal $p(X)F[X]$ of multiples of $p(X)$ in $F[X]$. Then the quotient ring $F[X]/J$ is a field.

PROOF Since $F[X]$ is commutative and unital, so is its quotient $F[X]/J$. It remains to be shown that each nonzero element $f(X) + J$ of the quotient actually has an inverse $g(X) + J$, so that $\big(f(X) + J \big) \cdot \big(g(X) + J \big) = 1 + J$. Now $f(X) + p(X)F[X]$ nonzero means that

$$f(X) \text{ is not a multiple of } p(X)\,. \tag{7.21}$$

Consider the ideal $f(X)F[X] + p(X)F[X]$ of $F[X]$. By Theorem 7.27, $F[X]$ is a principal ideal domain. Thus there is some element $a(X)$ of $F[X]$ with

$$f(X)F[X] + p(X)F[X] = a(X)F[X]\,. \tag{7.22}$$

In particular, the element $p(X) = f(X) \cdot 0 + p(X) \cdot 1$ of the left-hand side of (7.22), as an element of the right-hand side of (7.22), is some multiple

$$p(X) = a(X)b(X)$$

of $a(X)$. Since $p(X)$ is irreducible, one of $a(X)$ or $b(X)$ is a nonzero constant. If $b(X)$ is the nonzero constant b_0, then $a(X) = p(X)b_0^{-1}$. But we would then have $f(X) = f(X) \cdot 1 + p(X) \cdot 0$ as an element of

$$f(X)F[X] + p(X)F[X] = a(X)F[X] = p(X)b_0^{-1}F[X] = p(X)F[X].$$

This would give $f(X)$ as a multiple of $p(X)$, contradicting (7.21). Thus it is actually $a(X)$ which is the nonzero constant, say a_0. Now the identity polynomial $1 = a_0 \cdot a_0^{-1} = a(X) \cdot a_0^{-1}$ is an element of (7.22), so there are polynomials $g(X)$ and $q(X)$ with $1 = f(X)g(X) + p(X)q(X)$. Then, recalling the definition of the multiplication in quotient rings, we have

$$\bigl(f(X) + J\bigr) \cdot \bigl(g(X) + J\bigr) = f(X)g(X) + J$$
$$= f(X)g(X) + p(X)F[X] = 1 + p(X)F[X]$$

as required. □

Example 7.40 (An 8-element field.)

By Example 7.37, there is an irreducible polynomial $p(X) = X^3 + X + 1$ over the field $\mathbb{Z}/_2$. Theorem 7.39 then shows that a field is given by the quotient $\mathbb{Z}/_2[X]\big/J$ of the polynomial ring $\mathbb{Z}/_2[X]$ over the ideal $J = p(X)\mathbb{Z}/_2[X]$ of multiples of $p(X)$. By Proposition 7.31, this field has 8 elements, and each coset may be represented by a polynomial of degree at most 2. It is instructive to compute the successive powers of the coset $X + J$:

$$X^3 + J = X^3 + (X^3 + X + 1) + J = X + 1 + J\,;$$
$$X^4 + J = X \cdot X^3 + J = X(X + 1) + J = X^2 + X + J\,;$$
$$X^5 + J = X \cdot (X^2 + X) + J = X^3 + X^2 + J = X^2 + X + 1 + J\,;$$
$$X^6 + J = X \cdot (X^2 + X + 1) + J = X^3 + X^2 + X + J = X^2 + 1 + J\,;$$
$$X^7 + J = X \cdot (X^2 + 1) + J = X^3 + X + J = 1 + J\,.$$

Note that the multiplicative group of nonzero elements of the field is cyclic, generated by the coset $X + J$. Thus the table of powers of $X + J$ may be used to compute products and inverses in the field, while Proposition 7.31 indicates how to compute sums. For example, we have

$$X^5 + X^6 = (X^2 + X + 1) + (X^2 + 1) = X$$

and

$$(X + 1) \cdot (X^2 + X + 1) = X^3 \cdot X^5 = X^8 = X^{7+1} = X\,,$$

using the list of powers (and omitting explicit mention of the ideal J). □

7.7   Lagrange interpolation

Field structures on a set are very useful in a wide variety of contexts. In this section, we discuss one typical application, to the specification of functions. Suppose that F is a field. For a natural number n and distinct elements x_0, x_1, ..., x_n of F, it is desired to construct a function $f : F \to F$ taking a particular function value

$$f(x_i) = y_i$$

in F for each $0 \le i \le n$ (compare Figure 7.2). The function should be easy to specify, and easy to compute, even when the set F is too large for a simple table look-up to be feasible.

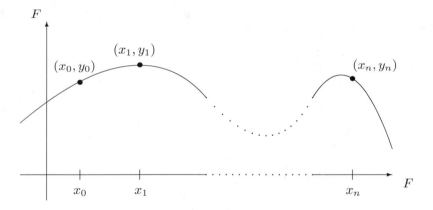

FIGURE 7.2:   Specifying a function.

To design a function with the desired properties, fix an index $0 \le j \le n$, and consider the simpler problem of finding a function δ_{x_j} satsifying the following specification:

$$\delta_{x_j}(x_i) = \begin{cases} 1 & \text{if } i = j\,; \\ 0 & \text{if } i \ne j\,. \end{cases}$$

Such a function is known as a (*Kronecker*) *delta function*. It is implemented as the polynomial

$$\delta_{x_j}(X) = \frac{(X - x_0)\ldots(X - x_{j-1})(X - x_{j+1})\ldots(X - x_n)}{(x_j - x_0)\ldots(x_j - x_{j-1})(x_j - x_{j+1})\ldots(x_j - x_n)}\,. \qquad (7.23)$$

of degree n. Note that, since x_j is distinct from all the other elements x_0, ..., x_{j-1}, x_{j+1}, ..., x_n, the denominator of (7.23) represents a nonzero field element, which thus has an inverse. Note further that for each index i distinct from j, the factor $(X - x_i)$ in the numerator of (7.23) vanishes at x_i, so $\delta_{x_j}(x_i) = 0$ in this case. Finally, note that each factor

$$\frac{(X - x_i)}{(x_j - x_i)}$$

(with $i \neq j$) takes the value $(x_j - x_i)(x_j - x_i)^{-1} = 1$ at x_j, so $\delta_{x_j}(x_j) = 1$, as required.

The function $f : F \to F$ can now be implemented by evaluating the polynomial

$$f(X) = \sum_{j=0}^{n} y_j \cdot \delta_{x_j}(X), \qquad (7.24)$$

a sum of constant multiples of the polynomials (7.23). The polynomial (7.24) is known as the *Lagrange interpolant* for the specified function values. Note that, for each $1 \leq i \leq n$, we have

$$
\begin{aligned}
f(x_i) &= \sum_{j=0}^{n} y_j \cdot \delta_{x_j}(x_i) \\
&= \sum_{j=0}^{i-1} y_j \cdot \delta_{x_j}(x_i) + y_i \cdot \delta_{x_i}(x_i) + \sum_{j=i+1}^{n} y_j \cdot \delta_{x_j}(x_i) \\
&= 0 + y_i \cdot 1 + 0 = y_i
\end{aligned}
$$

as required. The result may be summarized as follows.

THEOREM 7.41 (Lagrange interpolation.)
Suppose that x_0, x_1, ..., x_n are $n + 1$ distinct elements of a field F. Then for any elements y_0, y_1, ..., y_n of F, there is a polynomial $f(X)$ of degree at most n over F such that $f(x_i) = y_i$ for $0 \leq i \leq n$.

Example 7.42 (Interpolants of unexpectedly low degree.)
A Lagrange interpolant for $n + 1$ function values may have a degree less than n. For instance, the Lagrange interpolant to the identity function on the 3-element field $\mathbb{Z}/3$ is X (Exercise 23). 　　　　　　　　　　　　　　　　　　⬚

COROLLARY 7.43 (Self-maps of finite fields are polynomials.)
Let F be a finite field. Then each function $f : F \to F$ may be implemented as the evaluation of a polynomial $f(X)$ over F.

The property enunciated in Corollary 7.43 is sometimes expressed by the statement that finite fields are *polynomially complete*.

7.8   Fields of fractions

A final method of constructing new fields, this time infinite, mimics the construction of the rationals from the integers (Section 3.3). Let D be an integral domain. Consider the set

$$X = \left\{ \begin{bmatrix} n & m \\ 0 & n \end{bmatrix} \;\middle|\; n, m \text{ in } D, \ n \neq 0 \right\}$$

of matrices over D. Note that if $D = \mathbb{Z}$, the top row of a matrix from X corresponds to a solid dot in Figure 3.1.

A relation R is defined on the set X by

$$\begin{bmatrix} n_1 & m_1 \\ 0 & n_1 \end{bmatrix} R \begin{bmatrix} n_2 & m_2 \\ 0 & n_2 \end{bmatrix} \qquad \text{if and only if} \qquad \begin{vmatrix} n_1 & m_1 \\ n_2 & m_2 \end{vmatrix} = 0 \qquad (7.25)$$

— compare (3.6). As in Proposition 3.8, the relation R is seen to be an equivalence relation on the set X. (The properties of the ring of integers that were used in the proof of Proposition 3.8 are all properties that have been abstracted into the concept of an integral domain.) Define an addition operation on the set X_R of equivalence classes by

$$\begin{bmatrix} n_1 & m_1 \\ 0 & n_1 \end{bmatrix}_R + \begin{bmatrix} n_2 & m_2 \\ 0 & n_2 \end{bmatrix}_R = \left(\begin{bmatrix} n_1 & m_1 \\ 0 & n_1 \end{bmatrix} \cdot \begin{bmatrix} n_2 & m_2 \\ 0 & n_2 \end{bmatrix} \right)_R \qquad (7.26)$$

(using the usual multiplication of matrices). This addition operation is well defined (compare Exercise 7 in Chapter 3). Since matrix multiplication is associative, it follows directly that the addition on X_R is associative. Since X forms a commutative subsemigroup of the semigroup of matrices under the usual multiplication, it again follows directly that the addition on X_R is commutative. Furthermore, the equivalence class

$$0 = \begin{bmatrix} 1 & 0 \\ 0 & 1 \end{bmatrix}_R$$

of the identity matrix is an additive identity for the semigroup $(X_R, +)$, so we obtain a commutative monoid $(X_R, +, 0)$. Finally, note that

$$\begin{bmatrix} n & m \\ 0 & n \end{bmatrix}_R + \begin{bmatrix} n & -m \\ 0 & n \end{bmatrix}_R = \left(\begin{bmatrix} n & m \\ 0 & n \end{bmatrix} \cdot \begin{bmatrix} n & -m \\ 0 & n \end{bmatrix} \right)_R = \begin{bmatrix} n^2 & 0 \\ 0 & n^2 \end{bmatrix}_R = \begin{bmatrix} 1 & 0 \\ 0 & 1 \end{bmatrix}_R$$

— the latter equality holding directly by (7.25) — so that $(X_R, +, 0)$ is seen to form an abelian group.

PROPOSITION 7.44

There is an injective abelian group homomorphism

$$j_D : (D, +, 0) \rightarrow (X_R, +, 0); m \mapsto \begin{bmatrix} 1 & m \\ 0 & 1 \end{bmatrix}_R .$$

PROOF   The map j_D is an abelian group homomorphism, since

$$\begin{bmatrix} 1 & m_1 \\ 0 & 1 \end{bmatrix} \cdot \begin{bmatrix} 1 & m_2 \\ 0 & 1 \end{bmatrix} = \begin{bmatrix} 1 & m_1 + m_2 \\ 0 & 1 \end{bmatrix}$$

for elements m_1, m_2 of D. By (7.25), the group kernel of j_D is the set $\{0\}$ of elements m in D with

$$0 = \begin{vmatrix} 1 & m \\ 1 & 0 \end{vmatrix} = -m,$$

so j_D is injective.   □

Now define a commutative operation of multiplication on X_R by

$$\begin{bmatrix} n_1 & m_1 \\ 0 & n_1 \end{bmatrix}_R \cdot \begin{bmatrix} n_2 & m_2 \\ 0 & n_2 \end{bmatrix}_R = \left(\begin{bmatrix} n_1 n_2 & m_1 m_2 \\ 0 & n_1 n_2 \end{bmatrix} \right)_R. \qquad (7.27)$$

Note that the right-hand side of (7.27) may be written as the equivalence class

$$\left(\begin{bmatrix} n_1 & m_1 \\ 0 & n_1 \end{bmatrix} \circ \begin{bmatrix} n_2 & m_2 \\ 0 & n_2 \end{bmatrix} \right)_R$$

of the componentwise or Hadamard product of the representatives of the classes on the left-hand side of (7.27) — compare (4.18). The operation (7.27) is well defined, as shown by the argument on page 55. Since the Hadamard multiplication of matrices is associative, X_R forms a commutative semigroup under the multiplication (7.27). The equivalence class

$$1 = \begin{bmatrix} 1 & 1 \\ 0 & 1 \end{bmatrix}_R$$

forms an identity element for the multiplication, yielding a monoid $(X_R, \cdot, 1)$. Moreover, the map $j_D : (D, \cdot, 1) \to (X_R, \cdot, 1)$ of Proposition 7.44 is a monoid homomorphism. Finally, note that

$$\begin{bmatrix} n & m \\ 0 & n \end{bmatrix}_R \cdot \begin{bmatrix} m & n \\ 0 & m \end{bmatrix}_R = \begin{bmatrix} nm & nm \\ 0 & nm \end{bmatrix}_R = \begin{bmatrix} 1 & 1 \\ 0 & 1 \end{bmatrix}_R$$

— the latter equality holding directly by (7.25) — for each nonzero element

$$\begin{bmatrix} n & m \\ 0 & n \end{bmatrix}_R$$

of X_R. Thus the set of nonzero elements of X_R forms a commutative group under multiplication. A direct verification shows that the multiplication in X_R distributes over the addition (Exercise 26). Altogether, we obtain a field $(X_R, +, \cdot)$, and $j_D : D \to X_R$ becomes a unital ring homomorphism.

DEFINITION 7.45 (The field of fractions of an integral domain.)
Let D be an integral domain. The field X_R is called the field of fractions F_D
of the integral domain D. Each equivalence class

$$\begin{bmatrix} n & m \\ 0 & n \end{bmatrix}_R$$

is written in the forms

$$\frac{m}{n} = n^{-1}m = m/n.$$

The addition becomes

$$\frac{a}{b} + \frac{c}{d} = \frac{ad + bc}{bd}.$$

The multiplication becomes

$$\frac{a}{b} \cdot \frac{c}{d} = \frac{ac}{bd}.$$

The inversion of nonzero elements becomes

$$\left(\frac{a}{b}\right)^{-1} = \frac{b}{a}.$$

Then the map j_D takes the form

$$j_D : D \to F_D; m \mapsto \frac{m}{1}.$$

*The integral domain D is usually identified with its image under the injective
map j_D.*

Example 7.46 (The field of fractions of the integers.)
The field of fractions of $\mathbb{Z}$ is the field $\mathbb{Q}$ of rational numbers. □

Example 7.47 (Rational functions.)
Let D be an integral domain, and let X be an indeterminate over D. By
Corollary 7.14, the ring $D[X]$ of polynomials in X over D forms an integral
domain. The field of fractions of the integral domain $D[X]$ is called the *field
$D(X)$ of rational functions* in X over D. Thus an element of $D(X)$ is of the
form

$$\frac{f(X)}{g(X)}$$

in which $f(X)$ and $g(X)$ are polynomials over D, the polynomial $g(X)$ being
nonzero. Such elements are known as *rational functions* over D. □

7.9   Exercises

1. Show that the ring $\mathbb{Z}/5$ of integers modulo 5 forms an integral domain.

2. Show that the ring $\mathbb{Z}/6$ of integers modulo 6 does not form an integral domain.

3. Let n be a prime number. Show that the ring $\mathbb{Z}/n$ of integers modulo n forms an integral domain.

4. Let n be a composite number. Show that the ring $\mathbb{Z}/n$ of integers modulo n does not form an integral domain.

5. Show that the ring $\mathbb{Z}[i]$ of Gaussian integers forms an integral domain.

6. Let $(R, +, \cdot)$ be a commutative ring. Let Z be the subset of R consisting of 0 and the zero divisors. Show that $(Z, \cdot)$ is a subsemigroup of $(R, \cdot)$.

7. (a) Show that with the definition (7.5), the union

$$\mathbb{N}_{-\infty} = \{-\infty\} \cup \mathbb{N}$$

of the singleton $\{-\infty\}$ with the set of natural numbers forms a commutative semigroup under addition.

(b) Is $\mathbb{N}_{-\infty}$ a monoid?

(c) Setting

$$2^{-\infty} = 0,$$

show that

$$(\mathbb{N}_{-\infty}, +) \to (\mathbb{N}, \cdot); d \mapsto 2^d$$

is a semigroup homomorphism.

8. What is the inverse of the nonzero element 25 in the field $\mathbb{Z}/41$?

9. Consider the subset

$$S = \{0, 2, 4, 6, 8\}$$

of $\mathbb{Z}/10\mathbb{Z}$ (compare Exercise 17 in Chapter 6). Show that S forms a field.

10. Let R be a simple, unital commutative ring.

(a) Suppose that a is a nonzero element of R. Show that the set $aR = \{ar \mid r \text{ in } R\}$ of multiples of a is a nontrivial ideal of R.

(b) Show that each nonzero element of R is invertible.

(c) Conclude that a simple, unital commutative ring is a field.

(d) Give an example of an integral domain which is not simple.

(e) Give an example of a simple, commutative ring which is not a field.

11. Show that the ring $\mathbb{Z}/5[i]$ has a zero divisor.

12. Consider the group $\big((\mathbb{Z}/2)^2, +, 00\big)$ of length 2 bit strings discussed in Example 7.20, with the multiplication $*$ displayed in Figure 7.1.

(a) Show that the left multiplication by 11 (using $*$) is inverse to the left multiplication by 10.

(b) Show that the left multiplication by 10 gives a homomorphism from the additive group $\big((\mathbb{Z}/2)^2, +, 00\big)$ to itself.

13. In the ring $\mathbb{Z}/3[X]$ of polynomials over the field $\mathbb{Z}/3$ of integers modulo 3, find the quotient and remainder when the polynomial $X^4 - 1$ is divided by $X^2 + X + 1$.

14. Let $f(X)$ be a polynomial over a field F. For a general linear polynomial $a_1 X + a_0$ (with $a_1 \neq 0$) over the field F, show that the remainder left after dividing $f(X)$ by $a_1 X + a_0$ is $f(-a_1^{-1}a_0)$.

15. **(Fermat's Little Theorem.)** Let p be a prime number.

(a) Show that each nonzero element of the field $\mathbb{Z}/p$ is a root of the polynomial $X^{p-1} - 1$ over $\mathbb{Z}/p$. (Hint: Apply Exercise 36 from Chapter 5 to the group G of nonzero elements of $\mathbb{Z}/p$.)

(b) Show that each element of the field $\mathbb{Z}/p$ is a root of the polynomial $X^p - X$ over $\mathbb{Z}/p$.

16. Consider the polynomial $p(X) = X^r - 1$, for a positive integer r.

(a) Show that $p(X)$ has r roots in the field of complex numbers. (Hint: Compare Exercise 35 in Chapter 5.)

(b) Show that $p(X)$ may have less than r roots in the field of real numbers.

(c) For which positive integers r does $p(X)$ have r roots in the field of real numbers?

17. Let J be an ideal in the ring $F[X]$ of polynomials over a field F. Consider the set M of nonzero polynomials in J whose degree is minimal in the set S of (7.14).

(a) Given two polynomials in M, show that each is a multiple of the other.

(b) Given two polynomials in M, show that each is a constant multiple of the other.

(c) Given a polynomial in M, show that each of its nonzero constant multiples also lies in M.

(d) Show that the set M contains a unique monic polynomial.

18. Consider the elements 2 and X in the ring $\mathbb{Z}[X]$ of polynomials over the integers.

   (a) Show that $2\mathbb{Z}[X] + X\mathbb{Z}[X]$ is an ideal of $\mathbb{Z}[X]$.

   (b) Show that there can be no polynomial $d(X)$ in $\mathbb{Z}[X]$ for which

$$2\mathbb{Z}[X] + X\mathbb{Z}[X] = d(X)\mathbb{Z}[X].$$

   (c) Conclude that Theorem 7.27 cannot be generalized, replacing the field F with an integral domain.

19. Let R be the set of all the 2×2 matrices of the form

$$\begin{bmatrix} x & 3y \\ y & x \end{bmatrix}$$

with x and y from the ring $\mathbb{Z}/5$ of integers modulo 5.

   (a) Show that R is a subring of the ring $(\mathbb{Z}/5)^2_2$ of all 2×2 matrices over $\mathbb{Z}/5$.

   (b) Show that R is a field.

   (c) Find the minimal polynomial of the kernel ideal of the evaluation homomorphism

$$\mathbb{Z}/5[X] \to R; p(X) \mapsto p\left(\begin{bmatrix} 0 & 3 \\ 1 & 0 \end{bmatrix}\right).$$

20. Show that the polynomial $X^3 + X + 1$ is an irreducible element of $\mathbb{Z}/7[X]$.

21. Let $p(X)$ be a polynomial over a field F, and let c be an element of F. Show that $p(X)$ is irreducible if and only if $p(X + c)$ is irreducible.

22. Consider the polynomial $p(X) = X^2 + 1$ over the field $\mathbb{Z}/3$ of integers modulo 3. Let J be the ideal $p(X)\mathbb{Z}/3[X]$ of multiples of $p(X)$.

   (a) Show that $p(X)$ is irreducible over $\mathbb{Z}/3[X]$.

   (b) Compute the powers of the coset $X + J$ in the field $\mathbb{Z}/3[X]\big/J$.

   (c) Find the unique representative $r(X)$ of the coset $X^3 + X^6 + J$ with $\deg r(X) < 2$.

(d) Compute the product

$$(X + 1 + J)(2X + 1 + J)$$

in the field $\mathbb{Z}/3[X]\big/J$.

(e) Show that $\mathbb{Z}/3[X]\big/J$ is isomorphic to the field $\mathbb{Z}/3[i]$ discussed in Example 7.18.

23. Verify that the Lagrange interpolant

$$0 \cdot \delta_0(X) + 1 \cdot \delta_1(X) + (-1) \cdot \delta_{-1}(X)$$

to the identity function id : $\mathbb{Z}/3 \to \mathbb{Z}/3$ on the 3-element field of integers modulo 3 is X (compare Example 7.42).

24. Show that the Lagrange interpolant passing through two specified points (x_0, y_0) and (x_1, y_1), with $x_0 \neq x_1$, is

$$\frac{1}{x_1 - x_0}\left\{(y_1 - y_0) \cdot X - \begin{vmatrix} x_0 & y_0 \\ x_1 & y_1 \end{vmatrix}\right\}.$$

25. Show that the infinite field of real numbers is not polynomially complete: Give an example of a function $f : \mathbb{R} \to \mathbb{R}$ which cannot be implemented as the evaluation of a polynomial $p(X)$ over R. Justify your claim.

26. Let D be an integral domain. Show that the multiplication (7.27) in X_R distributes over the addition (7.26).

27. Suppose that an integral domain D is actually a field. Show that the embedding $j_D : D \to F_D$ of D into its field of fractions is an isomorphism of fields.

28. Show that the field of fractions of the integral domain $\mathbb{Z}[i]$ of Gaussian integers (compare Exercise 5) is $\mathbb{Q}[i]$.

29. Let D be an integral domain, and let $\theta : D \to K$ be an injective unital ring homomorphism from D to a field K. Show that there is a unique field homomorphism $F_\theta : F_D \to K$ such that $F_\theta \circ j_D = \theta$.

30. Let D be an integral domain, and let X be an indeterminate over D. Show that the field of fractions F_D of D is a subfield of the field $D(X)$ of rational functions in X over D.

31. Let D be an integral domain, and let X be an indeterminate over D. Show that

$$\frac{X^{n+1} - 1}{X - 1} = X^n + X^{n-1} + \ldots + X + 1$$

for each natural number n.

7.10 Study projects

1. **Permutation polynomials.** Let F be a finite field. A polynomial $p(X)$ in $F[X]$ is a *permutation polynomial* if its polynomial function

$$F \to F; c \mapsto p(c)$$

— compare Definition 6.40(b) — is a permutation of the finite set F.

   (a) Show that a polynomial of degree 1 in $F[X]$ is a permutation polynomial.

   (b) Show that the symmetric group S_3 is the full set of polynomial functions of degree 1 polynomials in $\mathbb{Z}/3[X]$.

   (c) Show that the degree 1 polynomial functions over the field $\mathbb{Z}/5$ form a proper subgroup of the symmetric group S_5.

   (d) Show that the degree 1 polynomial functions, together with the degree 3 permutation polynomial functions, combine to give the full symmetric group S_5.

   (e) Give an example of a degree 3 polynomial over the field $\mathbb{Z}/5$ which is not a permutation polynomial.

2. **Simpson's Rule.** Suppose that

$$f : [-1, 1] \to \mathbb{R}$$

is a continuous, real-valued function defined on the interval $[-1, 1]$ of real numbers r with $-1 \le r \le 1$. Simpson's Rule approximates the Riemann integral

$$\int_{-1}^{1} f(x)dx$$

as the integral of the Lagrange interpolant polynomial function $p(X)$ that is determined by the specifications

$$-1 \mapsto f(-1), \qquad 0 \mapsto f(0), \qquad 1 \mapsto f(1). \qquad (7.28)$$

   (a) Find the quadratic interpolant $p(X)$ specified by (7.28).

   (b) Evaluate the Riemann integral

$$\int_{-1}^{1} p(x)dx \qquad (7.29)$$

   of the quadratic interpolant.

(c) Consider the function $\cos\left(\pi x/2\right)$ on the interval $[-1, 1]$. Compute the true value of the Riemann integral

$$\int_{-1}^{1} \cos\left(\frac{\pi x}{2}\right) dx,$$

and compare it with the value (7.29) given by Simpson's Rule in this case.

3. **Primitive elements.** Given a prime number p, and an irreducible polynomial $p(X)$ of degree n in the ring $\mathbb{Z}/p[X]$ of polynomials over the p-element field $\mathbb{Z}/p$, Theorem 7.39 shows that the quotient ring $\mathbb{Z}/p[X]/J$ by the ideal $p(X)\mathbb{Z}/p[X]$ of multiples of $p(X)$ forms a field K. Proposition 7.31 shows that this field has p^n elements, and it also reduces the additive group structure $(K, +, 0)$ to the power $(\mathbb{Z}/p, +, 0)^n$. The problem of computing products and inverses in the field K still remains. This problem is solved by the use of primitive elements.

An element e of a finite field K is said to be *primitive* if the set $\langle e \rangle$ of powers of e is the full set K^* of nonzero elements of the field K.

(a) Show that $X + J$ is a primitive element of the 8-element field in Example 7.40.

(b) Show that the polynomial $p(X) = X^2 + X + 1$ is irreducible over the field $\mathbb{Z}/5$.

(c) Define $J = (X^2 + X + 1)\mathbb{Z}/5[X]$ and $K = \mathbb{Z}/5[X]/J$. Show that $X + J$ is not a primitive element of K.

(d) Show that $(X + 1) + J$ is not a primitive element of K.

(e) Show that $(X + 2) + J$ is a primitive element of K.

4. **Discrete logarithms.** Given a primitive element e for a finite field K, each nonzero element x of the field appears as a unique power e^l of e with $0 \le l < |K| - 1$. This power is defined as the *discrete logarithm* $\log_e x$ of x to the base e. A table of the discrete logarithms to base $X + J$ for the 8-element field of Example 7.40 is displayed in Figure 7.3. The coefficients of the unique coset representatives of degree less than 3 are written as bit strings.

x	001	010	011	100	101	110	111
$\log_e x$	0	1	3	2	6	4	5

FIGURE 7.3: Discrete logs to base X in $\mathbb{Z}/2[X]/(X^2 + X + 1)\mathbb{Z}/2[X]$.

n	0	1	2	3	4	5	6
e^n	001	010	100	011	110	111	101

FIGURE 7.4: Discrete antilogs to base X in $\mathbb{Z}/_2[X]/(X^2 + X + 1)\mathbb{Z}/_2[X]$.

Figure 7.4 displays the corresponding "antilogarithms," the successive powers of the primitive element e.

Compute similar tables of discrete logarithms and antilogarithms for the primitive element $e = (X + 2) + J$ in the field $K = \mathbb{Z}/_5[X]/J$ with $J = (X^2 + X + 1)\mathbb{Z}/_5[X]$. Field elements can be represented as pairs $r_1 r_0$ of elements of $\mathbb{Z}/_5$, the coefficients of the representative polynomial.

7.11 Notes

Section 7.1

Some authors use the absence of zero divisors in nontrivial, commutative unital rings to define integral domains (compare Proposition 7.8). However, Definition 7.1 is more direct and positive.

Section 7.7

L. Kronecker was a German mathematician who lived from 1823 to 1891.

Section 7.8

The use of matrices in the construction of the field of fractions helps to minimize the number of formal verifications required, leaving only the proof of transitivity of the equivalence relation R, and confirmation of the distributive law.

Section 7.9

P. Fermat was a French mathematician who lived from 1601 to 1665. His "Little Theorem" (Exercise 15) is not to be confused with his "Last Theorem," the nonexistence of positive integers x, y, z with $x^n + y^n = z^n$ when $n > 2$.

Section 7.10

T. Simpson was an English mathematician who lived from 1710 to 1761.

Chapter 8

FACTORIZATION

The Fundamental Theorem of Arithmetic governs factorization of integers. Factorization of a polynomial is the key step to location of its roots. In this chapter, we consider the factorization of elements in various rings, obtaining a deeper insight into the structure of finite fields.

8.1   Factorization in integral domains

Integral domains follow the model of the ring $\mathbb{Z}$ of integers. Factorization in the ring of integers was studied in Chapter 1. However, the discussion there focussed mainly on positive integers. Now the group of units of $\mathbb{Z}$ is $\mathbb{Z}^* = \{\pm 1\}$, and each nonzero integer n is related to the positive integer $|n|$ by the equation $|n| = u \cdot n$ for a unit u of $\mathbb{Z}$. When discussing factorization in general integral domains, consideration of units (invertible elements) is a key feature. In an integral domain D, a factorization

$$a = b \cdot c$$

of an element a of D as a product of elements b and c of D is said to be *proper* if neither b nor c is a unit.

The following observation is very useful.

PROPOSITION 8.1

Let d be a nonzero element of an integral domain D. Then for an element u of D, the equality $dD = duD$ holds if and only if u is a unit.

PROOF   Each multiple of du is certainly a multiple of d, so the set duD is always a subset of dD.

If u is a unit, say $uv = 1$ for v in D, then each multiple dx of d is the multiple $duvx$ of du, so dD is a subset of duD, and the equality $dD = duD$ holds.

Conversely, suppose $dD = duD$, so the element $d1$ of dD is some multiple duv of du. Cancellation (Proposition 7.10, page 159) in the equation $d1 = duv$ implies that $1 = uv$, so u is a unit. $\Box$

In Section 1.7, two properties of positive integers were introduced:

- the internal property of being *irreducible*, and

- the external property of being *prime*.

Here are the corresponding properties in general integral domains:

DEFINITION 8.2 (Irreducibles and primes.)   *Let D be an ID.*

(a) *An element a of D is said to be* irreducible *in D if it is not zero, not a unit, and if it has no proper factorization $a = b \cdot c$ in D.*

(b) *An element a of D is said to be* prime *in D if it is not zero, not a unit, and if*

$$a \mid bc \qquad implies \qquad (\, a \mid b \quad or \quad a \mid c \,)$$

for elements b and c of D.

REMARK 8.3   (a) For a unit u, an element a of an integral domain D is irreducible if and only if ua is irreducible (Exercise 2).

(b) A nonzero, noninvertible element a of an integral domain D is *reducible* if it is not irreducible. A reducible element does have a proper factorization in D.

(c) The clause "in D" is an important rider on the terms "irreducible" and "prime" in Definition 8.2. For example, even if a nonzero element a of an integral domain D is prime in D, it is not prime in the field of fractions F_D, since there it is invertible. Also, compare Example 7.35 (page 170).   ▯

Example 8.4 **(Integers.)**
An integer n is irreducible in $\mathbb{Z}$ according to Definition 8.2(a) if and only if $|n|$ is a prime positive integer. This condition, in turn, is equivalent to n being prime in the sense of Definition 8.2(b).   ▯

Example 8.5 **(Polynomials.)**
Let F be a field. Consider the integral domain $F[X]$ of polynomials over F, and an element $f(X)$ of $F[X]$. The units in the ring $F[X]$ are the nonzero constants, the polynomials of degree 0. Now $f(X)$ is nonzero and not a unit if it is not a constant, so $\deg f(X) > 0$. Thus $f(X)$ is irreducible in $F[X]$ by Definition 8.2(a) if and only if it is irreducible by Definition 7.34.   ▯

For positive integers, the two properties: being irreducible, and being prime, turned out to be equivalent (Proposition 1.11, page 13), so the single term "prime" sufficed in that case. An example shows that in a general integral domain, the two concepts may differ.

Example 8.6 (Irreducibles need not be prime.)
Consider the commutative unital subring

$$\mathbb{Z}[\sqrt{-3}] = \left\{ \begin{bmatrix} x & -3y \\ y & x \end{bmatrix} \middle| x, y \text{ in } \mathbb{Z} \right\}$$

of the ring $\mathbb{Z}_2^2$ of 2×2 matrices over $\mathbb{Z}$. Note that

$$\begin{bmatrix} 0 & -3 \\ 1 & 0 \end{bmatrix} \begin{bmatrix} 0 & -3 \\ 1 & 0 \end{bmatrix} = \begin{bmatrix} -3 & 0 \\ 0 & -3 \end{bmatrix} = (-3)I_2,$$

so the matrix $\begin{bmatrix} 0 & -3 \\ 1 & 0 \end{bmatrix}$ serves as a square root of -3, and each element of $\mathbb{Z}[\sqrt{-3}]$ may be written in the form

$$\begin{bmatrix} x & -3y \\ y & x \end{bmatrix} = x + y\sqrt{-3}$$

with integers x and y. Now

$$\det \begin{bmatrix} x & -3y \\ y & x \end{bmatrix} = x^2 + 3y^2 = 0$$

if and only if $x = y = 0$. Recalling

$$\det(AB) = \det(A)\det(B)$$

for matrices A and B (Exercise 7 in Chapter 5), it is apparent that $\mathbb{Z}[\sqrt{-3}]$ is an integral domain. Moreover, the determinants of elements of $\mathbb{Z}[\sqrt{-3}]$ are

$$0, 1, 3, 4, 7, 9, 12, 13, 16, \ldots \tag{8.1}$$

in increasing order, an element being invertible in $\mathbb{Z}[\sqrt{-3}]$ if and only if its determinant is 1. Now

$$\begin{bmatrix} 1 & -3 \\ 1 & 1 \end{bmatrix} \begin{bmatrix} 1 & 3 \\ -1 & 1 \end{bmatrix} = \begin{bmatrix} 4 & 0 \\ 0 & 4 \end{bmatrix} = \begin{bmatrix} 2 & 0 \\ 0 & 2 \end{bmatrix} \begin{bmatrix} 2 & 0 \\ 0 & 2 \end{bmatrix}. \tag{8.2}$$

The factors

$$\begin{bmatrix} 1 & -3 \\ 1 & 1 \end{bmatrix}, \begin{bmatrix} 1 & 3 \\ -1 & 1 \end{bmatrix}, \text{ and } \begin{bmatrix} 2 & 0 \\ 0 & 2 \end{bmatrix}$$

of $4I_2$ are irreducible in $\mathbb{Z}[\sqrt{-3}]$, since their respective determinants are all equal to 4, an integer which does not factor nontrivially into a product of members of the ordered list (8.1). Thus elements of $\mathbb{Z}[\sqrt{-3}]$ may not have a unique factorization into products of irreducibles. Moreover, since the irreducible element $2I_2$ divides the product on the left-hand side of (8.2), but does not divide either of the two individual factors in that product, it is not a prime element of $\mathbb{Z}[\sqrt{-3}]$. □

188   *Introduction to Abstract Algebra*

The following result clarifies the general relationship between the concepts of Definition 8.2 within an integral domain D:

---

prime (external property)
implies
irreducible (internal property)

PROPOSITION 8.7 (Primes are irreducible.)
Let D be an integral domain. If an element a of D is prime in D, then it is also irreducible in D.

PROOF Suppose that the prime element a factorizes as $a = bc$ in D. Then certainly $a \mid bc$. Since a is prime, it follows that a divides at least one of b and c, say $c = ua$ for some element u of D. Then $1c = c = ua = ubc$, so $1 = ub$ by cancellation (Proposition 7.10, page 159). It follows that b is a unit in D: the factorization $a = bc$ is improper. $\square$

8.2   Noetherian domains

The Fundamental Theorem of Arithmetic has two parts. The "existence part" of the theorem, Theorem 1.13 (page 15), shows that each integer $n > 1$ may be factorized as some product of irreducibles. Then the "uniqueness part" of the theorem, Theorem 1.14 (page 15), shows that any such factorization is essentially unique. Noetherian domains, the topic of this section, form a general class of integral domains having an analogue of the "existence part" of the Fundamental Theorem of Arithmetic. (For a complete characterization of those integral domains in which factorizations exist, without necessarily existing uniquely, see Exercise 7.)

Let R be a ring. An *ascending chain of ideals* in R is a sequence

$$J_0 \hookrightarrow J_1 \hookrightarrow \ldots \hookrightarrow J_n \hookrightarrow J_{n+1} \hookrightarrow \ldots \tag{8.3}$$

of ideals J_n of R, for natural numbers n, such that each ideal J_n has an embedding $J_n \hookrightarrow J_{n+1}$ as a subset of its successor. (Compare Example 6.21, page 135.) The union

$$J = \bigcup_{n \in \mathbb{N}} J_n = \{r \mid r \text{ lies in } J_n \text{ for some } n \text{ in } \mathbb{N}\} \tag{8.4}$$

of the members of the sequence is again an ideal of R (Exercise 5).

DEFINITION 8.8 (ACC, Noetherian domains.)   *Let R be a ring.*

(a) *The ring R is said to satisfy the* ascending chain condition, *or* ACC, *if for each ascending chain (8.3) of ideals J_n of R, there is a natural number N such that $J_m = J_N$ for all $m \geq N$.*

(b) *The ring R is said to be a* Noetherian domain *if it is an integral domain that satisfies the ascending chain condition.*

The ascending chain condition may be paraphrased as saying that there are no infinite ascending chains (8.3): All eventually stabilize at some point J_N.

PROPOSITION 8.9 (PID implies ACC.)
Each principal ideal domain D is Noetherian.

PROOF   Consider an ascending chain (8.3) of ideals J_n in D. Since the union (8.4) is again an ideal of the principal ideal domain D, it may be written in the form

$$J = aD$$

as the set of multiples of a certain element a of D. By the definition (8.4) of J, there is an element J_N of the sequence (8.3) such that a lies in J_N. For each integer $m \geq N$, the ideal J_N is a subset of J_m, since the sequence (8.3) is ascending. But J_m is a subset of $J = aD$. By the absorption property of the ideal J_N, the set aD is contained inside J_N, so J_m is a subset of J_N. Thus $J_m = J_N$ for all $m \geq N$.   ▯

PROPOSITION 8.10 (Factors in Noetherian domains.)
Let D be a Noetherian domain. Then each nonzero, noninvertible element of D has an irreducible factor.

PROOF   Assume that the proposition is false, and that a_0 is a nonzero, noninvertible element of D that does not have an irreducible factor. As an induction hypothesis, suppose that a_n is a nonzero, noninvertible element of D that does not have an irreducible factor. Since a_n is not itself irreducible, it has a proper factorization $a_n = a_{n+1}b_{n+1}$, in which the factor a_{n+1} is a nonzero, nonunit element of D that does not have an irreducible factor. By induction, we obtain such elements a_n for each natural number n, with $a_n D = a_{n+1}b_{n+1}D$ contained in $a_{n+1}D$.

Now consider the ascending chain

$$a_0 D \hookrightarrow a_1 D \hookrightarrow \ldots \hookrightarrow a_n D \hookrightarrow a_{n+1} D \hookrightarrow \ldots$$

of ideals in D. By the ascending chain condition, there is a natural number N with $a_N D = a_{N+1}b_{N+1}D = a_{N+1}D$. Proposition 8.1 then implies that b_{N+1} is a unit, contradicting the properness of the factorization $a_N = a_{N+1}b_{N+1}$.   ▯

PROPOSITION 8.11 (Factorization in Noetherian domains.)
Let D be a Noetherian domain. Then each nonzero, noninvertible element of D is a product of irreducible factors.

PROOF   Let $a = a_0$ be a nonzero, noninvertible element of D. As an induction hypothesis, suppose that a_i is a nonzero, noninvertible element of D. If a_i is not irreducible, it has a proper factorization $a_i = p_{i+1}a_{i+1}$ with an irreducible factor p_{i+1}, by Proposition 8.10. Thus

$$a_0 = p_1 a_1$$
$$= p_1 p_2 a_2$$
$$\vdots$$
$$= p_1 p_2 \ldots p_i a_i$$
$$= p_1 p_2 \ldots p_i p_{i+1} a_{i+1}$$
$$= \ldots$$

Now $a_i D = p_{i+1} a_{i+1} D$ is contained in $a_{i+1} D$ for $i = 0, 1, \ldots$, and so on. Consider the ascending chain

$$a_0 D \hookrightarrow a_1 D \hookrightarrow \ldots \hookrightarrow a_i D \hookrightarrow a_{i+1} D \hookrightarrow \ldots$$

of ideals in D. Since D is Noetherian, the chain stabilizes as $a_{n-1} D = a_n D$ for some natural number n. Thus a_n is an irreducible element p_n of D, and a factorizes as $a = p_1 \ldots p_n$.   ∎

COROLLARY 8.12 (Factorization in principal ideal domains.)
In a principal ideal domain D, each nonzero, noninvertible element is a product of irreducible factors.

PROOF   By Proposition 8.9, the principal ideal domain D is Noetherian. Proposition 8.11 then gives the desired result.   ∎

8.3   Unique factorization domains

The full conclusion of the Fundamental Theorem of Arithmetic, existence and uniqueness, is formalized in the concept of a unique factorization domain. In order to capture the correct level of uniqueness, two elements p and q of a general integral domain D are defined to be *associates* if there is a unit u from D^* such that $p = u \cdot q$. The relation of being associate is an equivalence relation (Exercise 10). By Remark 8.3, associates of irreducibles are irreducible.

DEFINITION 8.13 (UFD.)   *A ring D is a* unique factorization domain, *or* UFD, *if it is an integral domain in which the following properties hold:*

(a) (**Existence of factorizations.**) *Each nonzero, nonunit element a of D has a factorization*

$$a = p_1 p_2 \ldots p_m \qquad\qquad (8.5)$$

as a (nonempty) product of irreducible elements $p_1, p_2, \ldots, p_m$ of D;

(b) (**Uniqueness of factorizations.**) *For each nonzero, nonunit element a of D, the factorization (8.5) is unique to within reordering and passage to associate irreducibles. In other words, if*

$$a = p_1 p_2 \ldots p_m = q_1 q_2 \ldots q_n \qquad\qquad (8.6)$$

with irreducibles $p_1, p_2, \ldots, p_m, q_1, q_2, \ldots, q_n$ in D, then $m = n$, and there is a permutation π of $\{1, 2, \ldots, n\}$ such that p_j is an associate of $q_{\pi(j)}$ for each $1 \le j \le n$.

In considering the uniqueness part of Definition 8.13, it is useful to extend the defining property Definition 8.2(b) of a prime element in an integral domain.

PROPOSITION 8.14 ("Divide and conquer.")
Let p be a prime element of an integral domain D. If p divides a product $a_1 \ldots a_n$ in D, then it divides one of the factors a_i for $1 \le i \le n$.

PROOF   By induction on n, the induction basis $n = 2$ being a direct application of Definition 8.2(b). Suppose the proposition holds for all products with less than n factors, for an integer $n \ge 3$. Then by Definition 8.2(b), $p \mid a_1 \ldots a_{n-1} a_n$ implies $p \mid a_1 \ldots a_{n-1}$ or $p \mid a_n$. The second case immediately gives the desired result. In the first case, $p \mid a_i$ for some $1 \le i < n$ by the induction hypothesis. ▯

PROPOSITION 8.15 (Irreducible $\equiv$ prime in a UFD.)
Let D be an integral domain satisfying the property of Definition 8.13(a). Then the following two conditions are equivalent:

(a) *D is a unique factorization domain;*

(b) *Each irreducible element of D is prime.*

PROOF   (a) implies (b): Suppose that p_1 is an irreducible element of a unique factorization domain D. Suppose that a product bc in D is a multiple $p_1 d$ of p_1. Suppose that the elements b, c, and d have respective factorizations

$$b = q_1 \ldots q_r, \qquad c = q_{r+1} \ldots q_n, \qquad \text{and} \qquad d = p_2 \ldots p_m$$

as products of irreducibles in D. The element

$$a = p_1 d = b \cdot c$$

of D then has the factorizations

$$a = p_1 p_2 \ldots p_m = q_1 \ldots q_r \cdot q_{r+1} \ldots q_n ,$$

so p_1 is the associate of $q_{\pi(1)}$, say $p_1 = v q_{\pi(1)}$ or $q_{\pi(1)} = u p_1$ for $uv = 1$. If $\pi(1) \leq r$, we have $p_1 \mid b$, while if $\pi(1) > r$, we have $p_1 \mid c$.

(b) implies (a): We will prove, by induction on m, that the uniqueness property of Definition 8.13(b) holds for each nonzero, nonunit element a of D. The induction basis is the case $m = 1$ of (8.6). In this case, the factorization $q_1 q_2 \ldots q_n$ of the irreducible element p_1 must be trivial, so $n = 1$ and $p_1 = q_1$. As the induction hypothesis, suppose that the uniqueness property holds for all factorizations with less than m irreducible factors p_i. Consider (8.6) with $m > 1$. By the assumption (b), the irreducible element p_m of D is a prime divisor of $q_1 \ldots q_n$. The "divide and conquer" Proposition 8.14 implies that p_m divides some factor q_r with $1 \leq r \leq n$. Since q_r is irreducible, it can only be a multiple $q_r = u p_m$ of p_m with a unit u. In particular, p_m is an associate of q_r. Now

$$p_1 \ldots p_{m-1} p_m = q_1 \ldots q_{r-1} u p_m q_{r+1} q_{r+2} \ldots q_n$$

implies

$$p_1 \ldots p_{m-1} = q_1 \ldots q_{r-1} (u q_{r+1}) q_{r+2} \ldots q_n \qquad (8.7)$$

by cancellation. Set

$$q_i' = \begin{cases} q_i & \text{for } 1 \leq i < r ; \\ u q_{i+1} & \text{for } i = r ; \\ q_{i+1} & \text{for } r < i < n . \end{cases}$$

The equality (8.7) reads as

$$p_1 \ldots p_{m-1} = q_1' \ldots q_{n-1}'$$

with irreducibles $q_1', \ldots, q_{n-1}'$. By the induction hypothesis, $n - 1 = m - 1$, and there is a permutation π' of $\{1, \ldots, m-1\}$ such that each p_j for $1 \leq j < m$ is an associate of $q_{\pi'(j)}'$. Now define

$$\pi(j) = \begin{cases} \pi'(j) & \text{for } 1 \leq \pi'(j) < r ; \\ \pi'(j) + 1 & \text{for } r \leq \pi'(j) < m ; \\ r & \text{for } j = m . \end{cases}$$

For $1 \leq j < m$, the irreducible p_j is an associate of $q_{\pi'(j)}'$, which in turn is an associate of $q_{\pi(j)}$. Also, the irreducible p_m is an associate of $q_r = q_{\pi(m)}$. Thus the uniqueness property holds for factorizations with m irreducible factors p_i, as required to complete the inductive proof. $\square$

THEOREM 8.16 (Every PID is a UFD.)
Let D be a principal ideal domain.

(a) *Each irreducible element p of D is prime.*

(b) *The integral domain D is a unique factorization domain.*

PROOF (a): Suppose $p \mid ab$, say $ab = pc$ with c in D. Consider the set

$$J = \{xp + ya \mid x, y \text{ in } D\} \tag{8.8}$$

of "D-linear combinations" of p and a. Now J is an ideal of D (Exercise 14). Since D is a principal ideal domain, J is just the set dD of multiples of some element d of D. As $1p + 0a$, the irreducible element p of D lies in J, and is thus expressible as a multiple $p = dz$ of D. Since p is irreducible, the factorization $p = dz$ is improper.

> (i) If d is a unit, then $J = dD = D$ contains 1, say $1 = x_1 p + y_1 a$. Now $b = 1b = x_1 pb + y_1 ab = x_1 bp + y_1 cp = (x_1 b + y_1 c)p$, so p divides b in this case.

> (ii) If z is a unit, $J = dD = dzD = pD$ by Proposition 8.1, so p divides the element $a = 0p + 1a$ of J in this case.

(b): By Corollary 8.12, D satisfies the existence property of Definition 8.13(a). Using part (a) of the theorem, part (b) then follows from Proposition 8.15. $\square$

8.4 Roots of polynomials

Let F be a field. By Theorem 7.27 (page 167), the ring $F[X]$ of polynomials over F is a principal ideal domain. Theorem 8.16(b) then shows that $F[X]$ is a unique factorization domain. The remainder of the chapter will rely heavily (and implicitly) on this property of such rings $F[X]$. Note that the group of units $F[X]^*$ of the ring $F[X]$ is the group F^* of nonzero constant polynomials (elements of F). In particular, each polynomial of positive degree is associate to a unique monic polynomial (Exercise 12).

A field F is said to be a *subfield* of a field E if F is a subset of E, and the inclusion $j : F \hookrightarrow E$ is a ring homomorphism. In this context, the field E is also described as an *extension (field)* of F. If there is a further field K with inclusions $F \hookrightarrow K \hookrightarrow E$ that are ring homomorphisms, then K is described as an *intermediate field* between F and E. If $a_1, \ldots, a_n$ are elements of E, then $F(a_1, \ldots, a_n)$ will denote the smallest intermediate field containing the set $\{a_1, \ldots, a_n\}$. For example, $\mathbb{C}$ is an extension field of $\mathbb{Q}$, with intermediate field $\mathbb{R}$. Then $\mathbb{Q}(i) = \{x + iy \mid x, y \text{ in } \mathbb{Q}\}$.

In its most basic application, the field $\mathbb{C}$ of complex numbers is used to furnish a root $i = \sqrt{-1}$ of the polynomial $X^2 + 1$ from $\mathbb{R}[X]$. The polynomial $X^2 + 1$ then factorizes in $\mathbb{C}[X]$ as the product

$$X^2 + 1 = (X + i)(X - i)$$

of linear factors. The following result, known as *Kronecker's Theorem*, shows that each nonconstant polynomial over a field has a root in some extension field.

THEOREM 8.17 (Kronecker's Theorem.)
Let F be a field. If $f(X)$ is a nonconstant polynomial in $F[X]$, then $f(X)$ has a root in some extension field E of F.

PROOF As a nonconstant polynomial, $f(X)$ is nonzero and noninvertible in the Noetherian domain $F[X]$. Proposition 8.10 shows that $f(X)$ has an irreducible factor $p(X)$. If $p(X)$ has a root x in an extension field E, then $f(x) = 0$ in E.

Let J be the ideal $p(X)F[X]$ of $F[X]$. Since $p(X)$ is irreducible, the quotient ring $E = F[X]/J$ is a field (Theorem 7.39, page 171). Let x be the element $X + J$ of E. Then $p(x) = p(X) + J = p(X) + p(X)F[X] = p(X)F[X] = J$, so x is a root of $p(X)$ in E. ⧠

Example 8.18 (The square root of −1.)
Kronecker's Theorem builds $\mathbb{R}[X]/(X^2+1)\mathbb{R}[X]$ as the extension field E of $\mathbb{R}$ in which the real polynomial $X^2 + 1$ has a root. As discussed in Example 7.29 (page 168), this field E is isomorphic to the field $\mathbb{C}$ of complex numbers. Moreover, the root $x = X + (X^2 + 1)\mathbb{R}[X]$ of $X^2 + 1$ in E maps to the complex number i under the isomorphism (7.16). ⧠

The real polynomial

$$f(X) = X^4 + 2X^2 + 1 = (X^2 + 1)^2 \tag{8.9}$$

has no real roots. However, in the extension field $\mathbb{C}$ of $\mathbb{R}$, it has i as a double root, since it factorizes as $X^4 + 2X^2 + 1 = (X + i)^2(X - i)^2$ in $\mathbb{C}[X]$. There is a way to recognize that the real polynomial $f(X)$ has a repeated root in the extension field $\mathbb{C}$, without leaving the field of real numbers. Consider the derivative

$$f'(X) = 4X^3 + 4X = 4X(X^2 + 1),$$

computed as usual in calculus. It is then apparent that $f(X)$ and $f'(X)$ share the common, nonconstant factor $X^2 + 1$ in $\mathbb{R}[X]$.

If F is an abstract field, there is no analytical method to differentiate a polynomial $f(X)$ in $F[X]$. However, one may define the *derivative* $Df(X)$ in

$F[X]$ of a polynomial $f(X)$ in $F[X]$ using the formal rules

$$DX^n = nX^{n-1}$$

and

$$D\big(af(X) + bg(X)\big) = aDf(X) + bDg(X)$$

for positive integers n, for constants a, b in F, and for polynomials $f(X)$ and $g(X)$ in $F[X]$. The following properties are readily verified, just as in calculus (Exercise 22).

PROPOSITION 8.19 (Rules for differentiation.)
Let F be a field.

(a) *For a polynomial*

$$p(X) = p_n X^n + p_{n-1} X^{n-1} + \cdots + p_1 X + p_0$$

in $F[X]$, the derivative is given as

$$Dp(X) = np_n X^{n-1} + (n-1)p_{n-1} X^{n-2} + \cdots + p_1 .$$

(b) *The* product rule *holds:*

$$D\big(f(X) \cdot g(X)\big) = Df(X) \cdot g(X) + f(X) \cdot Dg(X)$$

for polynomials $f(X)$ and $g(X)$ in $F[X]$.

The formal derivative may then be used to spot when a polynomial will have repeated roots in some extension field.

THEOREM 8.20 (Derivatives and repeated roots.)
Let F be a field, and let $f(X)$ be a nonconstant polynomial in $F[X]$. The following conditions are equivalent:

(a) *There is an extension field E of F in which $f(X)$ has a repeated root.*

(b) *The polynomials $f(X)$ and $Df(X)$ share a common factor of positive degree in the unique factorization domain $F[X]$.*

PROOF **(a) implies (b):** Suppose that there is a factorization

$$f(X) = (X - a)^2 g(X)$$

in the ring $E(X)$ of polynomials over an extension field E of F. By the product rule,

$$Df(X) = 2(X - a)g(X) + (X - a)^2 Dg(X) .$$

Thus $f(a) = 0$ and $Df(a) = 0$. Let $j : F \hookrightarrow E$ be the inclusion of F in E. The Substitution Principle (Theorem 6.39, page 145) gives a unique ring homomorphism $j_a : F[X] \to E; h(X) \mapsto h(a)$ with $j_a(X) = a$. The ring kernel $\operatorname{Ker} j_a$ in the principal ideal domain $F[X]$ is the set $p(X)F[X]$ of multiples of a certain nonconstant polynomial $p(X)$. Since $f(a) = 0$ and $Df(a) = 0$, the polynomial $p(X)$ is a common factor of $f(X)$ and $DF(X)$ in $F[X]$.

(b) implies (a): Suppose that $p(X)$ is a nonconstant factor of both $f(X)$ and $Df(X)$ in $F[X]$. By Kronecker's Theorem, there is an extension field E of F in which $p(X)$ has a root a, so $f(a) = 0 = Df(a)$ in E. Then

$$f(X) = (X - a)h(X) \tag{8.10}$$

in $E[X]$, and

$$Df(X) = h(X) + (X - a)Dh(X)$$

by the product rule. Now $Df(a) = 0$ in E implies $h(a) = 0$ in E. From (8.10), it is apparent that $f(X)$ has a as a repeated root in E. $\qquad\qquad\square$

8.5   Splitting fields

Let $f(X)$ be a polynomial over a field F. The polynomial $f(X)$ is said to *split* over an extension field E of F if $f(X)$ factorizes as a product of linear factors in $E[X]$. For example, the polynomial $X^4 + 2X^2 + 1$ of (8.9), considered as a polynomial over the field $\mathbb{Q}$ of rationals, splits over the field of complex numbers. The splitting of nonconstant polynomials is a consequence of Kronecker's Theorem.

PROPOSITION 8.21 (Every nonconstant polynomial splits.)
Let $f(X)$ be a nonconstant polynomial over a field F. Then $f(X)$ splits over some extension field E of F.

PROOF   As an induction hypothesis, assume that the proposition is true for all fields F and for all nonconstant polynomials of degree less than some positive integer n. Note that the proposition is trivial if $\deg f = 1$. Suppose $\deg f = n > 1$.

Case A: If $f(X)$ has a root a in F, then it factorizes as $(X - a)g(X)$ in $F[X]$, with $\deg g = n - 1$. By induction, $g(X)$ splits over some extension field E of F. Then so does $f(X)$.

Case B: If $f(X)$ has no root in F, it has an irreducible factor $p(X)$ in $F[X]$. By Kronecker's Theorem, there is an extension field K of F such that $p(X)$ has a root a in K. Then $f(X)$, as a nonconstant polynomial of degree n in $K[X]$, has a root a in K. Case A now applies, with F replaced by K.

It follows that $f(X)$ splits over some extension field E of K. Since F is a subfield of K, the field E is an extension field of F. ⬚

The proof of Proposition 8.21 embodies a process for finding an extension field E over which a polynomial $f(X)$ from $F[X]$ will split. Consider the example of $f(X) = X^4 + 2X^2 + 1$ in $\mathbb{Q}[X]$. Since $f(X)$ has no root in $\mathbb{Q}$, Case B becomes relevant. Now $f(X)$ has the irreducible factor $X^2 + 1$ in $\mathbb{Q}[X]$. Kronecker's Theorem builds

$$\mathbb{Q}(i) = \mathbb{Q}[X]/(X^2+1)\mathbb{Q}[X] = \{x + iy \mid x, y \text{ in } \mathbb{Q}\} \qquad (8.11)$$

as an extension field of $\mathbb{Q}$ in which $X^2 + 1$ has a root, namely

$$i = X + (X^2 + 1)\mathbb{Q}[X].$$

At this point, we are transferred to Case A, working with the polynomial $f(X)$ as an element of $\mathbb{Q}(i)[X]$. Factorization in $\mathbb{Q}(i)[X]$ yields $f(X) = (X-i)g(X)$ with $g(X) = (X+i)^2(X-i)$. Since $g(X)$ already splits over $\mathbb{Q}(i)$, this is the extension field over which $f(X)$ will split, as $f(X) = (X-i)^2(X+i)^2$.

Let F be a field. Let $f(X)$ be a nonconstant polynomial in $F[X]$. Suppose that $f(X)$ splits over an extension field L of F, say as a product

$$f(X) = c(X - l_1)(X - l_2)\ldots(X - l_n)$$

of a nonzero constant c from F and (not necessarily distinct) linear factors $(X - l_i)$ with l_i in L. The smallest subfield $F(l_1, \ldots, l_n)$ of L containing F and the set $\{l_1, \ldots, l_n\}$ of roots is called a *splitting field* of $f(X)$ over F. In other words, a field K is a splitting field for a polynomial $f(X)$ in $F[X]$ if

- $f(X) = c_0(X - c_1)\ldots(X - c_n)$ in $K[X]$ (with $c_0, c_1, \ldots, c_n$ in K), and

- $f(X)$ doesn't split over any proper intermediate field between F and K.

Note the dependence on the choice of coefficient field F for the polynomial $f(X)$, just as with the definition of irreducibility.

Example 8.22 (Dependence of the splitting field.)
With $F = \mathbb{Q}$ and $f(X) = (X^2 + 1)^2$, we obtain (8.11) as a splitting field. On the other hand, $\mathbb{C}$ is a splitting field for $f(X) = (X^2 + 1)^2$ over $F = \mathbb{R}$. ⬚

Example 8.23 (Use of the quadratic formula.)
Consider the problem of specifying a splitting field for the quadratic equation

$$f(X) = X^2 - X + 1$$

over $\mathbb{Q}$. Recall that for complex numbers a, b, and c, the quadratic polynomial $aX^2 + bX + c$ has roots given by the formula

$$\frac{-b \pm \sqrt{b^2 - 4ac}}{2a}.$$

Applying that formula to $f(X)$ in $\mathbb{C}[X]$, we obtain the two roots $\frac{1}{2}(1 \pm i\sqrt{3})$. These roots lie in the field $\mathbb{Q}(i\sqrt{3})$ or $\mathbb{Q}[X]/(X^2+3)\mathbb{Q}[X]$, giving

$$f(X) = \left(X - \frac{1}{2}(1 + i\sqrt{3}) \right) \cdot \left(X - \frac{1}{2}(1 - i\sqrt{3}) \right)$$

as the factorization in $\mathbb{Q}(i\sqrt{3})[X]$. Since $i\sqrt{3}$ is not rational, no smaller field will split $f(X)$. Thus $\mathbb{Q}(i\sqrt{3})$ is the splitting field. ⬜

Example 8.24 (Roots of unity.)

Let $n > 2$ be an integer. To specify a splitting field for $X^n - 1$ over $\mathbb{Q}$, consider the complex number

$$\omega = \begin{bmatrix} \cos(2\pi/n) & -\sin(2\pi/n) \\ \sin(2\pi/n) & \cos(2\pi/n) \end{bmatrix} = \cos\left(\frac{2\pi}{n}\right) + i\sin\left(\frac{2\pi}{n}\right).$$

The powers ω^r, for $0 \le r < n$, are n roots of $X^n - 1$. They are known as *roots of unity*. Then

$$X^n - 1 = (X - 1)(X - \omega)(X - \omega^2)\ldots(X - \omega^{n-1})$$

in $\mathbb{Q}(\omega)$, and $\mathbb{Q}(\omega)$ is seen to be a splitting field for $X^n - 1$ over $\mathbb{Q}$. ⬜

8.6   Uniqueness of splitting fields

The proof of Proposition 8.21 provides a technique for obtaining a splitting field of a nonconstant polynomial $f(X)$ in the ring of polynomials over a field F. However, the technique involves some arbitrary choices: choice of a particular root a in Case A, or choice of a particular irreducible factor $p(X)$ in Case B. Different choices at these points cause a branching in the construction process that could potentially lead to different splitting fields. In this section, it will be shown that the divergent paths eventually land back in the same place: Any two splitting fields for $f(X)$ over F are isomorphic.

We begin with the case of an irreducible polynomial.

PROPOSITION 8.25

Let F be a field, and let $p(X)$ be an irreducible polynomial over F. Suppose that a_i is a root of $p(X)$ in an extension field E_i of F, for $i = 1, 2$. Then the fields $F(a_1)$ and $F(a_2)$ are each isomorphic to $F[X]/p(X)F[X]$.

PROOF   For $i = 1, 2$, let $j^i : F \hookrightarrow E_i$ be the inclusion of F in the extension field E_i. The Substitution Principle (Theorem 6.39, page 145) gives a unique ring homomorphism $j_{a_i}^i : F[X] \rightarrow E_i; h(X) \mapsto h(a_i)$ with $j_{a_i}^i(X) = a_i$. The ring kernel $\text{Ker} \, j_{a_i}^i$ in the principal ideal domain $F[X]$ is the set $q_i(X)F[X]$ of multiples of a certain nonconstant polynomial $q_i(X)$. Now $p(a_i) = 0$ implies that the irreducible polynomial $p(X)$ is a multiple of $q_i(X)$. Since $p(X)$ is prime in the principal ideal domain $F[X]$, the polynomials $p(X)$ and $q_i(X)$ are associates, so $\text{Ker} \, j_{a_i}^i = q_i(X)F[X] = p(X)F[X]$. The image of $j_{a_i}^i$ is $F(a_i)$, so the First Isomorphism Theorem for rings gives a well-defined isomorphism

$$b_i : F[X]/p(X)F[X] \rightarrow F(a_i); h(X) + p(X)F[X] \mapsto h(a_i) \qquad (8.12)$$

for $i = 1, 2$. $\qquad\qquad\qquad\qquad\qquad\qquad\qquad\qquad\qquad\qquad\qquad\qquad$ ▯

A slight refinement of Proposition 8.25 is needed for the treatment of the general case.

PROPOSITION 8.26
Let F be a field, and let $p(X)$ be an irreducible polynomial over F. Suppose that a_1 is a root of $p(X)$ in an extension field of F. Let $\theta : F \rightarrow G; x \mapsto \overline{x}$ be a field isomorphism, with corresponding ring homomorphism

$$\theta_X : F[X] \rightarrow G[X]; h(X) \mapsto \overline{h}(X)$$

given by the Substitution Principle. Let a_2 be a root of $\overline{p}(X)$ in an extension field of G. Then there is a field isomorphism $\overline{\theta} : F(a_1) \rightarrow G(a_2)$, restricting to $\theta : F \rightarrow G$, with $\overline{\theta}(a_1) = a_2$.

PROOF The ring homomorphism $\theta_X : F[X] \rightarrow G[X]$ is an isomorphism, with two-sided inverse $G[X] \rightarrow F[X]; \overline{h}(X) \mapsto h(X)$. Thus $\overline{p}(X)$ is irreducible in $G[X]$. By Proposition 8.25, there is an isomorphism

$$b_1 : F[X]/p(X)F[X] \rightarrow F(a_1); h(X) + p(X)F[X] \mapsto h(a_1)$$

as in (8.12), restricting to the identity map on F in the form

$$F \rightarrow F; x + p(X)F[X] \mapsto x \, .$$

Similarly, there is an isomorphism

$$b_2 : G[X]/\overline{p}(X)G[X] \rightarrow G(a_2); \overline{h}(X) + \overline{p}(X)G[X] \mapsto \overline{h}(a_2)$$

as in (8.12), restricting to the identity map on G in the form

$$G \rightarrow G; \overline{x} + \overline{p}(X)G[X] \mapsto \overline{x} \, .$$

Moreover, the First Isomorphism Theorem for rings, applied to

$$F[X] \rightarrow G[X]/\overline{p}(X)G[X]; h(X) \mapsto \overline{h}(X) + \overline{p}(X)G[X] \, ,$$

yields a well-defined isomorphism

$$b : F[X]/p(X)F[X] \to G[X]/\overline{p}(X)G[X];$$
$$h(X) + p(X)F[X] \mapsto \overline{h}(X) + \overline{p}(X)G[X].$$

The desired isomorphism $\overline{\theta} : F(a_1) \to G(a_2)$ is then realized by $b_2 \circ b \circ b_1^{-1}$. □

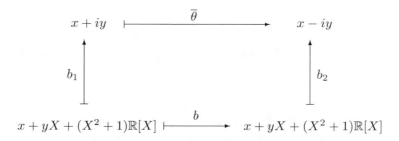

FIGURE 8.1:   Complex conjugation from Proposition 8.26.

Example 8.27 (Complex conjugation.)
For the irreducible polynomial $p(X) = X^2 + 1$ in $\mathbb{R}[X]$, consider the root $a_1 = i$ in the extension field $\mathbb{C} = \mathbb{R}(i)$. Take $\theta : \mathbb{R} \to \mathbb{R}$ to be the identity map $\mathrm{id}_{\mathbb{R}}$ on the reals. Take a_2 to be the root $-i$ of $\overline{p}(X) = X^2 + 1$. Then the field isomorphism $\overline{\theta} : \mathbb{R}(i) \to \mathbb{R}(-i)$ with $\overline{\theta}(i) = -i$ is the complex conjugation

$$\mathbb{C} \to \mathbb{C}; z = x + iy \mapsto \overline{z} = x - iy$$

(Figure 8.1). Compare Study Project 1 in Chapter 6.   □

It will now be shown that splitting fields are unique up to isomorphism.

PROPOSITION 8.28
Let F be a field. Let $f(X)$ be a nonconstant polynomial in $F[X]$. Let

$$\theta : F \to G; x \mapsto \overline{x}$$

be a field isomorphism, with corresponding ring homomorphism

$$\theta_X : F[X] \to G[X]; h(X) \mapsto \overline{h}(X)$$

given by the Substitution Principle. Suppose that K is a splitting field for $f(X)$ over $F[X]$, and that L is a splitting field for $\overline{f}(X)$ over G. Then there is a field isomorphism $\overline{\theta} : K \to L$ extending $\theta : F \to G$.

PROOF As an induction hypothesis, suppose that the proposition is true for all fields F and for all nonconstant polynomials of degree less than some positive integer n. (Note that the proposition is trivial if $\deg f = 1$, since $K = F$, $L = G$, and $\overline{\theta} : K \to L$ is just $\theta : F \to G$ in that case.) Suppose $\deg f = n > 1$.

Let $p(X)$ be an irreducible factor of $f(X)$. Let a_1 be a root of $p(X)$ in K, and let a_2 be a root of $\overline{p}(X)$ in L. By Proposition 8.26, there is a field isomorphism $\widetilde{\theta} : F(a_1) \to G(a_2)$ extending $\theta : F \to G$, with $\widetilde{\theta}(a_1) = a_2$. Within the polynomial ring $F(a_1)[X]$, the polynomial $f(X)$ factorizes as $f(X) = (X - a_1)b(X)$.

Since $\deg f > 1$, the polynomial $b(X)$ in $F(a_1)[X]$ is nonconstant, of degree $n - 1$. Now K is a splitting field for $b(X)$ over $F(a_1)$. The field isomorphism

$$\widetilde{\theta} : F(a_1) \to G(a_2)$$

furnishes a ring isomorphism

$$\widetilde{\theta}_X : F(a_1)[X] \to G(a_2)[X]; h(X) \mapsto \widetilde{h}(X)$$

by the Substitution Principle. Note that $\widetilde{\theta}_X$ extends θ_X. Under the ring isomorphism $\widetilde{\theta}_X$, the factorization $f(X) = (X - a_1)b(X)$ in $F(a_1)[X]$ maps to $\overline{f}(X) = (X - a_2)\widetilde{b}(X)$ in $G(a_2)[X]$. Thus L is a splitting field for $\widetilde{b}(X)$ over $G(a_2)$. By the induction hypothesis, there is a field isomorphism $\overline{\theta} : K \to L$ extending $\widetilde{\theta} : F(a_1) \to G(a_2)$. Since $\widetilde{\theta} : F(a_1) \to G(a_2)$ extends $\theta : F \to G$, it follows that $\overline{\theta} : K \to L$ extends $\theta : F \to G$. This completes the inductive proof. $\square$

THEOREM 8.29 (Uniqueness of splitting fields.)
Let F be a field. Let $f(X)$ be a nonconstant polynomial in $F[X]$. Suppose that K and L are splitting fields for $f(X)$ over F. Then K and L are isomorphic.

PROOF In Proposition 8.28, take $\theta : F \to G$ to be the identity map $\mathrm{id}_F : F \to F$ on F. $\square$

Example 8.30 (Splitting $X^n - a$ over the rationals.)
Let $n > 2$ be an integer. Let a be a positive real number. Consider the problem of determining the splitting field for $X^n - a$ over $\mathbb{Q}$. Since $x \mapsto x^n$ is a strictly increasing, continuous function on the set of positive reals, there is a unique positive real number r with $r^n = a$. This number is denoted by $a^{1/n}$. The field $\mathbb{Q}(a^{1/n})$ is a subfield of $\mathbb{R}$. In $\mathbb{Q}(a^{1/n})[X]$, the polynomial $X^n - a$ has a linear factor $X - a^{1/n}$, but does not factorize further. However, in $\mathbb{Q}(a^{1/n})(\omega)$, with ω as in Example 8.24, the polynomial $X^n - a$ splits, with roots of the form $a^{1/n}\omega^k$ for $0 \le k < n$. Thus the splitting field for $X^n - a$ over $\mathbb{Q}$ is $\mathbb{Q}(a^{1/n}, \omega)$. $\square$

8.7   Structure of finite fields

Some finite fields made their appearance in Chapter 7. In the final two sections of this chapter, we undertake a more comprehensive study, starting with an observation about finite abelian groups.

PROPOSITION 8.31
Let A be a finite abelian group. Let x be an element of A whose order d_x is maximal. Then for each element y of A, the order d_y of y divides d_x.

PROOF   If d_y does not divide d_x, then d_y has a prime factor p which does not divide d_x. Since $\gcd(p, d_x) = 1$, there are integers l and m with $lp + md_x = 1$.

Now $z = y^{d_y/p}$ has order p. Consider the subgroup $\langle xz \rangle$ of A. It contains

$$\langle (xz)^{lp} \rangle = \langle x^{lp} z^{lp} \rangle = \langle x^{1-md_x} \rangle = \langle x \rangle$$

as a subgroup of order d_x, and

$$\langle (xz)^{md_x} \rangle = \langle x^{md_x} z^{md_x} \rangle = \langle z^{1-lp} \rangle = \langle z \rangle$$

as a subgroup of order p. By Lagrange's Theorem, the order of xz is a multiple of pd_x. This contradicts the maximality of d_x.   ☐

PROPOSITION 8.32 (Multiplicative groups of finite fields.)
Let K be a finite field. Then the group K^* of nonzero elements of K is cyclic.

PROOF   Let e be an element of K^* of maximal (multiplicative) order d. By Proposition 8.31, each element of K^* is a root of the polynomial $X^d - 1$. By Proposition 7.24 (page 166), $|K^*| \le d$. On the other hand, $d = |\langle e \rangle|$ is a divisor of $|K^*|$, by Lagrange's Theorem. Thus K^* is the cyclic group $\langle e \rangle$.   ☐

An element e of a finite field K is called a *primitive element* if $K^* = \langle e \rangle$ — compare Study Project 3 in Chapter 7. Now if F is a subfield of a finite field E, the additive group $(F, +, 0)$ of F is a subgroup of the additive group $(E, +, 0)$ of E. Lagrange's Theorem then shows that the order $|F|$ of F is a divisor of the order $|E|$ of E. In fact, a much stronger statement is true.

PROPOSITION 8.33 (Additive groups of finite fields.)
Let E be a finite field, with a subfield F. Then the additive group $(E, +, 0)$ of E is isomorphic to a power of the additive group $(F, +, 0)$ of F. In particular, the order $|E|$ of E is a power of the order $|F|$ of F.

PROOF Let e be a primitive element of E. Let $j : F \hookrightarrow E$ be the inclusion of F in E. The Substitution Principle (Theorem 6.39, page 145) gives a unique ring homomorphism $j_e : F[X] \to E$ with $j_e(X) = e$. The ring kernel of j_e in the principal ideal domain $F[X]$ is the set $p(X)F[X]$ of multiples of a certain polynomial $p(X)$. If the degree of $p(X)$ is m, Proposition 7.31 (page 169) shows that the additive group of $F[X]/p(X)F[X]$ is isomorphic to $(F,+,0)^m$.

Each nonzero element of E, as a power e^r of e, appears as $j_e(X^r)$ in the image of j_e. Thus $j_e(F[X]) = E$. The First Isomorphism Theorem for Rings (Theorem 6.34, page 140) applied to $j_e : F[X] \to E$ yields the ring isomorphism

$$F[X]/p(X)F[X] \cong E .$$

It follows that $(E, +, 0)$ is isomorphic to the power $(F, +, 0)^m$ of $(F, +, 0)$. $\quad\square$

PROPOSITION 8.34 (Prime subrings of finite fields.)
Let K be a finite field. Then the prime subring of K is the field $\mathbb{Z}/_p$ of integers modulo a prime number p.

PROOF Since K is finite, its prime subring is finite, and thus of the form $\mathbb{Z}/_d$ for a positive integer d. If d is composite, say $d = ab$ with $1 < a, b < d$, then $a \cdot b = 0$ in K. This cannot happen in the integral domain K. $\quad\square$

COROLLARY 8.35
The additive group $(K, +, 0)$ of the finite field K is isomorphic to a power $(\mathbb{Z}/_p, +, 0)^n$ of the group $(\mathbb{Z}/_p, +, 0)$ of integers modulo p under addition. In particular, the order $|K|$ is a power p^n of the prime number p.

PROOF Set $E = K$ and $F = \mathbb{Z}/_p$ in Proposition 8.33. $\quad\square$

Proposition 8.34 shows that the characteristic of a finite field K is a prime number p. Now in a ring R of characteristic p, the Frobenius map

$$\varphi : R \to R; x \mapsto x^p$$

is a ring homomorphism (Exercise 45 in Chapter 6). In an integral domain of characteristic p, the Frobenius map is injective (Exercise 35).

PROPOSITION 8.36 (Subfields and the Frobenius map.)
Let K be a field of characteristic p. Then for each natural number r, the subset

$$L = \{x \mid \varphi^r(x) = x\}$$

of K forms a subfield of K.

PROOF   The set L is nonempty, indeed $\varphi^r(1) = 1$. Now suppose that x and y are elements of L. Then

$$\varphi^r(x - y) = \varphi^r(x) - \varphi^r(y) = x - y$$

and

$$\varphi^r(x \cdot y) = \varphi^r(x) \cdot \varphi^r(y) = x \cdot y\,,$$

so L is a unital subring of the integral domain K. As such, L is an integral domain. Each element of L is a root of the polynomial $X^m - X$, with $m = p^r$, over the field $\mathbb{Z}/_p$. Proposition 7.24 (page 166) shows that the set L is finite. By Proposition 7.16 (page 162), it follows that the integral domain L is actually a subfield of K.   ⬚

8.8   Galois fields

Corollary 8.35 shows that the only possible orders for a finite field are the powers of a prime number. The following theorem shows that for each power $q = p^n$ of a prime number p (with $n > 1$), there is a field of order q. To within isomorphism, this field is unique. It is known as the *Galois field* GF(q) of order q. In this notation, the field $\mathbb{Z}/_p$ of integers modulo p is written as GF(p).

THEOREM 8.37 (Classification of finite fields.)
Let $q = p^n$ be a power of a prime number p, with positive index n.

(a) *There is a field* GF(q) *of order q.*

(b) *Each field K of order q is isomorphic to* GF(q).

PROOF   (a): Consider the splitting field E of the polynomial $X^q - X$ in $\mathbb{Z}/_p[X]$. The set of roots of $X^q - X$ in E is the subset

$$\mathrm{GF}(q) = \{x \mid \varphi^n(x) = x\}$$

of E. By Proposition 8.36, GF(q) is a subfield of E. In fact, since $X^q - X$ splits over GF(q), we have $E = \mathrm{GF}(q)$. Now the derivative of $X^q - X$ is

$$qX^{q-1} - 1 = p^n X^{q-1} - 1 = -1\,,$$

which does not have any factor of positive degree. Theorem 8.20 shows that $X^q - X$, as a polynomial of degree q, has no repeated roots in any extension field. Thus its set GF(q) of roots has exactly q elements.

(b): If K is a field of order q, Proposition 8.34 shows that K contains $\mathbb{Z}/_p$ as a subfield. Since K^* is a cyclic group of order $q-1$, each nonzero element of K is a root of the polynomial $X^q - X$ in $\mathbb{Z}/_p[X]$, while 0 is certainly a root. Thus K is a splitting field of the polynomial $X^q - X$ in $\mathbb{Z}/_p[X]$. As such, it is isomorphic to the splitting field $\mathrm{GF}(q)$ for $X^q - X$ over $\mathbb{Z}/_p[X]$, by Theorem 8.29. ⬜

For a prime power $q = p^n$, the final question concerns the possible subfields of $\mathrm{GF}(q)$. By Proposition 8.33, the only possible orders for subfields are the powers p^r in which r divides n. Conversely, for such a power $m = p^r$, Proposition 8.36 shows that $\mathrm{GF}(q)$ does have a subfield

$$L = \{x \mid \varphi^r(x) = x\}$$

consisting of the roots of the polynomial $X^m - X$ in $\mathrm{GF}(q)$. By Theorem 8.20, this polynomial has no repeated roots, so L is a subfield of order $m = p^r$. Furthermore, it is the unique subfield of this order. Indeed, the elements of any subfield K of order p^r would consist of roots of $X^m - X$, and would thus lie in L. We summarize as follows.

PROPOSITION 8.38 (Subfields of finite fields.)
Let $q = p^n$ be a power of a prime number p, with positive index n. The only possible orders of subfields of $\mathrm{GF}(q)$ are the powers p^r for a divisor r of n. For each such power $m = p^r$, there is a unique subfield $\mathrm{GF}(m)$ of $\mathrm{GF}(q)$.

For a prime number p, the subfields of $\mathrm{GF}(p^{72})$ are displayed in Figure 8.2. Note that this figure is essentially just a relabelled version of Figure 1.3, which displayed the positive divisors of 72.

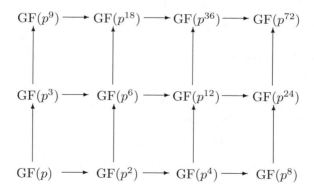

FIGURE 8.2: The subfields of $\mathrm{GF}(p^{72})$.

8.9 Exercises

1. Let a be a nonzero element of a zero ring A (compare Example 6.6, page 128). Show that a does not factorize as the product $a = b \cdot c$ of any pair of elements b and c of A.

2. Let u be a unit of an integral domain D. Show that a nonzero, nonunit element a of D is irreducible if and only if ua is irreducible.

3. Let A be a matrix in the ring $\mathbb{Z}[\sqrt{-3}\,]$ of Example 8.6. If $\det(A)$ is a prime number, show that A is irreducible.

4. Give an example of an irreducible element A of $\mathbb{Z}[\sqrt{-3}\,]$ for which $\det(A)$ is a composite number.

5. Show that the union (8.4) of the chain (8.3) is an ideal of the ring R.

6. Consider the ring $\mathbb{R}^{\mathbb{R}}$ of all functions $f : \mathbb{R} \to \mathbb{R}$, with componentwise unital ring structure (compare Example 6.12, page 130). For each natural number n, define
$$J_n = \{f : \mathbb{R} \to \mathbb{R} \mid f(x) = 0 \text{ if } x > n\}.$$

   (a) Show that J_n is an ideal of $\mathbb{R}^{\mathbb{R}}$ for each natural number n.

   (b) Show that
$$J_0 \hookrightarrow J_1 \hookrightarrow \ldots \hookrightarrow J_n \hookrightarrow J_{n+1} \hookrightarrow \ldots$$
   is an ascending chain of ideals in $\mathbb{R}^{\mathbb{R}}$.

   (c) Show that the ideal J_n is a proper subset of the ideal J_{n+1} for each natural number n.

   (d) Conclude that $\mathbb{R}^{\mathbb{R}}$ does not satisfy the ascending chain condition.

7. Let D be an integral domain. Show that the following two conditions are equivalent:

   (a) Each nonzero, noninvertible element a of D factorizes as a product of irreducibles in D;

   (b) D has no ascending chain
$$a_0 D \hookrightarrow a_1 D \hookrightarrow \ldots \hookrightarrow a_n D \hookrightarrow a_{n+1} D \hookrightarrow \ldots$$
   of principal ideals with $a_n D$ properly contained in $a_{n+1} D$ for each natural number n.

8. A ring R is said to satisfy the *descending chain condition*, or *DCC*, if for each descending chain

$$\ldots J_n \hookrightarrow J_{n-1} \hookrightarrow \ldots \hookrightarrow J_1 \hookrightarrow J_0 = R$$

of ideals J_n of R, there is a natural number N such that $J_m = J_N$ for all $m \geq N$. Suppose that an integral domain D satisfies the descending chain condition. Show that D is a field.

9. Show that the integer 12 may be factored as a product of irreducibles in $\mathbb{Z}$ in 12 different ways (ordering the factors).

10. Define a relation A on an integral domain D by setting x A y if and only if $x = u \cdot y$ for some u in D^*.

   (a) Show that A is an equivalence relation on D.

   (b) Show that the group D^* of units of D is the equivalence class of 1 under the equivalence relation A.

   (c) Show that $\{0\}$ is the equivalence class of 0 under the equivalence relation A.

11. Show that (3.2) gives the associate classes in $\mathbb{Z}$.

12. Let F be a field, and let $p(X)$ be a nonzero, nonunit element of the integral domain $F[X]$ of polynomials over F. Show that $p(X)$ is the associate of a unique monic polynomial in $F[X]$.

13. Let D be an integral domain, and let $\operatorname{Id} D$ be the set of ideals of D. Show that the relation of being associate is the kernel relation $\ker \alpha$ of the function

$$\alpha : D \to \operatorname{Id} D; a \mapsto aD.$$

14. Verify that the set J of (8.8) is an ideal of the integral domain D.

15. Show that the ring $\mathbb{Z}[\sqrt{-3}]$ of Example 8.6 is not a principal ideal domain.

16. Show that each field F is a subfield of the field $F(X)$ of rational functions over F — compare Example 7.47, page 177.

17. If E is an extension field of a field F, show that $E(X)$ is an extension field of $F(X)$.

18. Find the inverse of a nonzero element $x + iy$ of $\mathbb{Q}(i)$.

19. Find the inverse of a nonzero element $x + y\sqrt{5}$ of the field $\mathbb{Q}(\sqrt{5}) = \{x + y\sqrt{5} \mid x, y \text{ in } Q\}$.

20. Find a monic polynomial $p(X)$ in $\mathbb{Q}[X]$ such that the quotient field $\mathbb{Q}[X]/p(X)\mathbb{Q}[X]$ is isomorphic to the field $\mathbb{Q}\left(\sqrt{1+\sqrt{5}}\right)$.

21. Show that $\mathbb{Q}(\sqrt{2}, \sqrt{3}) = \mathbb{Q}(\sqrt{2}+\sqrt{3})$.

22. Verify the claims of Proposition 8.19.

23. Let $f(X) = X^4 + X^2 + 1$ in $\mathbb{Z}/_2[X]$. Show that $f(X)$ is a solution of the differential equation $Df(X) = 0$.

24. Let F be a field of characteristic zero. Show that the only solutions $f(X)$ to the differential equation $Df(X) = 0$ in $F[X]$ are the constant polynomials $f(X) = c$ for c in F.

25. Specify a splitting field for the quadratic equation
$$f(X) = X^2 + X + 1$$
over $\mathbb{Q}$.

26. Specify a splitting field for the quadratic equation
$$f(X) = X^2 + X + 1$$
over $\mathbb{Z}/_2$.

27. Specify the splitting field for $X^3 - 2$ over $\mathbb{Q}$.

28. Find the smallest positive integer n for which the splitting field of the polynomial
$$X^2 + 4X + 2$$
over $\mathbb{Q}$ is $\mathbb{Q}\left(\sqrt{n}\right)$.

29. Determine the splitting field for
$$X^3 + 2X^2 + 4X + 8$$
over $\mathbb{Q}$.

30. Determine the splitting field for
$$X^4 + 10X^3 + 100X^2 + 1000X + 10000$$
over $\mathbb{Q}$. (Hint: Set $n = 5$ and $a = 10^5$ in Example 8.30.)

31. Show that the fields $\mathbb{Q}(\sqrt{2})$ and $\mathbb{Q}(\sqrt{3})$ are not isomorphic.

32. Give an example of a finite group G with two elements x and y, such that x has maximal order d_x in G, while the order d_y of y is not a divisor of d_x. Why does this not contradict Proposition 8.31?

FACTORIZATION

33. In the proof of Proposition 8.33, show that the polynomial $p(X)$ is irreducible in $F[X]$.

34. (a) Show that there are no nontrivial finite subgroups in an infinite cyclic group.

   (b) Give an example of an infinite field K whose group K^* of nonzero elements is not cyclic.

35. Let p be a prime number.

   (a) Let R be an integral domain of characteristic p. Show that the Frobenius map

   $$\varphi : R \to R; x \mapsto x^p$$

   is injective.

   (b) Let R be a finite field of characteristic p. Show that the Frobenius map

   $$\varphi : R \to R; x \mapsto x^p$$

   is surjective.

   (c) Show that a polynomial $f(X)$ in $\mathbb{Z}/p[X]$ is the image $\varphi(g(X))$ of some polynomial $g(X)$ under the Frobenius map

   $$\varphi : \mathbb{Z}/p[X] \to \mathbb{Z}/p[X]$$

   if and only if it is a solution of the differential equation $Df(X) = 0$.

   (d) Show that the field $\mathbb{Z}/p(X)$ of rational functions over the integral domain $\mathbb{Z}/p$ has characteristic p.

   (e) Show that the Frobenius map

   $$\varphi : \mathbb{Z}/p(X) \to \mathbb{Z}/p(X)$$

   fails to be surjective.

36. Consider the following ideals in the ring $R = \mathbb{Z}/2[X]$ of polynomials over $\mathbb{Z}/2$:

$$J_1 = (X^3 + X + 1)\mathbb{Z}/2[X],$$
$$J_2 = (X^3 + X^2 + 1)\mathbb{Z}/2[X],$$
$$J_3 = (X^3 + X^2 + X + 1)\mathbb{Z}/2[X].$$

   (a) Show that the quotient rings R/J_1 and R/J_2 are isomorphic.

   (b) Show that the quotient rings R/J_1 and R/J_3 are not isomorphic.

37. Let F be a finite field, and let n be a positive integer. Show that there is an irreducible polynomial of degree n in $F[X]$.

38. Let p be a prime number, and let m and n be positive integers.

    (a) Show that $p^m - 1$ is a factor of $p^{mn} - 1$. (Hint: Use Lagrange's Theorem, or Exercise 31 from Chapter 7.)

    (b) Let e be a primitive element of $GF(p^{mn})$. Let r be the integer

    $$\frac{p^{mn} - 1}{p^m - 1} \, .$$

    Show that the subfield $GF(p^m)$ of $GF(p^{mn})$ consists of 0 and the powers of e^r.

39. Find the integer q such that $GF(q)$ is the splitting field of the polynomial $X^4 + X + 1$ over $\mathbb{Z}/2$.

40. Let $p(X)$ be an irreducible polynomial of degree 5 in $\mathbb{Z}/2[X]$. Let J be the ideal $p(X)\mathbb{Z}/2[X]$ of $\mathbb{Z}/2[X]$. Explain why $X + J$ is a primitive element of $\mathbb{Z}/2[X]/J$.

41. For

    $$\omega = \cos \frac{2\pi}{13} + i \sin \frac{2\pi}{13} \, ,$$

    show that $\mathbb{Q}(-\omega)$ is the splitting field of $X^{27} - X$ over $\mathbb{Q}$.

42. What is the splitting field of $X^{27} - X$ over $GF(3)$?

8.10 Study projects

1. **The Sieve of Eratosthenes** is a method for generating a partial listing of the irreducible elements in an integral domain whose associate classes may be ordered systematically. In its classical version, it produces a list of prime numbers, as illustrated in Figure 8.3.

2	3	4	5	6	7	8	9	10	11	12	13	14	15	16	17	18	19	20	21	22	23	24	25
2	3		5		7		9		11		13		15		17		19		21		23		25
2	3		5		7				11		13				17		19				23		25
2	3		5		7				11		13				17		19				23		

FIGURE 8.3: The Sieve of Eratosthenes.

The first row lists the integers bigger than 1 in increasing order. In the second row, all the multiples of 2, except for 2 itself, are deleted. The element to the right of 2 is 3, so at the next step all the proper multiples of 3 are deleted. Continuing in this fashion, the sieve leaves a list of the prime numbers. In Figure 8.3, the last row has just had the proper multiples of 5 deleted. It already exhibits the list of primes less than 25.

The sieve may be used to produce a list of irreducible polynomials over $\mathbb{Z}/2$. For convenience during the calculation, a polynomial

$$f(X) = a_n X^n + a_{n-1} X^{n-1} + \cdots + a_1 X + a_0$$

in $\mathbb{Z}/2[X]$ may be represented by the binary expansion $a_n a_{n-1} \ldots a_1 a_0$ of the integer obtained by interpreting $f(X)$ in $\mathbb{Z}[X]$, and then computing $f(2)$. For example, the polynomial $X^4 + X^3 + 1$ is represented as

$$2^4 + 2^3 + 1 = 11001$$

(i.e., the binary expansion of 25). The elements $f(X)$ are then listed in increasing order of the corresponding integers $f(2)$.

Use the sieve to show that the following is a complete list of all the irreducible polynomials over $\mathbb{Z}/2$ of degree less than 5:

Degree 1 : X, $X + 1$;

Degree 2 : $X^2 + X + 1$;

Degree 3 : $X^3 + X + 1$, $X^3 + X^2 + 1$;

Degree 4 : $X^4 + X + 1$, $X^4 + X^3 + 1$, $X^4 + X^3 + X^2 + X + 1$.

2. **The ring $\mathbb{Z}[X]$ is not a principal ideal domain** (compare Exercise 18 in Chapter 7). The goal of this project is to show that $\mathbb{Z}[X]$ is a unique factorization domain. In particular, the converse of Theorem 8.16(b) is false. Recall that $\mathbb{Z}[X]$ is an integral domain (Corollary 7.14, page 161).

   (a) Show that the group of units $\mathbb{Z}[X]^*$ of $\mathbb{Z}[X]$ is $\{\pm 1\}$.

   (b) Show that each prime number is an irreducible element of $\mathbb{Z}[X]$.

   (c) Define a polynomial

   $$p(X) = p_n X^n + p_{n-1} X^{n-1} + \cdots + p_1 X + p_0$$

   of positive degree n in $\mathbb{Z}[X]$ to be *primitive* if $p_n > 0$ and

   $$\gcd(p_n, \ldots, p_0) = 1.$$

   Show that each element $f(X)$ of the ring $\mathbb{Q}[X]$ of polynomials over the rationals has a unique expression of the form

   $$f(X) = q_f p_f(X)$$

   with q_f in $\mathbb{Q}$ and with $p_f(X)$ as a primitive polynomial in $\mathbb{Z}[X]$. The rational number q_f is known as the *content* of $f(X)$.

(d) Show that $\mathbb{Z}[X] = \{f(X) \text{ in } \mathbb{Q}[X] \mid q_f \text{ in } \mathbb{Z}\}$.

(e) For $f(X)$ in $\mathbb{Z}[X]$, show that $f(X)$ is primitive if and only if $q_f = 1$.

(f) Let p be a prime number. Let $\theta : \mathbb{Z} \to \mathbb{Z}/p; n \mapsto n + p\mathbb{Z}$ denote reduction modulo p. Use the Substitution Principle to define a homomorphism $\theta_X : \mathbb{Z}[X] \to \mathbb{Z}/p[X]; f(X) \mapsto \overline{f}(X)$. Suppose that a polynomial $f(X)$ in $\mathbb{Z}[X]$ has positive leading coefficient. Show that $f(X)$ is a primitive polynomial if and only if $\overline{f}(X)$ is nonzero for each prime p.

(g) Prove *Gauss' Lemma*: the product of two primitive polynomials is primitive.

(h) Let $f(X)$ and $p(X)$ be elements of $\mathbb{Z}[X]$, with $p(X)$ primitive. If $p(X)$ divides $f(X)$ in $\mathbb{Q}[X]$, show that $p(X)$ divides $f(X)$ in $\mathbb{Z}[X]$.

(i) Let $f(X)$ and $g(X)$ be elements of $\mathbb{Q}[X]$. If $f(X)$ divides $g(X)$ in $\mathbb{Q}[X]$, show that $p_f(X)$ divides $p_g(X)$ in $\mathbb{Z}[X]$.

(j) Let $f(X)$ and $g(X)$ be elements of $\mathbb{Z}[X]$. If $f(X)$ and $g(X)$ have a common nonconstant factor in $\mathbb{Q}[X]$, show that they also have a common nonconstant factor in $\mathbb{Z}[X]$.

(k) Show that if a nonconstant polynomial is irreducible in $\mathbb{Z}[X]$, it is also irreducible in $\mathbb{Q}[X]$.

(l) Let $f(X)$ be an element of $\mathbb{Z}[X]$ with positive leading coefficient. Show that $f(X)$ is irreducible in $\mathbb{Z}[X]$ if and only if it is a prime number, or is a primitive polynomial that is irreducible in $\mathbb{Q}[X]$.

(m) Show that each irreducible element of $\mathbb{Z}[X]$ is prime. Conclude that $\mathbb{Z}[X]$ is a unique factorization domain.

3. **Eisenstein's Criterion and cyclotomic polynomials.** Section 7.6 noted that it may be very tricky to decide whether a given polynomial is irreducible or not. One of the numerous tricks available is *Eisenstein's Criterion* for the irreducibility over $\mathbb{Q}[X]$ of a polynomial $f(X)$ from $\mathbb{Z}[X]$. A polynomial

$$f(X) = f_n X^n + f_{n-1} X^{n-1} + \cdots + f_1 X + f_0 \qquad (8.13)$$

with integral coefficients $f_n, \ldots, f_0$ satisfies the criterion if there is a prime number p for which

$$p \nmid f_n, \quad p \mid f_{n-1}, \quad \ldots, \quad p \mid f_0, \quad p^2 \nmid f_0. \qquad (8.14)$$

(a) Show that no constant integral polynomial can satisfy Eisenstein's Criterion (8.14).

(b) Suppose that an integral polynomial (8.13) satisfies Eisenstein's Criterion (8.14). Suppose that there is a proper factorization

$$f(X) = g(X)h(X) \qquad (8.15)$$

in $\mathbb{Z}[X]$. Consider the corresponding factorization

$$\overline{f}(X) = \overline{g}(X)\overline{h}(X) \qquad (8.16)$$

in $\mathbb{Z}/_p[X]$ — compare (f) in Study Project 2. Show that (8.16) reduces to

$$\overline{f}(X) = \overline{f_n}X^n = \overline{g}(X)\overline{h}(X)$$

in the unique factorization domain $\mathbb{Z}/_p[X]$. Since

$$X \mid \overline{g}(X) \text{ and } X \mid \overline{h}(X)$$

in $\mathbb{Z}/_p[X]$, conclude that

$$\overline{g}(0) = \overline{h}(0) = 0$$

in $\mathbb{Z}/_p$ or

$$p \mid g(0) \text{ and } p \mid h(0)$$

in $\mathbb{Z}$. Obtain a contradiction to the final condition of Eisenstein's Criterion (8.14).

(c) Using (k) in Study Project 2, show that an integral polynomial (8.13) satisfying Eisenstein's Criterion is irreducible over $\mathbb{Q}[X]$.

(d) For a prime number p, the p-th *cyclotomic polynomial* is

$$\Phi_p(X) = \frac{X^p - 1}{X - 1} = X^{p-1} + X^{p-2} + \cdots + X + 1$$

— compare Exercise 31 in Chapter 7. Use Eisenstein's Criterion, and Exercise 44 in Chapter 6, to show that $\Phi_p(1+X)$ is irreducible over $\mathbb{Q}[X]$. Conclude that $\Phi_p(X)$ is irreducible over $\mathbb{Q}[X]$.

8.11 Notes

Section 8.2

A. "Emmy" Noether was a German mathematician who lived from 1882 to 1935.

Section 8.3

Many treatments of factorization introduce the concept of a "Euclidean domain," as an integral domain in which a version of the Division Algorithm is available. Elements of the domain are weighted, so that remainders have lesser weight than divisors. However, it appears that there is no complete agreement between different sources about the multiplicativity properties that should be demanded of the weighting function. For this reason, we refrain from a discussion of Euclidean domains, merely noting the use of the Division Algorithm to show that $\mathbb{Z}$, and rings of polynomials over fields, are principal ideal domains.

Section 8.7

Proposition 8.32 could have been derived from the structure theorem for finitely generated abelian groups (Theorem 9.56), while Proposition 8.33 may be proved by regarding E as a finite-dimensional vector space over F (compare Section 9.7). The proofs given here are designed to depend only on concepts that have already been encountered in the text.

Section 8.8

E. Galois was a French mathematician who lived from 1811 to 1832.

Section 8.10

Eratosthenes ($E\rho\alpha\tau o\sigma\theta\varepsilon\nu\eta\varsigma$) was a Greek mathematician and geographer who lived from around 276 to 194 B.C.

F.M.G. Eisenstein was a German mathematician who lived from 1823 to 1852.

Chapter 9

MODULES

In Chapter 6, rings were introduced axiomatically (Definition 6.1, page 127), based on the examples of the integers $\mathbb{Z}$, the reals $\mathbb{R}$, and the ring of 2×2 real matrices. On the other hand, the axiomatic definition of an abstract group (Definition 4.14, page 71) was founded on the general class of concrete groups of permutations, while Cayley's Theorem (Section 5.6) showed that each abstract group is isomorphic to a group of permutations. In this chapter, we investigate the corresponding general class of concrete rings, namely rings of endomorphisms (of abelian groups), and the accompanying concept of a module. For fields, modules are just vector spaces. For the ring of integers, modules are just abelian groups. Thus modules capture the features that are common to vector spaces and abelian groups, providing a general context for the pervasive phenomenon of linearity.

9.1 Endomorphisms

Suppose that $(A, +, 0)$ is an abelian group. An *endomorphism* θ of A is a group homomorphism $\theta : A \to A$ from A to A. (The prefix "endo-" suggests that after leaving the domain A, the homomorphism θ goes back **in to** A again, as a codomain.) Endomorphisms are not required to be injective or surjective. In particular, the constant map

$$0_A : A \to A; a \mapsto 0 \qquad (9.1)$$

sending each element of A to 0 is an endomorphism of A, known as the *zero map*. Another endomorphism of A is the identity map id_A on A, often written as

$$1_A : A \to A; a \mapsto a \qquad (9.2)$$

using the symbol 1, in parallel with the use of 0 in (9.1).

Example 9.1 (Endomorphisms of integers modulo 2.)
Consider the abelian group $(\mathbb{Z}/_2, +, 0)$ of integers modulo 2 under addition. The only endomorphisms of $(\mathbb{Z}/_2, +, 0)$ are the zero map (9.1) and the identity

map (9.2). Indeed, an endomorphism θ of $\mathbb{Z}/_2$ is the zero map if $\theta(1) = 0$, and is the identity map if $\theta(1) = 1$. $\quad\square$

Example 9.2 (Endomorphisms of the trivial group.)
Consider the trivial abelian group $\{0\}$. The only endomorphism of $\{0\}$ is the zero map (9.1), which coincides with the identity map (9.2) in this case. $\quad\square$

Example 9.3 (Multiples.)
Let $(A, +, 0)$ be an abelian group. For each integer n, the map

$$\mu_n : A \to A; a \mapsto na, \tag{9.3}$$

sending a group element a to its multiple na, is an endomorphism of A. Taking the integer $n = 0$ gives the zero map (9.1). Taking the integer $n = 1$ gives the identity map (9.2). $\quad\square$

Example 9.4 (Left multiplication in rings.)
Let $(R, +, \cdot)$ be a ring, with additive group $(R, +, 0)$. Let r be an element of R. As discussed in Section 6.2, the left multiplication

$$R \to R; x \mapsto r \cdot x \tag{9.4}$$

by r is an endomorphism of $(R, +, 0)$.

(a) For $r = 0$, the left multiplication (9.4) is the zero map (9.1).

(b) If R is unital, and $r = 1$, then the left multiplication (9.4) is the identity map (9.2).

$\quad\square$

Example 9.5 (Scalar multiplication.)
Consider the abelian group $\mathbb{R}_2^1$ of 2-dimensional real column vectors with componentwise addition. For each real number λ, the *scalar multiplication*

$$\lambda : \mathbb{R}_2^1 \to \mathbb{R}_2^1; \begin{bmatrix} x_1 \\ x_2 \end{bmatrix} \mapsto \begin{bmatrix} \lambda x_1 \\ \lambda x_2 \end{bmatrix}$$

is an endomorphism of $\mathbb{R}_2^1$. $\quad\square$

Given two real scalars λ and μ in Example 9.5, we have the equality

$$(\lambda - \mu) \begin{bmatrix} x_1 \\ x_2 \end{bmatrix} = \lambda \begin{bmatrix} x_1 \\ x_2 \end{bmatrix} - \mu \begin{bmatrix} x_1 \\ x_2 \end{bmatrix}$$

for vectors

$$\begin{bmatrix} x_1 \\ x_2 \end{bmatrix}$$

in $\mathbb{R}_2^1$. More generally, the *difference* $\theta - \varphi$ of two endomorphisms θ and φ of an abelian group A is given by

$$(\theta - \varphi)(a) = \theta(a) - \varphi(a) \tag{9.5}$$

for each element a of A.

PROPOSITION 9.6 (The difference of two endomorphisms.)

Let A be an abelian group. Then the difference (9.5) of two endomorphisms θ and φ of A is again an endomorphism of A.

PROOF　For elements a and b of A, we have

$$\begin{aligned}
(\theta - \varphi)(a + b) &= \theta(a + b) - \varphi(a + b) \\
&= \theta(a) + \theta(b) - \varphi(a) - \varphi(b) \\
&= \theta(a) - \varphi(a) + \theta(b) - \varphi(b) \\
&= (\theta - \varphi)(a) + (\theta - \varphi)(b).
\end{aligned}$$

The first and last equalities use the definition (9.5) of the difference. The second equality holds since θ and φ are group homomorphisms. The third equality holds since A is abelian. □

COROLLARY 9.7 (The additive group of endomorphisms.)

Consider the group A^A of all maps from A to $(A, +, 0)$, equipped with the componentwise abelian group structure $(A^A, +, 0)$ given by Definition 4.34(c). Then the set of all endomorphisms of the abelian group A forms a subgroup of A^A.

PROOF　The set of endomorphisms of the abelian group A is nonempty, since it contains the zero endomorphism (9.1). Proposition 9.6 shows that the set of endomorphisms satsifies the closure property required by Proposition 4.43 (page 80) for a subgroup — compare Remark 4.44. □

DEFINITION 9.8 (The endomorphism group End A.)　*The set of endomorphisms of an abelian group $(A, +, 0)$ is denoted by $\mathrm{End}(A, +, 0)$, or just $\mathrm{End}\, A$. As a group with the structure given by Corollary 9.7, it is known as the* endomorphism group $(\mathrm{End}\, A, +, 0)$ *of the abelian group A.*

Along with its closure under the subtraction (9.5), the endomorphism set $\mathrm{End}\, A$ of an abelian group A is also closed under functional composition.

Indeed, if θ and φ are endomorphisms of A, then so is $\theta \circ \varphi$ — compare Exercise 4 in Chapter 5. The full structure is described as follows.

THEOREM 9.9 (The endomorphism ring.)

Let $(A, +, 0)$ be an abelian group. Then the set $\operatorname{End} A$ of endomorphisms of A forms a unital ring $(\operatorname{End} A, +, \circ)$. The additive group structure $(\operatorname{End} A, +, 0)$ follows Definition 9.8:

$$(\theta + \varphi)(a) = \theta(a) + \varphi(a)$$

for θ, φ in $\operatorname{End} A$ and a in A. The monoid structure $(\operatorname{End} A, \circ, 1)$ is given by functional composition:

$$(\theta \circ \varphi)(a) = \theta\big(\varphi(a)\big)$$

for θ, φ in $\operatorname{End} A$ and a in A. The zero element 0 is the zero map (9.1), while the unit element is the identity map (9.2).

PROOF In order to obtain $(\operatorname{End} A, +, \circ)$ as a ring, the distributive laws must be verified. Consider endomorphisms θ, φ, and ψ of A. Then for each element a of A, we have

$$\begin{aligned}
\theta \circ (\varphi + \psi)(a) &= \theta\big(\varphi(a) + \psi(a)\big) \\
&= \theta \circ \varphi(a) + \theta \circ \psi(a) \\
&= (\theta \circ \varphi + \theta \circ \psi)(a)
\end{aligned}$$

and

$$\begin{aligned}
(\varphi + \psi) \circ \theta(a) &= (\varphi + \psi)\big(\theta(a)\big) \\
&= \varphi \circ \theta(a) + \psi \circ \theta(a) \\
&= (\varphi \circ \theta + \psi \circ \theta)(a) \, .
\end{aligned}$$

Thus the left distributive law $\theta \circ (\varphi + \psi) = \theta \circ \varphi + \theta \circ \psi$ and right distributive law $(\varphi + \psi) \circ \theta = \varphi \circ \theta + \psi \circ \theta$ hold in $\operatorname{End} A$, as required. ⬜

For an abelian group A, the ring $\operatorname{End} A$ or $(\operatorname{End} A, +, \circ)$ of Theorem 9.9 is known as the *endomorphism ring* of A.

Example 9.10 (Groups of order 2.)

Let A be an abelian group of order 2. Then the endomorphism ring $\operatorname{End} A$ is isomorphic to the ring of integers modulo 2 (Exercise 7). ⬜

9.2 Representing a ring

Cayley's Theorem shows that an abstract group $(G, \cdot, 1)$ is isomorphic to a subgroup of the symmetric group $G!$ on the underlying set G. Here are the analogous results for rings.

THEOREM 9.11

Let $(R, +, \cdot)$ be a ring, with additive group $(R, +, 0)$. For each element r of R, consider the left multiplication

$$\lambda_r : R \to R; x \mapsto r \cdot x$$

by r. Then there is a ring homomorphism

$$\Lambda : R \to \mathrm{End}(R, +, 0); r \mapsto \lambda_r \tag{9.6}$$

from R to the endomorphism ring of the additive group $(R, +, 0)$ of R.

PROOF For elements r, s, and x of R, we have

$$(\lambda_r + \lambda_s)(x) = \lambda_r(x) + \lambda_s(x) = r \cdot x + s \cdot x = (r + s) \cdot x = \lambda_{r+s}(x)$$

by the left distributive law and

$$\lambda_r \circ \lambda_s(x) = r \cdot (s \cdot x) = (r \cdot s) \cdot x = \lambda_{r \cdot s}(x)$$

by the associative law, so that Λ is a ring homomorphism. $\quad\Box$

COROLLARY 9.12 (Cayley's Theorem for unital rings.)
Let $(R, +, \cdot)$ be a unital ring. Then the map

$$\Lambda : R \to \mathrm{End}(R, +, 0)$$

of (9.6) is an injective homomorphism of unital rings.

PROOF For elements r and s of R, the equation $\lambda_r = \lambda_s$ implies

$$r = r \cdot 1 = \lambda_r(1) = \lambda_s(1) = s \cdot 1 = s,$$

so the map

$$\Lambda : R \to \mathrm{End}\, R; r \mapsto \lambda_r$$

is injective. Moreover, λ_1 is the identity map id_R, so that Λ is a unital ring homomorphism. $\quad\Box$

Example 9.13

If $(R, +, \cdot)$ is not unital, it may happen that the ring homomorphism

$$\Lambda : R \to \text{End}(R, +, \cdot); r \mapsto \lambda_r$$

of (9.6) is not injective. Consider the subring $R = \{0, 2\}$ of the ring of integers modulo 4. Since $2 \cdot 2 = 0 = 2 \cdot 0$, the left multiplications λ_2 and λ_0 coincide. In this case, the image of (9.6) is trivial, consisting only of the zero endomorphism. □

9.3 Modules

Let R be a unital ring. According to Corollary 9.12, there is a unital ring homomorphism

$$R \to \text{End}(R, +, 0)$$

from R to the endomorphism ring $\text{End}(R, +, 0)$ of the additive group $(R, +, 0)$ of R. In this case the abelian group $(R, +, 0)$ is directly obtained as part of the ring structure on R. However, we may just as well consider a unital ring homomorphism

$$R \to \text{End}(A, +, 0)$$

from R to the endomorphism ring $\text{End}(A, +, 0)$ of an arbitrary abelian group $(A, +, 0)$.

DEFINITION 9.14 (Modules with explicit structure map.) *Let R be a unital ring, and let $(A, +, 0)$ be an abelian group. Suppose that there is a unital ring homomorphism*

$$\sigma : R \to \text{End}(A, +, 0); r \mapsto \sigma_r \qquad (9.7)$$

from R to the endomorphism ring $\text{End}\, A$ of the group A.

(a) *The abelian group A is described as a (unital) (left) module over R, or a (unital) (left) R-module $(A, +, \sigma)$, with structure map (9.7).*

(b) *For each element r of R, the endomorphism*

$$\sigma_r : A \to A; a \mapsto \sigma_r(a)$$

is described as the action of the ring element r on A.

REMARK 9.15 Since $\sigma_1 = \text{id}_A$ in (9.7), we have

$$\sigma_1(a) = a$$

for each element a of A. □

Example 9.16 (Scalar action.)
Let F be a field. Consider the abelian group F_2^1 of 2-dimensional column vectors

$$\begin{bmatrix} x_1 \\ x_2 \end{bmatrix}$$

with componentwise structure. For each element λ of F, consider the action

$$\sigma_\lambda : F_2^1 \to F_2^1; \begin{bmatrix} x_1 \\ x_2 \end{bmatrix} \mapsto \begin{bmatrix} \lambda x_1 \\ \lambda x_2 \end{bmatrix}$$

of λ on F_2^1 by scalar multiplication (compare Example 9.5 for the real case). Then F_2^1 becomes a module $(F_2^1, +, \sigma)$ over F with structure map

$$\sigma : F \to \operatorname{End} F_2^1; \lambda \mapsto \sigma_\lambda$$

(Exercise 8). $\quad\square$

Example 9.17 (Conjugated action.)
Consider the field $\mathbb{C}$ of complex numbers. For each element λ of $\mathbb{C}$, define the action

$$\delta_\lambda : \mathbb{C}_2^1 \to \mathbb{C}_2^1; \begin{bmatrix} x_1 \\ x_2 \end{bmatrix} \mapsto \begin{bmatrix} \overline{\lambda} x_1 \\ \overline{\lambda} x_2 \end{bmatrix}$$

of λ on $\mathbb{C}_2^1$ by scalar multiplication with the complex conjugate $\overline{\lambda}$ of λ. Then $\mathbb{C}_2^1$ becomes a module $(\mathbb{C}_2^1, +, \delta)$ over $\mathbb{C}$ with structure map

$$\delta : \mathbb{C} \to \operatorname{End} \mathbb{C}_2^1; \lambda \mapsto \delta_\lambda$$

(Exercise 9). $\quad\square$

Example 9.18 (Abelian groups as $\mathbb{Z}$-modules.)
Let $(A, +, 0)$ be an abelian group. Example 9.3 noted that for each integer n, the map $\mu_n : A \to A; a \mapsto na$, sending an element a of A to the multiple na, is an endomorphism of A. Now consider the map

$$\mu : \mathbb{Z} \to \operatorname{End}(A, +, 0); n \mapsto \mu_n . \tag{9.8}$$

Since

$$\mu_{m+n}(a) = (m+n)a = ma + na = \mu_m(a) + \mu_n(a)$$

by the Law of Exponents (5.13), and

$$\mu_{mn}(a) = (mn)a = m(na) = \mu_m \circ \mu_n(a)$$

for elements a of A and integers m, n, with $\mu_1(a) = a = \operatorname{id}_A(a)$, the map (9.8) gives a unital ring homomorphism. Thus the abelian group $(A, +, 0)$ is a left $\mathbb{Z}$-module $(A, +, \mu)$. In fact, (9.8) represents the only possible choice for a structure map making A into a $\mathbb{Z}$-module (Exercise 10). $\quad\square$

Definition 9.14 specifies modules using the structure map (9.7) as a unital ring homomorphism. However, it is often easier to recognize the presence of a module directly.

PROPOSITION 9.19 (Characterizing modules.)

Let R be a unital ring, and let $(A, +, 0)$ be an abelian group. Then A is a left R-module if and only if there is a map

$$\sigma_r : A \to A; a \mapsto \sigma_r(a) \, ,$$

for each element r of R, such that the following properties are satisfied:

(a) $\sigma_r(a + b) = \sigma_r(a) + \sigma_r(b)$;

(b) $\sigma_{r+s}(a) = \sigma_r(a) + \sigma_s(a)$;

(c) $\sigma_{rs}(a) = \sigma_r \circ \sigma_s(a)$;

(d) $\sigma_1(a) = a$

for r, s in R and a, b in A.

PROOF   First, suppose that A is a left R-module, by virtue of a unital ring homomorphism (9.7). Then (a) holds since σ_r is an endomorphism of $(A, +, 0)$, while (b) and (c) hold since σ is a ring homomorphism, and (d) holds since the ring homomorphism σ is unital (Remark 9.15).

Conversely, suppose that the conditions (a) – (d) are satisfied. By (a), each map σ_r, for r in R, is an endomorphism of $(A, +, 0)$. Thus a function

$$\sigma : R \to \text{End}(A, +, 0); r \mapsto \sigma_r$$

is specified. By (b) and (c), the function σ is a ring homomorphism. By (d), the ring homomorphism σ is unital.   ⬚

Example 9.20 (Rings as modules over subrings.)

Let R be a unital subring of a unital ring A. For each element r of R, and for each element a of A, define $\sigma_r(a) = ra$. Condition (a) of Proposition 9.19 is satisfied by the left distributive law in the ring A. Condition (b) follows from the right distributive law. Condition (c) follows by the associative law, and (d) holds since A is unital. Thus A is a left R-module, denoted by $_RA$. ⬚

Consider the scalar action of Example 9.16 for the field $F = \mathbb{C}$ of complex numbers, and contrast it with the conjugated action of Example 9.17. In each of these two cases, the abelian group $\mathbb{C}_2^1$ of 2-dimensional complex column vectors becomes a module over $\mathbb{C}$. Example 9.16 yields the module $(\mathbb{C}_2^1, +, \sigma)$, while Example 9.17 yields the module $(\mathbb{C}_2^1, +, \delta)$. When comparing the two examples, it is critical to distinguish the structure maps. However, in most cases there is only one structure map in question, and there is no need to keep mentioning it explicitly all the time. Proposition 9.19 allows us to simplify the notation for an R-module. Instead of writing $\sigma_r(a)$, we just write $r \cdot a$ or ra for a ring element r and module element a. The conditions (a) through (d) of Proposition 9.19 become the axioms (a) through (d) of the following definition. Proposition 9.19 is then interpreted as saying that the new definition is consistent with the previous Definition 9.14.

DEFINITION 9.21 (Modules with implicit structure map.)   *Let R be a unital ring, and let $(A, +, 0)$ be an abelian group. Suppose that for each element r of R, and for each element a of A, an element $r \cdot a$ or ra of A is defined. Suppose that the following properties are satisfied:*

(a) $r \cdot (a + b) = r \cdot a + r \cdot b$;

(b) $(r + s) \cdot a = r \cdot a + s \cdot a$;

(c) $(rs) \cdot a = r \cdot (s \cdot a)$;

(d) $1 \cdot a = a$

for r, s in R and a, b in A. Then the abelian group A is described as a (unital) (left) *module* over R, *or a* (unital) (left) R-module $(A, +, R)$.

9.4   Submodules

Let R be a unital ring. Modules over the ring R are abstract algebras in their own right, just like groups, rings, or any of the other algebras we have been studying. Thus there is a concept of a submodule, analogous to subgroups or subrings. We use the notation of Definition 9.21.

DEFINITION 9.22 (Submodules.)   *Let R be a unital ring. Let $(A, +, R)$ be a left R-module. Then a subset B of A is said to be an R-submodule of A, or just a* submodule *of M, if B is a subgroup of the abelian group $(A, +, 0)$, and $r \cdot b$ lies in B for each r in R and b in B.*

Using Remark 4.44 (page 80), we see that a subset B of an R-module $(A, +, R)$ is a submodule if:

- B is nonempty;

- $x - y$ lies in B if x, y lie in B;

- $r \cdot x$ lies in B for r in R and x in B.

Example 9.23 (Ideals as submodules.)

Let R be a unital subring of a unital ring A. According to Example 9.20, A becomes a left R-module $_RA$, with the action $r \cdot a$ defined by the multiplication $\cdot$ in the ring $(A, +, \cdot)$ for r in R and a in A. Each ideal J of the ring A forms an R-submodule of $_RA$, since J is a subgroup of $(A, +, 0)$, while $r \cdot j$ lies in J (for r in R and j in J) by the absorptive property (6.16) of the ideal J.   ⬜

In the theory of modules, one of the most fundamental concepts is the notion of generating a submodule. In order to understand this concept, we begin with a simple proposition. (Compare Exercise 30 in Chapter 4.)

PROPOSITION 9.24 (The intersection of submodules.)

Let $(A, +, R)$ be a module over a ring R. Suppose that I is a set, and that for each element i of I, there is a submodule B_i of A. Then the intersection

$$\bigcap_{i \in I} B_i = \{a \mid a \text{ lies in } B_i \text{ for all } i \text{ in } I\} \tag{9.9}$$

is again a submodule of A.

PROOF   Write B for the intersection (9.9).

- B is nonempty, since each submodule B_i contains 0.

- If a_1 and a_2 lie in each B_i, so does $a_1 - a_2$.

- If a lies in each B_i, so does $r \cdot a$ for each element r of R.

                                                                        ⬜

DEFINITION 9.25   *Let X be a subset of a module A over a ring R.*

(a) *The submodule RX generated or spanned by X is the intersection of all the submodules of A that contain X.*

(b) *An R-linear combination of elements of X is an element of A of the form*

$$r_1 \cdot x_1 + r_2 \cdot x_2 + \ldots + r_n \cdot x_n \tag{9.10}$$

for a natural number n, elements $x_1, \cdots x_n$ of X, and elements $r_1, \ldots r_n$ of R. The elements $r_1, \ldots r_n$ of R are known as the coefficients of the linear combination. If $n = 0$, the linear combination (9.10) is understood to be the zero element 0 of A.

Rephrasing Definition 9.25(a), the submodule RX generated by a subset X is the smallest submodule containing X. The following result provides an explicit description of the submodule generated by a subset of a module, in terms of linear combinations.

THEOREM 9.26 (The submodule generated by a subset.)
Let X be a subset of a unital module $(A, +, R)$ over a ring R. Then the submodule RX generated by X is the set

$$\{r_1 \cdot x_1 + \ldots + r_n \cdot x_n \mid n \text{ in } \mathbb{N}, r_i \text{ in } R, x_i \text{ in } X\} \tag{9.11}$$

of all R-linear combinations of elements of X.

PROOF   Write L for the set (9.11) of R-linear combinations of elements of X. We must show that $L = RX$.

- Each submodule B of A that contains X also contains each R-linear combination

$$r_1 \cdot x_1 + \ldots + r_n \cdot x_n$$

of elements x_1, ..., x_n of X. (A formal proof proceeds by induction on the natural number n, using the closure properties for submodules given in Definition 9.22 — see Exercise 16.) Thus each such submodule B contains L. It follows that their intersection RX also contains L.

- Conversely, we must show that L contains RX. Now since A is unital, each element x of X appears in L as the R-linear combination $1 \cdot x$. Thus L contains X. Furthermore, L is a submodule of A:

  - L is nonempty, since it contains 0 as the linear combination (9.10) with $n = 0$;
  - If L contains

  $$x = r_1 \cdot x_1 + \ldots + r_n \cdot x_n$$

  and

  $$x' = r_1' \cdot x_1' + \ldots + r_{n'}' \cdot x_{n'}' ,$$

  then it contains $x - x'$ as

  $$r_1 \cdot x_1 + \ldots + r_n \cdot x_n + (-r_1') \cdot x_1' + \ldots + (-r_{n'}') \cdot x_{n'}' ;$$

  - If L contains

  $$x = r_1 \cdot x_1 + \ldots + r_n \cdot x_n ,$$

  then it contains $r \cdot x$ (for each r in R) as

  $$(rr_1) \cdot x_1 + \ldots + (rr_n) \cdot x_n .$$

Since L is a submodule of A that contains X, it contains RX.

◻

COROLLARY 9.27 (**Closure under linear combinations.**)
A subset B of a module $(A, +, R)$ over a ring R is a submodule of A if and only if each linear combination

$$r_1 \cdot x_1 + \ldots + r_n \cdot x_n$$

(with n in $\mathbb{N}$, elements $r_1, \ldots, r_n$ in R and $x_1, \ldots, x_n$ in B) lies in B.

PROOF   Set $X = B$ in Theorem 9.26.   ▢

Example 9.28
Let R be a unital ring, and let A be the ring $R[Y]$ of polynomials over the ring R in an indeterminate Y. Taking R as the subring of constant polynomials, consider A as an R-module ${}_R A$ or $(A, +, R)$ according to Example 9.20. Then for each natural number n, the R-submodule of A generated by the set

$$\{Y^n, Y^{n-1}, \ldots, Y^2, Y, 1\}$$

is the set of all polynomials

$$r_n Y^n + r_{n-1} Y^{n-1} + \ldots + r_1 Y + r_0 \tag{9.12}$$

of degree at most n. Note that in this example, the coefficients r_i of (9.12) as a linear combination — in the sense of Definition 9.25(b) — are just the coefficients of (9.12) as a polynomial.   ▢

Attention will soon focus on modules that are generated by a finite subset.

DEFINITION 9.29 (**Finitely generated modules.**)   *A module $(A, +, R)$ over a ring R is said to be finitely generated if there is a finite subset X of A such that A is generated by X.*

Example 9.30 (**Column vectors.**)
For a field F, the F-module F_2^1 of 2-dimensional column vectors (compare Example 9.16) is finitely generated, say by the subset

$$\left\{ \begin{bmatrix} 1 \\ 0 \end{bmatrix}, \begin{bmatrix} 0 \\ 1 \end{bmatrix} \right\}.$$

If the field F is infinite, note that F_2^1 itself is an infinite set.   ▢

Any finite module $(A, +, R)$ is finitely generated (by its underlying set A). On the other hand, the module A defined in Example 9.28 is not finitely generated — see Exercise 19.

9.5   Direct sums

Let R be a unital ring. We continue with the theme that R-modules form another kind of abstract algebra, analogous to groups and rings. Accordingly, given two R-modules $(A, +, R)$ and $(B, +, R)$, a function $f : A \to B$ is defined to be an R-*module homomorphism* (or just a *module homomorphism*) if

$$f(x + y) = f(x) + f(y) \qquad (9.13)$$

and

$$f(r \cdot x) = r \cdot f(x) \qquad (9.14)$$

for r in R and x, y in A. The image $f(A)$ of an R-module homomorphism $F : A \to B$ is a submodule of B (Exercise 20). By (9.13), each R-module homomorphism

$$f : (A, +, R) \to (B, +, R)$$

is an abelian group homomorphism

$$f : (A, +, 0) \to (B, +, 0) \, .$$

If R is the ring $\mathbb{Z}$ of integers, then (9.13) implies (9.14) — Exercise 21.

For a unital ring R, bijective R-module homomorphisms are described as R-*isomorphisms*, or just *isomorphisms* if the context is clear. Two R-modules A and B are said to be *isomorphic*, a relation that is written as $A \cong B$, if there is an R-module isomorphism $b : X \to Y$. In particular, two $\mathbb{Z}$-modules A and B are isomorphic if and only if they are isomorphic as abelian groups.

For groups and rings, the idea of imposing componentwise structure was used to define direct products. For R-modules, the same idea may also be applied. However, since there are special properties that hold in the case of modules, different terminology is used. (Compare Exercises 24 and 25.)

DEFINITION 9.31 **(Direct sums.)**   *Let R be a unital ring. Let n be a positive integer. Let A_i be an R-module, for $1 \le i \le n$. Then the* (external) *direct sum $A_1 \oplus \cdots \oplus A_n$ or $\bigoplus_{1 \le i \le n} A_i$ is the set*

$$\{(a_1, \dots, a_n) \mid a_i \text{ in } A_i \text{ for } 1 \le i \le n\}$$

equipped with the componentwise abelian group structure

$$(a_1, \dots, a_n) + (a_1', \dots, a_n') = (a_1 + a_n', \dots, a_n + a_n')$$

and componentwise actions

$$r \cdot (a_1, \dots, a_n) = (r \cdot a_1, \dots, r \cdot a_n)$$

for r in R.

Example 9.32 (Direct products of abelian groups.)
If $A_1, \ldots, A_n$ are abelian groups, then the direct sum

$$A_1 \oplus \cdots \oplus A_n$$

of the $\mathbb{Z}$-modules $A_1, \ldots, A_n$ is just the direct product

$$A_1 \times \cdots \times A_n$$

of the abelian groups $A_1, \ldots, A_n$ (compare Section 4.4). We thus speak of a *direct sum* of abelian groups, as a synonym for the direct product. ☐

Example 9.33 (Column vectors.)
Let F be a field. Consider F as a left F-module $_F F$ with the structure map given by the ring multiplication, as in Example 9.20. Recall the F-module F_2^1 of 2-dimensional column vectors with scalar action, as in Example 9.16. Then the map

$$_F F \oplus {}_F F \to F_2^1; (x_1, x_2) \mapsto \begin{bmatrix} x_1 \\ x_2 \end{bmatrix}$$

is an F-module isomorphism (Exercise 22). ☐

The following result characterizes modules (such as F_2^1 in Example 9.33) that are isomorphic to direct sums.

THEOREM 9.34 (Recognizing direct sums.)
Let B be a unital module over a unital ring R. Let n be a positive integer. Then the following conditions on B are equivalent:

(a) *B is isomorphic to a direct sum*

$$A_1 \oplus A_2 \oplus \ldots \oplus A_n$$

of modules $A_1, \ldots, A_n$;

(b) *There are submodules $B_1, \ldots, B_n$ of B such that each element b of B has a unique expression as a sum of the form*

$$b = b_1 + b_2 + \ldots + b_n \tag{9.15}$$

with b_i in B_i for $1 \le i \le n$;

(c) *There are submodules $B_1, \ldots, B_n$ of B such that B is isomorphic to the direct sum $B_1 \oplus B_2 \oplus \ldots \oplus B_n$.*

PROOF **(c) implies (a):** This is immediate.

(a) implies (b): For $1 \le i \le n$, let $\overline{A}_i$ be the submodule

$$\overline{A}_i = \left\{ (0,\ldots,0, \overset{\text{slot } i}{\overbrace{a_i}}, 0,\ldots,0) \ \middle|\ a_i \text{ in } A_i \right\}$$

of $A_1 \oplus \ldots \oplus A_n$. Each element $a = (a_1,\ldots,a_n)$ of $A_1 \oplus \ldots \oplus A_n$ has a unique expression $(a_1,\ldots,a_n) = (a_1,0\ldots,0) + \ldots + (0,\ldots,0,a_n)$ as a sum of the form

$$a = \overline{a}_1 + \ldots + \overline{a}_n \tag{9.16}$$

with $\overline{a}_i = (0,\ldots,0,a_i,0,\ldots,0)$ in $\overline{A}_i$ for $1 \le i \le n$. Consider the given isomorphism as

$$\beta : A_1 \oplus \ldots \oplus A_n \to B. \tag{9.17}$$

For $1 \le i \le n$, define the submodule B_i of B to be the image $\beta(\overline{A}_i)$ of the submodule $\overline{A}_i$ of $A_1 \oplus \ldots \oplus A_n$. Since (9.17) is an isomorphism, each element of B has a unique expression of the form $\beta(a)$ for a in $A_1 \oplus \ldots \oplus A_n$. Applying the isomorphism β to the unique expression (9.16) for a as a sum of respective elements $\overline{a}_1, \ldots, \overline{a}_n$ of $\overline{A}_1, \ldots, \overline{A}_n$ we obtain the unique expression

$$\beta(a) = \beta(\overline{a}_1) + \ldots + \beta(\overline{a}_n)$$

of $\beta(a)$ as a sum of elements of $B_1, \ldots, B_n$.

(b) implies (c): If (b) holds, there is a well-defined isomorphism

$$B \to B_1 \oplus B_2 \oplus \ldots \oplus B_n; b_1 + b_2 + \ldots + b_n \mapsto (b_1, b_2, \ldots, b_n)$$

of R-modules. ∎

COROLLARY 9.35

If B is isomorphic to the direct sum $A_1 \oplus \ldots \oplus A_n$ of modules $A_1, \ldots, A_n$, then the submodules B_i of B, for which B is isomorphic to the direct sum $B_1 \oplus \ldots \oplus B_n$, may be chosen so that $A_i \cong B_i$ for $1 \le i \le n$.

PROOF For $1 \le i \le n$, define $\alpha_i : A_i \to \overline{A}_i; a_i \mapsto (0,\ldots,0,a_i,0,\ldots,0)$. Then the desired isomorphism is $\beta \circ \alpha_i : A_i \to B_i$. ∎

DEFINITION 9.36 (Internal direct sums.) *A module B over a ring R is described as an internal direct sum (of submodules $B_1, \ldots, B_n$) if it satisfies the equivalent conditions of Theorem 9.34(b),(c) for some positive integer n.*

Example 9.37 (Column vectors.)
Let F be a field. Consider the F-module F_2^1 of column vectors discussed in Example 9.33. The module F_2^1 is the internal direct sum of the submodules

$$B_1 = \left\{ \begin{bmatrix} x \\ 0 \end{bmatrix} \;\middle|\; x \text{ in } F \right\}$$

and

$$B_2 = \left\{ \begin{bmatrix} 0 \\ x \end{bmatrix} \;\middle|\; x \text{ in } F \right\}$$

(Exercise 23). □

Suppose that $B_1, \ldots, B_n$ are submodules of a module B over a unital ring R. The condition for B to be an internal direct sum of $B_1, \ldots, B_n$ is the existence and uniqueness of the expression (9.15) for elements b of B. It is helpful to consider the existence and uniqueness questions separately. Write

$$B_1 + B_2 + \ldots + B_n \tag{9.18}$$

for the submodule of B that is generated by the union of the subsets $B_1, \ldots, B_n$ of B. By Theorem 9.26, the set (9.18) comprises all the elements b of B that have a (not necessarily unique) expression of the form

$$b = b_1 + b_2 + \ldots + b_n \tag{9.19}$$

with b_i in B_i for $1 \le i \le n$ (Exercise 26). Given the existence of expressions of the form (9.19), their uniqueness is characterized as follows.

PROPOSITION 9.38
For a positive integer n, let $B_1, \ldots, B_n$ be submodules of a module B. The following conditions are equivalent:

(a) *The submodule $B_1 + B_2 + \ldots + B_n$ of B is the internal direct sum of its submodules $B_1, \ldots, B_n$;*

(b) *Each element b of $B_1 + \ldots + B_n$ has a unique expression of the form*

$$b = b_1 + b_2 + \ldots + b_n$$

with b_i in B_i for $1 \le i \le n$;

(c) *The zero element 0 of B has a unique expression of the form*

$$0 = z_1 + z_2 + \ldots + z_n$$

with z_i in B_i for $1 \le i \le n$.

PROOF   The equivalence of (a) and (b) follows directly by Definition 9.36. Condition (c) is just a special case of condition (b). It remains to show that (c) implies (b). Suppose that (c) holds. In that case, $z_i = 0$ for $1 \leq i \leq n$. Suppose that an element b of $B_1 + \ldots + B_n$ has expressions

$$b = x_1 + \ldots + x_n = y_1 + \ldots + y_n$$

with x_i, y_i in B_i for $1 \leq i \leq n$. Then $0 = (x_1 - y_1) + \ldots + (x_n - y_n)$. By (c), we have $x_i - y_i = z_i = 0$ for $1 \leq i \leq n$, so the expression for b is unique. □

9.6   Free modules

We now begin the study of finitely generated modules. Let R be a unital ring. Using the construction of Example 9.20, the underlying abelian group $(R, +, 0)$ becomes a left R-module $_RR$, with action given by the multiplication in the ring. Direct sums of a finite number of copies of $_RR$, and modules isomorphic to such direct sums, are very important. They are recognized with the help of Proposition 9.38, using the concept of linear independence introduced in the following definition.

DEFINITION 9.39 (Free modules and linear independence.)   *Let A be a module over a unital ring R. Let l be a natural number, and let $X = \{x_1, \ldots, x_l\}$ be an l-element subset of A.*

(a) *An expression*
$$0 = r_1 x_1 + \ldots + r_l x_l \tag{9.20}$$

   of 0 as an R-linear combination of the elements of X is said to be nontrivial *if there is at least one nonzero coefficient r_1, ..., r_l.*

(b) *The set X is said to be* linearly independent *(over R) if there is no nontrivial expression (9.20) of 0 as a linear combination of the elements of X. (In particular, the empty set is linearly independent.)*

(c) *If X is a linearly independent generating set for the R-module A, then A is described as the* free R-module *over the generating set X. In this case, the set X or sequence x_1, ..., x_l is also known as a* basis *for A.*

Example 9.40 (Standard bases.)
Let l be a positive integer. For a nontrivial unital ring R, write R^l for the direct sum

$$\overbrace{_RR \oplus \ldots \oplus _RR}^{l \text{ summands}}$$

of l copies of the module $_RR$. For $1 \le i \le l$, consider the element

$$e_i = (0, \dots, 0, \overset{\text{slot } i}{1}, 0, \dots, 0)$$

of R^l. The sequence $e_1, \dots, e_l$, or the subset

$$E_l = \{e_i \mid 1 \le i \le l\}$$

of R^l, are both known as the *standard basis* of R^l. Now R-linear combinations of standard basis elements are easily computed:

$$r_1 e_1 + \dots + r_l e_l = (r_1, \dots, r_l)$$

for coefficients $r_1, \dots, r_l$ in R. Thus the set E_l is linearly independent, and R^l is the free R-module over the standard basis E_l. For the ring $\mathbb{R}$ of real numbers, the standard basis E_3 of $\mathbb{R}^3$ is illustrated in Figure 9.1. □

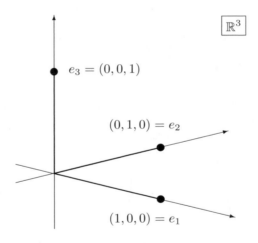

FIGURE 9.1:   The standard basis E_3 of $\mathbb{R}^3$.

The trivial R-module $\{0\}$ is the free R-module over the empty generating set. Expanding the notation of Example 9.40, it is convenient to denote $\{0\}$ by R^0. General free modules over nonempty generating sets are described as follows.

PROPOSITION 9.41 (Structure of free modules.)
Let R be a unital ring. Let A be a free R-module over a generating set $X = \{x_1, \dots, x_l\}$ with a positive number l of elements.

(a) *A is the internal direct sum of the submodules $R\{x_1\}, \ldots, R\{x_l\}$.*

(b) *A is isomorphic to the module R^l of Example 9.40.*

(c) *Each element a of A has a unique expression as an R-linear combination*

$$a = r_1 x_1 + \ldots + r_l x_l .$$

PROOF **(a):** For $1 \leq i \leq l$, each element of $R\{x_i\}$ is of the form $r_i x_i$ for some element r_i of R. Since X is linearly independent, the submodules $R\{x_1\}, \ldots, R\{x_l\}$ of A satisfy the condition (c) of Proposition 9.38. Since the module A is generated by X, it becomes the sum $A = R\{x_1\} + \ldots + R\{x_l\}$. By condition (a) of Proposition 9.38, it follows that A is the internal direct sum of the submodules $R\{x_1\}, \ldots, R\{x_l\}$.

(b): For each $1 \leq i \leq l$, there is an isomorphism

$$g_i : {}_R R \to R\{x_i\}; r \mapsto r x_i$$

from the module ${}_R R$ to the submodule $R\{x_i\}$ of A. Note that as an abelian group homomorphism, the map g_i is injective. Indeed $r x_i = 0$ implies $r = 0$, since the set X is linearly independent.

(c) now follows by condition (b) of Proposition 9.38 and the existence of the isomorphisms g_i for $1 \leq i \leq l$. $\quad\square$

The importance of free modules stems from the following.

THEOREM 9.42 (Universality property of free modules.)
Let A be a free R-module over a finite generating set X. Let $j : X \hookrightarrow A$ denote the inclusion of X as a subset of A. Then for each function $f : X \to B$ from X to the underlying set B of an R-module B, there is a unique R-module homomorphism $\overline{f} : A \to B$ with $f = \overline{f} \circ j$.

PROOF Suppose that X is the l-element subset $\{x_1, \ldots, x_l\}$ of A. For natural numbers k not exceeding l, we will prove by induction on k that an R-module homomorphism $\overline{f} : A \to B$ with $f = \overline{f} \circ j$ has to satisfy

$$\overline{f}(r_1 x_1 + \ldots + r_k x_k) = r_1 f(x_1) + \ldots r_k f(x_k) \tag{9.21}$$

for coefficients $r_1, \ldots, r_k$ in R. The induction basis is the case $k = 0$ of (9.21), namely $\overline{f}(0) = 0$. This must hold if $\overline{f} : (A, +, 0) \to (B, +, 0)$ is to be a group homomorphism. For the induction step, we have

$$\overline{f}(r_1 x_1 + \ldots + r_{k-1} x_{k-1} + r_k x_k)$$
$$= \overline{f}(r_1 x_1 + \ldots + r_{k-1} x_{k-1}) + \overline{f}(r_k x_k)$$
$$= \overline{f}(r_1 x_1 + \ldots + r_{k-1} x_{k-1}) + r_k \overline{f}(x_k)$$
$$= r_1 f(x_1) + \ldots + r_{k-1} f(x_{k-1}) + r_k f(x_k) .$$

The first two equalities hold since $\overline{f}$ is required to be a module homomorphism. The last equality holds by the induction hypothesis, and by the consequence $\overline{f}(x_k) = f(x_k)$ of the requirement $f = \overline{f} \circ j$.

Since A is generated by X, the case $k = l$ of (9.21) defines the function $\overline{f} : A \to B$ uniquely as

$$\overline{f}(r_1 x_1 + \ldots + r_l x_l) = r_1 f(x_1) + \ldots r_l f(x_l).$$

Note that $\overline{f}$ is well defined, by Proposition 9.41(c). It remains to check that the function $\overline{f}$ actually is a module homomorphism. But for elements r_i, s_i, r of R, we have

$$\overline{f}\left(\sum_{i=1}^{l} r_i x_i + \sum_{i=1}^{l} s_i x_i\right) = \overline{f}\left(\sum_{i=1}^{l}(r_i + s_i)x_i\right) = \sum_{i=1}^{l}(r_i + s_i)f(x_i)$$

$$= \sum_{i=1}^{l} r_i f(x_i) + \sum_{i=1}^{l} s_i f(x_i) = \overline{f}\left(\sum_{i=1}^{l} r_i x_i\right) + \overline{f}\left(\sum_{i=1}^{l} s_i x_i\right)$$

and

$$\overline{f}\left(r\sum_{i=1}^{l} r_i x_i\right) = \overline{f}\left(\sum_{i=1}^{l} r r_i x_i\right) = \sum_{i=1}^{l} r r_i f(x_i)$$

$$= r\sum_{i=1}^{l} r_i f(x_i) = r\overline{f}\left(\sum_{i=1}^{l} r_i x_i\right)$$

as required.   □

Example 9.43 (Freeness of the group of integers.)

Consider the abelian group $\mathbb{Z}$ as the free $\mathbb{Z}$-module $\mathbb{Z}^1$ over the standard basis $E_1 = \{1\}$. Theorem 9.42 states that for each abelian group G, and for each element x of G, i.e., for each function $f : \{1\} \to G; 1 \mapsto x$, there is a unique group homomorphism $\overline{f} : \mathbb{Z} \to G$ with $\overline{f}(1) = x$. This assertion is just the special case of Theorem 5.26 (page 107), the universality of the group of integers, in which the target group G is abelian. In the notation of Theorem 5.26, the group homomorphism $\overline{f}$ is $\exp_x$.   □

As an application of Theorem 9.42, we may compute the endomorphism ring of each finite cyclic group. Note the choice of the unital ring R here.

PROPOSITION 9.44 (The endomorphism ring of a cyclic group.)
Let n be a positive integer. Then the map

$$\mathrm{End}(\mathbb{Z}/_n, +, 0) \to (\mathbb{Z}/_n, +, \cdot); \theta \mapsto \theta(1) \tag{9.22}$$

is a ring isomorphism.

PROOF The abelian group $(\mathbb{Z}/_n, +, 0)$ is the free $\mathbb{Z}/_n$-module over the standard basis $\{1\}$. Each endomorphism θ of the abelian group $(\mathbb{Z}/_n, +, 0)$ is a $\mathbb{Z}/_n$-module homomorphism $\theta : \mathbb{Z}/_n \to \mathbb{Z}/_n$. By Theorem 9.42, each such module homomorphism is uniquely specified as the extension $\overline{f}$ of a function $f : \{1\} \to \mathbb{Z}/_n$. Thus the map (9.22) is a bijection. It is straightforward to verify that it is a ring homomorphism (Exercise 29). ☐

9.7 Vector spaces

Modules over a field are traditionally known as *vector spaces*. Module elements are known as *vectors*, and submodules are *subspaces*. Vector spaces enjoy special properties that are not shared by general modules. In particular, it will transpire that every (finitely generated) module over a field is free.

For a general unital ring R, Definition 9.39(b) introduced the concept of linear independence for an l-element subset $X = \{x_1, \dots, x_l\}$ of an R-module A. The stipulation that X has l elements means that there are no repeats in the sequence $x_1, x_2, \dots, x_l$. For the opposite concept of linear dependence, we do need to admit the possibility of repeats.

DEFINITION 9.45 (Linear dependence.) *Let A be a module over a unital ring R. Consider a sequence $x_1, x_2, \dots, x_k$ of elements of A. The sequence is said to be* linearly dependent *if there is an expression*

$$0 = r_1 x_1 + \dots + r_k x_k$$

of 0 as an R-linear combination of the elements of the sequence, with at least one of the coefficients $r_1, \dots, r_k$ in R being nonzero.

Suppose that A is a module over a nontrivial unital ring R. Let $x_1, \dots, x_k$ be a sequence of elements of A. There are three trivial ways for the sequence to be linearly dependent:

- If $x_i = 0$ for some $1 \le i \le k$, then the sequence is linearly dependent by virtue of the expression $0 = 1 \cdot x_i$;

- If the sequence contains a linearly dependent subsequence $x_{i_1}, \dots, x_{i_l}$, say
$$0 = r_{i_1} x_{i_1} + \dots + r_{i_l} x_{i_l}$$
with some $r_{i_j} \ne 0$, then the same expression serves to show that the original sequence $x_1, \dots, x_k$ is linearly dependent;

- If the sequence has a repeat, say $x_i = x_j$ for $1 \le i < j \le k$, then it is linearly dependent by virtue of the expression $0 = 1 \cdot x_i + (-1) \cdot x_j$.

A sequence $x_1, \ldots, x_l$ is *linearly independent* if it is not linearly dependent. Since there are then exactly l elements in the sequence, repeats being excluded, we recover the concept of linear independence of the set $\{x_1, \ldots, x_l\}$ from Definition 9.39(b).

The following result gives an indication of the special properties that hold for modules (vector spaces) over fields.

PROPOSITION 9.46 (Linear dependence and linear combinations.)
Let A be a vector space over a field F. Let $x_1, \ldots, x_k$ be a sequence of nonzero elements of A. Then the following conditions are equivalent:

(a) *The sequence $x_1, \ldots, x_k$ is linearly dependent;*

(b) *An element x_j of the sequence, with $j > 1$, is a linear combination of the earlier members $x_1, \ldots, x_{j-1}$ of the sequence.*

PROOF **(a) implies (b):** Since $x_1, \ldots, x_k$ is linearly dependent, there is a relation
$$0 = r_1 x_1 + \ldots + r_{j-1} x_{j-1} + r_j x_j$$
with $r_j \neq 0$. Then
$$x_j = -r_j^{-1} r_1 x_1 - \ldots - r_j^{-1} r_{j-1} x_{j-1}$$
expresses x_j as a linear combination of the earlier members of the sequence.
(b) implies (a): Suppose that
$$x_j = r_1 x_1 + \ldots + r_{j-1} x_{j-1} .$$
Then
$$0 = r_1 x_1 + \ldots + r_{j-1} x_{j-1} + (-1) x_j$$
expresses 0 as a linear combination involving the nonzero coefficient -1. ▯

Example 9.47 (Linear dependence without a linear combination.)
In the $\mathbb{Z}$-module $\mathbb{Z}/4$ of integers modulo 4, take $x_1 = 2$ and $x_2 = 1$. Then the sequence x_1, x_2 is linearly dependent, since $0 = 1 \cdot x_1 + 2 \cdot x_2$. Nevertheless, the element x_2 is not a linear combination (multiple) of x_1. ▯

For the remainder of this section, we will consider modules (vector spaces) over a field F. The following theorem uses Proposition 9.46 to show that every finitely generated vector space is free. The name comes from the way that a generating set may be pruned down to a linearly independent generating set, a basis.

THEOREM 9.48 (Pruning Theorem.)
Let V be a finitely generated vector space over a field F. Suppose that $Y = \{y_1, \ldots, y_m\}$ is a set of nonzero vectors in V that generates V. Then Y has a linearly independent subset $\{y_{i_1}, \ldots, y_{i_l}\}$ which forms a basis of V.

PROOF   Suppose that $X = \{x_1, \ldots, x_k\}$ is a k-element subset of Y which still generates V. If this subset X is already linearly independent, it forms the required basis. Otherwise, $x_1, \ldots, x_k$ is a linearly dependent sequence of nonzero vectors. According to Proposition 9.46, there is some vector x_j in the sequence, with $j > 1$, that is linearly dependent on its predecessors. Then $X' = \{x_1, \ldots, x_{j-1}, x_{j+1}, \ldots, x_k\}$ is a $(k-1)$-element subset of Y which still generates V. The pruning process, passing from a generating set X to a smaller generating set X', may be continued until a linearly independent generating subset is obtained.   ▯

Example 9.49 (Eighth roots of unity.)
Consider the set $Y = \{\omega^r \mid 1 \leq r < 8\}$ of 8-th roots of unity, the complex roots of the polynomial $X^8 - 1$ in $\mathbb{Q}[X]$ — Example 8.24, page 198 — with

$$\omega = \cos\left(\frac{\pi}{4}\right) + i\sin\left(\frac{\pi}{4}\right) = \frac{1}{\sqrt{2}} + \frac{i}{\sqrt{2}}.$$

These roots are displayed in Figure 9.2.

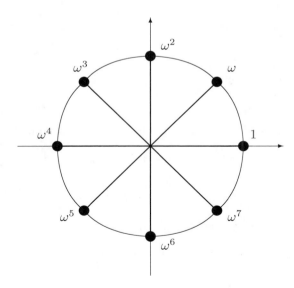

FIGURE 9.2:   Eighth roots of unity.

The field $\mathbb{Q}(\omega)$ is the splitting field for $X^8 - 1$ over $\mathbb{Q}$. Using the notation of Example 9.20, the extension field $\mathbb{Q}(\omega)$ of $\mathbb{Q}$ forms a module or vector space $_{\mathbb{Q}}\mathbb{Q}(\omega)$ over $\mathbb{Q}$. The set Y spans the $\mathbb{Q}$-vector space $\mathbb{Q}(\omega)$. (In fact, $\mathbb{Q}(\omega)$ may also be characterized as the subspace of $_{\mathbb{Q}}\mathbb{C}$ generated by Y.) By Theorem 9.48, Y may be pruned down to a basis for $\mathbb{Q}(\omega)$ over $\mathbb{Q}$. Starting with the sequence

$$1, \omega, \omega^2, \omega^3, \omega^4, \omega^5, \omega^6, \omega^7,$$

the elements $\omega^7 = -\omega^3$, $\omega^6 = -\omega^2$, $\omega^5 = -\omega$, $\omega^4 = -1$ are pruned in turn. This leaves the linearly independent set $\{1, \omega, \omega^2, \omega^3\}$ as a basis for the vector space $\mathbb{Q}(\omega)$ over $\mathbb{Q}$. ⬚

THEOREM 9.50 (Exchange Theorem.)

Let V be a vector space over a field F, and let l be a positive integer. Suppose:

(a) $X = \{x_1, \ldots, x_k\}$ *is an l-element basis of V, and*

(b) $Y = \{y_1, \ldots, y_m\}$ *is an l-element, linearly independent subset of V.*

Then V is generated by Y.

PROOF   For each $0 \le k \le l$, we will prove that there is an injective function $f : \{k+1, \ldots, l\} \to \{1, \ldots, l\}$ such that the set

$$\{y_1, \ldots, y_k, x_{f(k+1)}, \ldots, x_{f(l)}\} \tag{9.23}$$

generates V. The claim is true for $k = 0$, with f as the identity function on the set $\{1, \ldots, l\}$. Suppose

$$\{y_1, \ldots, y_{k-1}, x_{g(k)}, \ldots, x_{g(l)}\}$$

generates V, with an injective function

$$g : \{k, \ldots, l\} \to \{1, \ldots, l\}.$$

Then the vector y_k of V may be written as a linear combination

$$y_k = r_1 y_1 + \ldots + r_{k-1} y_{k-1} + r_k x_{g(k)} + \ldots + r_l x_{g(l)}. \tag{9.24}$$

Now $r_k = \cdots = r_l = 0$ would contradict the linear independence of Y, so there is an integer $k \le m \le l$ with $r_m \ne 0$. Rewriting (9.24) as

$$y_k = r_1 y_1 + \ldots + r_{k-1} y_{k-1} + r_k x_{g(k)} + \ldots + r_m x_{g(m)} + \ldots + r_l x_{g(l)},$$

we see that $x_{g(m)}$ is expressed as the linear combination

$$- r_m^{-1} r_1 y_1 - \ldots - r_m^{-1} r_{k-1} y_{k-1} + r_m^{-1} y_k$$
$$- r_m^{-1} r_k x_{g(k)} - \ldots - r_m^{-1} r_{m-1} x_{g(m-1)} - r_m^{-1} r_{m+1} x_{g(m+1)} - \ldots - r_m^{-1} r_l x_{g(l)}.$$

Since the set $\{y_1, \ldots, y_{k-1}, x_{g(k)}, \ldots, x_{g(l)}\}$ generates V, it is apparent that the set $\{y_1, \ldots, y_k, x_{g(k+1)}, \ldots, x_{g(m-1)}, x_{g(m+1)}, \ldots, x_{g(l)}\}$ also generates V. Define an injective function $h : \{k+1, \ldots, l\} \to \{1, \ldots, l\}$ by

$$h(i) = \begin{cases} g(i-1) & \text{if } i \leq m \, ; \\ g(i) & \text{if } i > m \, . \end{cases}$$

Then $\{y_1, \ldots, y_k, x_{h(k+1)}, \ldots, x_{h(l)}\}$ generates V, as required to complete the inductive proof. The statement of the theorem follows, since Y is just the case $k = l$ of (9.23), with $f : \varnothing \hookrightarrow \{1, \ldots, l\}$. □

COROLLARY 9.51

Let V be a finitely generated vector space over a field F. Then any two bases of V have the same number of elements.

PROOF Suppose that $X = \{x_1, \ldots, x_l\}$ and $Y' = \{y_1, \ldots, y_n\}$ are bases of V over F, with l and n elements respectively. If $l \neq n$, say $l < n$, then Theorem 9.50 shows that the linearly independent subset $Y = \{y_1, \ldots, y_l\}$ of Y' already generates V. The expression of the nonzero vector y_{l+1} as a linear combination of $y_1, \ldots, y_l$ would then violate the linear independence of Y'. □

DEFINITION 9.52 (Dimension of a vector space.) *Let V be a finitely generated vector space over a field F. Then the* dimension $\dim_F V$ *of V over F is the number of elements in a basis of V over F.*

Example 9.53
In Example 9.49, $\dim_{\mathbb{Q}} \mathbb{Q}(\omega) = 4$. □

Example 9.54
For a field F and positive integer n, the vector space F^n has n elements in its standard basis E_n (Example 9.40). Thus $\dim_F F^n = n$. □

Since finitely generated modules over a field F have finite dimension, they are usually known as *finite-dimensional* vector spaces. If a vector space is not finite-dimensional, it is described as being *infinite-dimensional*. For the proof of the following, see Exercise 34.

PROPOSITION 9.55

Let V and W be finite-dimensional vector spaces over a field F. Then V and W are isomorphic if and only if they have the same dimension.

9.8   Abelian groups

This section analyzes the structure of finitely generated abelian groups. Each such group is an internal direct sum of certain cyclic subgroups, given in terms of an ascending chain of ideals of $\mathbb{Z}$ — compare (8.3), page 188.

THEOREM 9.56 (Structure of finitely generated abelian groups.)
Let A be a finitely generated, nontrivial abelian group. Then there is a finite ascending chain

$$J_1 \hookrightarrow J_2 \hookrightarrow \ldots \hookrightarrow J_{l-1} \hookrightarrow J_l \tag{9.25}$$

of proper ideals of $\mathbb{Z}$ such that A is isomorphic to the direct sum

$$\mathbb{Z}/J_1 \oplus \mathbb{Z}/J_2 \oplus \ldots \oplus \mathbb{Z}/J_{l-1} \oplus \mathbb{Z}/J_l \tag{9.26}$$

of nontrivial cyclic groups.

PROOF   Consider the set

$$S_1 = \{ |X| < \infty \mid \mathbb{Z}X = A \}$$

of sizes of the finite subsets X of A that generate A. Since A is finitely generated, the set S_1 of positive integers is nonempty. By the Well-Ordering Principle, the set S_1 has a least element l. This integer l will become the number of ideals in the chain (9.25). The proof proceeds by induction on l. If $l = 1$, the group A is cyclic, and the theorem is immediate. Thus for the inductive proof, we proceed to the case $l > 1$, and assume that the theorem is true for all abelian groups generated by fewer than l elements.

Suppose that some generating set $X = \{x_1, \ldots, x_l\}$ of minimal size l is linearly independent, so that A is free. Proposition 9.41 then gives

$$A \cong \mathbb{Z}/J_1 \oplus \mathbb{Z}/J_2 \oplus \ldots \oplus \mathbb{Z}/J_l$$

with $J_1 = \cdots = J_l = \{0\}$, and the theorem is proved directly.

In the remaining cases, each generating set $X = \{x_1, \ldots, x_l\}$ of minimal size l satisfies a relation

$$0 = r_1 x_1 + \ldots + r_l x_l \tag{9.27}$$

with at least one nonzero coefficient integer r_i. Multiplying (9.27) by -1 if necessary, we may assume that there is a strictly positive coefficient. Let S_2 be the nonempty set of all such positive coefficients, taken over all the generating sets of minimal size l. By the Well-Ordering Principle, the set S_2 has a least element. This least element appears as the coefficient r_l in the relation (9.27) on a generating set $X = \{x_1, \ldots, x_l\}$.

For $1 \le i < l$, the Division Algorithm with divisor r_l gives $r_i = q_i r_l + r_i'$ with remainder $0 \le r_i' < r_l$. Rewriting the relation (9.27) as

$$0 = r_1 x_1 + \ldots + r_{i-1} x_{i-1} + r_i x_i + r_{i+1} x_{i+1} + \ldots + r_l x_l$$
$$= r_1 x_1 + \ldots + r_{i-1} x_{i-1} + (q_i r_l + r_i') x_i + r_{i+1} x_{i+1} + \ldots + r_l x_l$$
$$= r_1 x_1 + \ldots + r_{i-1} x_{i-1} + r_i' x_i + r_{i+1} x_{i+1} + \ldots + r_l (x_l + q_i x_i)$$

gives r_i' as a coefficient in a relation on the generating set

$$\{x_1, \ldots, x_i, \ldots, x_{l-1}, x_l + q_i x_i\}$$

— note that $x_l = (-q_i) x_i + (x_l + q_i x_i)$. By the choice of r_l as the least positive coefficient, we must have $r_i' = 0$, so that $r_i = q_i r_l$.

Now consider the minimal generating set $X' = \{x_1, \ldots, x_{l-1}, y_l\}$ with

$$y_l = q_1 x_1 + \ldots + q_{l-1} x_{l-1} + x_l .$$

The relation (9.27) is rewritten as

$$0 = r_1 x_1 \ldots + r_{l-1} x_{l-1} + r_l x_l$$
$$= q_1 r_l x_1 + \ldots + q_{l-1} r_l x_{l-1} + r_l x_l$$
$$= r_l (q_1 x_1 + \ldots + q_{l-1} x_{l-1} + x_l) = r_l y_l ,$$

so the element y_l has order r_l in A. Setting $J_l = r_l \mathbb{Z}$, we have $\mathbb{Z}/J_l \cong \mathbb{Z}\{y_l\}$.

Let B be the subgroup of A generated by $\{x_1, \ldots, x_{l-1}\}$. Note that $A = B + \mathbb{Z}\{y_l\}$. Suppose that

$$0 = b + s_l y_l \tag{9.28}$$

with $b = s_1 x_1 + \ldots + s_{l-1} x_{l-1}$ in B and $s_l y_l$ in $\mathbb{Z}\{y_l\}$. Use the Division Algorithm to write $s_l = p r_l + r$ with $0 \le r < r_l$. Recalling that $r_l y_l = 0$, the relation (9.28) takes the form

$$0 = b + s_l y_l = (s_1 x_1 + \ldots + s_{l-1} x_{l-1}) + r y_l .$$

By the choice of r_l as the least element of the set S_2 of positive coefficients, it follows that $r = 0$, so $b = 0 = s_l y_l$ in (9.28). By Proposition 9.38(c), the group A is the internal direct sum of B and $\mathbb{Z}\{y_l\}$.

Since B is generated by $l-1$ elements, the induction hypothesis yields an ascending chain

$$J_1 \hookrightarrow J_2 \hookrightarrow \ldots \hookrightarrow J_{l-1} \tag{9.29}$$

of ideals of $\mathbb{Z}$ such that B is isomorphic to the direct sum

$$\mathbb{Z}/J_1 \oplus \mathbb{Z}/J_2 \oplus \ldots \oplus \mathbb{Z}/J_{l-1}$$

of nontrivial cyclic groups. It follows that A, as the internal direct sum of B and $\mathbb{Z}\{y_l\}$, is isomorphic to the direct sum (9.26).

By Corollary 9.35, there is a generating set $\{y_1, \ldots, y_{l-1}\}$ of B such that $\mathbb{Z}/J_i \cong \mathbb{Z}\{y_i\}$ for $1 \leq i \leq l - 1$. If B is free, then $J_{l-1} = \{0\}$, so the chain (9.29) extends to (9.25). If B is not free, then $J_{l-1} = s\mathbb{Z}$ with $0 < s = qr_l + r'$ and $0 \leq r' < r_1$. Consider the generating set $Y = \{y_1, \ldots, y_{l-1}, qy_{l-1} + y_l\}$ for A — note $y_l = (-q)y_{l-1} + (qy_{l-1} + y_l)$. Since $0 = sy_{l-1} = r_l y_l$, we have

$$0 = (qr_l + r')y_{l-1} + r_l y_l = r'y_{l-1} + r_l(qy_{l-1} + y_l)$$

as a relation on the minimal generating set Y of A. By the choice of r_l as the least element of the set S_2 of positive coefficients, it follows that $r' = 0$. Thus $s = qr_l$ lies in J_l, and the chain (9.29) extends to (9.25) in this case as well. $\square$

Example 9.57
Consider the abelian group $A = C_3 \times C_4 \times C_5 \times C_6$. By the Chinese Remainder Theorem (Theorem 5.24, page 105), $C_3 \times C_4 \times C_5 \cong C_{60}$, so $A \cong C_{60} \times C_6$. Theorem 9.56 gives $A \cong \mathbb{Z}/J_1 \oplus \mathbb{Z}/J_2$ with $J_1 = 60\mathbb{Z}$ and $J_2 = 6\mathbb{Z}$. $\square$

Example 9.58
Consider the subgroup

$$N = \{(0, 10x, 2x) \mid x \text{ in } \mathbb{Z}\}$$

of $\mathbb{Z}^3$. Let A be the quotient group $\mathbb{Z}^3/N$. For the standard basis $E_3 = \{e_1, e_2, e_3\}$ of $\mathbb{Z}^3$, consider the cosets $x_i = e_i + N$ in A for $1 \leq i \leq 3$. The generating set $X = \{x_1, x_2, x_3\}$ of A is minimal, and satisfies the relation

$$10x_2 + 2x_3 = (0, 10, 2) + N = N = 0. \tag{9.30}$$

Using the notation from the proof of Theorem 9.56, the least element of the set S_2 of positive coefficients is $r_3 = 2$, the coefficient of x_3 in the relation (9.30). If B is the subgroup of A generated by $\{x_1, x_2\}$, we have $A \cong B \oplus \mathbb{Z}\{y_3\}$ with $y_3 = 5x_2 + x_3$ and $\mathbb{Z}\{y_3\} \cong \mathbb{Z}/2$. In turn, $B \cong \mathbb{Z} \oplus \mathbb{Z}$ (Exercise 36). Thus the chain of ideals in the description of A provided by Theorem 9.56 is $J_1 = \{0\}$, $J_2 = \{0\}$, and $J_3 = 2\mathbb{Z}$. $\square$

Example 9.59
Consider the subgroup

$$N = \{(0, 10x, 2y) \mid x, \ y \text{ in } \mathbb{Z}\}$$

of $\mathbb{Z}^3$. Let A be the quotient group $\mathbb{Z}^3/N$. Then the chain of ideals in the description of A provided by Theorem 9.56 is $J_1 = \{0\}$, $J_2 = 10\mathbb{Z}$, and $J_3 = 2\mathbb{Z}$ (Exercise 37). $\square$

9.9 Exercises

1. Determine all the endomorphisms of the group $(\mathbb{Z}/_3, +, 0)$ of integers modulo 3 under addition.

2. Let A and B be abelian groups. Show that the maps

$$\rho : A \times B \to A \times B; (a, b) \mapsto (a, 0)$$

and

$$\sigma : A \times B \to A \times B; (a, b) \mapsto (0, b)$$

are endomorphisms of $A \times B$.

3. Let A be an abelian group. Show that the map

$$\tau : A \times A \to A \times A; (a, b) \mapsto (b, a)$$

is an endomorphism of $A \times A$.

4. Determine all the injective endomorphisms of the Klein 4-group.

5. Let r be an element of a ring R. Which ring property in Definition 6.1 (page 127) ensures that (9.4) is an endomorphism of the additive group $(R, +, 0)$?

6. Let $(G, \cdot, e)$ be a group. Consider the group $\mathbb{Z}$ of integers. For two homomorphisms $\theta : \mathbb{Z} \to G$ and $\varphi : \mathbb{Z} \to G$, define the *componentwise product*

$$\theta \cdot \varphi : \mathbb{Z} \to G; n \mapsto \theta(n)\varphi(n).$$

Show that G is abelian if and only if the following condition is satisfied:

For all homomorphisms $\theta : \mathbb{Z} \to G$ and $\varphi : \mathbb{Z} \to G$, the componentwise product $\theta \cdot \varphi : \mathbb{Z} \to G$ is a homomorphism.

7. Let A be an abelian group of order 2.

   (a) Show that there are just 2 endomorphisms of A, the zero map 0 of (9.1) and the identity map 1 of (9.2).

   (b) Show that $1 + 1 = 0$ in the endomorphism ring $\mathrm{End}(A)$.

   (c) Conclude that the ring $\mathrm{End}(A)$ is isomorphic to the ring $(\mathbb{Z}/_2, +, \cdot)$ of integers modulo 2.

8. In Example 9.16, verify that the map

$$F \to \mathrm{End}\,(F_2^1, +); \lambda \mapsto \sigma_\lambda$$

is a unital ring homomorphism.

9. In Example 9.17, verify that the map

$$\mathbb{C} \to \text{End}\,(\mathbb{C}_2^1, +); \lambda \mapsto \delta_\lambda$$

is a unital ring homomorphism.

10. Let A be an abelian group. Suppose that A forms a $\mathbb{Z}$-module $(A, +, \sigma)$ with structure map $\sigma : \mathbb{Z} \to \text{End}\,A$. Prove that $\sigma_n(a)$ is the multiple na for each integer n and element a of A.

11. Let A be an abelian group, interpreted as a $\mathbb{Z}$-module according to Example 9.18. Show that a subset B of A is a subgroup of A if and only if it is a submodule of A.

12. For a field F, consider the module F_2^1 of 2-dimensional column vectors given by the scalar action of Example 9.16. For each 2-dimensional row vector $[f_1\ f_2]$ over F, show that the solution set

$$B = \left\{ \begin{bmatrix} x_1 \\ x_2 \end{bmatrix} \middle| [f_1\ f_2] \begin{bmatrix} x_1 \\ x_2 \end{bmatrix} = 0 \right\}$$

of the homogeneous linear equation

$$f_1 x_1 + f_2 x_2 = 0$$

is a submodule of F_2^1.

13. Continue the notation of Exercise 12. Let k be a nonzero element of the field F. Under what conditions is the solution set of the inhomogeneous linear equation

$$f_1 x_1 + f_2 x_2 = k$$

a submodule of F_2^1?

14. Let $(R, +, \cdot)$ be a ring. A subset J of R is said to be a *left ideal* of R if it is a subgroup of $(R, +, 0)$ satisfying the *left absorptive property*

$$j \text{ in } J \quad \text{implies} \quad r \cdot j \text{ in } J$$

for each r in R. Take R to be a left R-module, as in Example 9.20 with $A = R$. Show that a subset B of R is a submodule of R if and only if it is a left ideal of R.

15. Let R be the ring $\mathbb{R}_2^2$ of 2×2 real matrices, under the usual addition and multiplication of matrices. Consider the subset

$$J = \left\{ \begin{bmatrix} 0 & x \\ 0 & y \end{bmatrix} \middle| x,\ y \text{ in } \mathbb{R} \right\}$$

of R.

(a) Show that J is a left ideal of R.

(b) Show that J is not an ideal of R.

16. Suppose that X is a subset of a submodule B of a module A over a ring R. Write out a formal proof, by induction on the natural number n, that each R-linear combination

$$r_1 \cdot x_1 + \ldots + r_n \cdot x_n$$

of elements $x_1, \ldots, x_n$ of X, with coefficients $r_1, \ldots, r_n$ from R, lies in the submodule B.

17. Consider the abelian group $\mathbb{Z}$ of integers as a unital $\mathbb{Z}$-module. Show that

$$\mathbb{Z}\{a, b\} = \gcd(a, b)\mathbb{Z}$$

for nonzero integers a and b.

18. Give a direct proof of Corollary 9.27 (without using Theorem 9.26).

19. Let R be a nontrivial ring. Consider the polynomial ring A over R in an indeterminate Y as an R-module $(A, +, R)$, according to Example 9.28. Let

$$X = \{p_1(Y), \ldots, p_r(Y)\}$$

be a finite subset of A.

(a) Let M be the maximum of the degrees $\deg p_1, \ldots, \deg p_r$ of the polynomials in the set X. Consider the submodule RX of A that is generated by X. Show that no polynomial $p(X)$ in RX has a degree $\deg p$ which exceeds M.

(b) Conclude that the R-module A is not finitely generated.

20. Show that the image of a module homomorphism is a submodule of its codomain.

21. Let $(A, +, \mathbb{Z})$ and $(B, +, \mathbb{Z})$ be $\mathbb{Z}$-modules. Show that a given function $f : A \to B$ is a $\mathbb{Z}$-module homomorphism if and only if it is an abelian group homomorphism.

22. Verify the claim of Example 9.33.

23. Verify the claim of Example 9.37.

24. Let A and B be abelian groups. Define group homomorphisms

$$\alpha : A \to A \oplus B; a \mapsto (a, 0)$$

and

$$\beta : B \to A \oplus B; b \mapsto (0, b).$$

Let C be a third abelian group, the codomain of group homomorphisms $f : A \to C$ and $g : B \to C$. Show that there is a uniquely defined group homomorphism $h : A \oplus B \to C$ with $h \circ \alpha = f$ and $h \circ \beta = g$.

25. Define group homomorphisms

$$\alpha : \mathbb{Z} \to \mathbb{Z} \times \mathbb{Z}; n \mapsto (n, 0)$$

and

$$\beta : \mathbb{Z} \to \mathbb{Z} \times \mathbb{Z}; n \mapsto (0, n).$$

For the symmetric group $S_3 = \{0, 1, 2\}!$, define group homomorphisms

$$f : \mathbb{Z} \to S_3; n \mapsto (0\ 1)^n$$

and

$$g : \mathbb{Z} \to S_3; n \mapsto (1\ 2)^n.$$

Show that there is no group homomorphism $h : \mathbb{Z} \times \mathbb{Z} \to S_3$ with $h \circ \alpha = f$ and $h \circ \beta = g$. (Hint: In $\mathbb{Z} \times \mathbb{Z}$, the equation

$$(1, 0) + (0, 1) = (1, 1) = (0, 1) + (1, 0)$$

holds, while

$$(0\ 1) \circ (1\ 2) \neq (1\ 2) \circ (0\ 1)$$

in S_3.)

26. For a positive integer n, let $B_1, \ldots, B_n$ be submodules of a module B. Verify that the set of all elements b of B that have an expression of the form

$$b = b_1 + b_2 + \ldots + b_n$$

(with b_i in B_i for $1 \leq i \leq n$) is the submodule of B generated by the union of the subsets $B_1, \ldots, B_n$ of B.

27. Let $X = \{x_1, \ldots, x_l\}$ be a basis for a free module A over a unital ring R. For $1 \leq i \leq l$, define

$$y_i = x_1 + \ldots + x_i.$$

Show that $Y = \{y_1, \ldots, y_l\}$ is a basis for A.

28. Let R be a nontrivial unital ring. Let l and m be integers. Show that the two R-modules $R^l \oplus R^m$ and R^{l+m} are isomorphic.

29. Verify that the map (9.22) is a ring homomorphism.

30. Let n be a positive integer. Show that the abelian group $C_n \times C_n$ has n^4 endomorphisms.

31. In the $\mathbb{Z}$-module $\mathbb{Z}/6$ of integers modulo 6, take $x_1 = 2$ and $x_2 = 3$.

    (a) Show that the sequence x_1, x_2 is linearly dependent.

    (b) Show that neither of the elements x_1, x_2 is a linear combination of the other.

32. Consider
$$\omega = \cos\left(\frac{\pi}{3}\right) + i\sin\left(\frac{\pi}{3}\right)$$
as a sixth root of unity. Find a basis for the vector space $\mathbb{Q}(\omega)$ over $\mathbb{Q}$.

33. Consider
$$\omega = \cos\left(\frac{2\pi}{5}\right) + i\sin\left(\frac{2\pi}{5}\right)$$
as a fifth root of unity. Find a basis for the vector space $\mathbb{Q}(\omega)$ over $\mathbb{Q}$.

34. Prove Proposition 9.55: Let F be a field.

    (a) Let V be a finite-dimensional vector space over F. If $\dim_F V = l$, let $X = \{x_1, \ldots, x_l\}$ be a basis for V. Define
$$f : X \rightarrow E_l; x_i \mapsto e_i$$
and
$$g : E_l \rightarrow X; e_i \mapsto x_i.$$
Use the universality of free modules (Theorem 9.42) to build the respective extensions $\overline{f} : V \rightarrow F^n$ and $\overline{g} : F^n \rightarrow V$. Show that $\overline{f}$ and $\overline{g}$ are mutually inverse vector space isomorphisms. (Hint: The unique homomorphic extension $\overline{\mathrm{id}_X}$ of id_X is id_V.)

    (b) Let l and m be positive integers. If $h : F^l \rightarrow F^m$ is a vector space isomorphism, show that $h(E_l)$ is a basis of F^m. Use Corollary 9.51 to conclude that $l = m$.

35. Let A be the abelian group $C_2 \times C_3 \times C_4 \times C_6$. Determine the chain (9.25) of ideals in the description of A provided by Theorem 9.56.

36. In Example 9.58, show that each element $(a_1, a_2', a_3) + N$ of A has an expression of the form
$$(a_1, a_2', a_3) + N = (a_1, a_2, 0) + (0, 5a_3, a_3) + N$$
$$= a_1 x_1 + a_2 x_2 + a_3 y_3$$
with integers a_1 and a_2. Conclude that the subgroup B is isomorphic to $\mathbb{Z} \oplus \mathbb{Z}$.

37. In Example 9.59, show that the group A is isomorphic to $\mathbb{Z} \oplus \mathbb{Z}/10 \oplus \mathbb{Z}/2$.

38. Consider the subgroup

$$N = \{(0, 2x, 2y) \mid x, \, y \text{ in } \mathbb{Z}\}$$

of $\mathbb{Z}^3$. Let A be the quotient group $\mathbb{Z}^3/N$. Find the chain (9.25) of ideals in the description of A provided by Theorem 9.56.

39. (a) Show that the sets $\mathbb{Z}$ and $\mathbb{Z}^2$ are isomorphic.

(b) Show that the groups $\mathbb{Z}$ and $\mathbb{Z}^2$ are not isomorphic.

9.10 Study projects

1. **Partitions and p-groups.** A *partition* (or more explicitly, an *integer partition*) is an expression of a positive integer n as a sum

$$n = n_1 + n_2 + \ldots + n_l \tag{9.31}$$

of positive integers. The integers n_i are known as the *parts* or *summands* of the partition. Conventionally, the summands may be arranged in decreasing order:

$$n_1 \geq n_2 \geq \cdots \geq n_l \, . \tag{9.32}$$

The number l of summands is known as the *length* of the partition. The integer n is known as the *sum* of the partition.

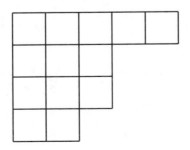

FIGURE 9.3: The partition $4 + 4 + 3 + 1 + 1$.

It is sometimes helpful to visualize a partition (9.31) as a series of l columns, of respective depths given by the parts of the partition, as illustrated in Figure 9.3.

For a prime number p, a finite group is said to be a *p-group* if its order is a power p^n of the prime number p. Here, we are concerned with abelian p-groups. Each partition (9.31) determines the p-group

$$\mathbb{Z}/p^{n_1}\mathbb{Z} \oplus \mathbb{Z}/p^{n_2}\mathbb{Z} \oplus \cdots \oplus \mathbb{Z}/p^{n_l}\mathbb{Z}.$$

The Structure Theorem 9.56 for abelian groups then yields the ascending sequence

$$p^{n_1}\mathbb{Z} \hookrightarrow p^{n_2}\mathbb{Z} \hookrightarrow \ldots \hookrightarrow p^{n_l}\mathbb{Z}$$

of ideals of $\mathbb{Z}$ corresponding to the decreasing sequence (9.32) of parts of the partition. A group A is said to be an *elementary abelian* group if it is isomorphic to a finite power C_p^l of a cyclic group of prime order p.

(a) Show that the underlying group of a finite-dimensional vector space over the field $\mathbb{Z}/_p$ is an elementary abelian p-group.

(b) Show that every elementary abelian p-group A of order p^l is the underlying group of an l-dimensional vector space over $\mathbb{Z}/_p$.

(c) Show that an elementary abelian p-group of order p^l has $p^l - 1$ elements of order p.

(d) Show that an abelian group of order p^l is elementary abelian if it has $p^l - 1$ elements of order p.

(e) Show that an elementary abelian p-group of order p^l corresponds to the partition $l = 1 + 1 + \ldots + 1$.

(f) Let A be an abelian p-group of order p^n. Using the notation of (9.3), consider the endomorphism $\mu_p : A \to A; a \mapsto pa$ of A. Show that A corresponds to a partition of n with length l if and only if the subgroup

$$\operatorname{Ker} \mu_p = \{a \mid pa = 0\}$$

is an elementary abelian p-group of order p^l.

(g) Use induction on n to show that two p-groups A and B of order p^n are isomorphic if and only if they correspond to the same partition (9.31) of n :

   (i). If A and B are elementary abelian, show that the claim results directly from (f).

  (ii). Otherwise, consider the subgroup

$$pA = \{pa \mid a \text{ in } A\}$$

  of A. If A corresponds to (9.32) with

$$n_1 \geq \cdots \geq n_m > n_{m+1} = \cdots = n_l = 1,$$

  show that pA corresponds to the partition

$$(n_1 - 1) + \ldots + (n_m - 1)$$

  of $n - l$.

(h) Determine all the integer partitions of 5.

(i) Determine representatives for each of the isomorphism classes of abelian groups of order 32.

(j) Determine the isomorphism class of the group of units $(\mathbb{Z}/51, \cdot, 1)^*$ of the monoid of integers modulo 51 under multiplication.

(k) Do the monoids $\mathbb{Z}/51$ and $\mathbb{Z}/15 \times \mathbb{Z}/5$ have isomorphic groups of units?

2. **Algebraic and transcendental numbers.** A complex number x is said to be *algebraic* if the field extension $\mathbb{Q}(x)$ is a finite-dimensional vector space over $\mathbb{Q}$. Otherwise, x is described as *transcendental*. For example, it is known that the real numbers e and π are transcendental.

(a) Show that $\sqrt{5}$ is algebraic.

(b) Show that x is algebraic if and only if it is the root of a certain polynomial $p(X)$ in $\mathbb{Q}[X]$.

(c) Show that $\sqrt{3} + \sqrt{5}$ is algebraic.

(d) Consider successive field extensions $F \hookrightarrow K \hookrightarrow L$. Suppose that K is a finite-dimensional vector space over F, with basis $\{k_1, \ldots, k_m\}$. Suppose that L is a finite-dimensional vector space over K, with basis $\{l_1, \ldots, l_n\}$. Show that L is a finite-dimensional vector space over F, with basis

$$\{k_i l_j \mid 1 \leq i \leq m,\ 1 \leq j \leq n\}.$$

(e) Show that the set of algebraic numbers forms a subfield $\mathbb{A}$ of $\mathbb{C}$. [Hint: If the dimensions $\dim_{\mathbb{Q}} \mathbb{Q}(x)$ and $\dim_{\mathbb{Q}} \mathbb{Q}(y)$ are finite, show that $\dim_{\mathbb{Q}(x)} \mathbb{Q}(x, y)$ and $\dim_{\mathbb{Q}} \mathbb{Q}(x, y)$ are finite.]

3. **Algebraically closed fields.** A field F is said to be *algebraically closed* if it satisfies one of the following conditions:

(a) Each nonconstant polynomial $f(X)$ in $F[X]$ has a root in F;

(b) Each nonconstant polynomial $f(X)$ in $F[X]$ splits over F;

(c) If E is an extension of F for which $\dim_F E$ is finite, then $E = F$.

Show that these conditions are equivalent.

Now, if you know a little complex analysis, recall Liouville's Theorem:

If a function $g : \mathbb{C} \to \mathbb{C}$ can be expanded as a power series

$$\sum_{n=0}^{\infty} a_n (z - z_0)^n$$

about each point z_0 of $\mathbb{C}$, and there is a real number M such that $|g(z)| < M$ for all complex numbers z, then $g : \mathbb{C} \to \mathbb{C}$ is a constant function.

Use Liouville's Theorem to derive the so-called *Fundamental Theorem of Algebra*: The field $\mathbb{C}$ is algebraically closed. [Hint: If $f(z) \neq 0$ for all complex numbers z, apply Liouville's Theorem to $g(z) = 1/f(z)$.]

9.11 Notes

Section 9.6

For certain noncommutative rings R, there may be an isomorphism

$$R^l \cong R^m$$

with $l \neq m$.

Section 9.7

The fundamental Corollary 9.51 may also be proved using a result from the elementary theory of linear equations, that a homogeneous system of l given equations in n unknowns has a nonzero solution if $l < n$. Traces of that method may be observed in the proof of Theorem 9.50, for example the selection of $x_{g(m)}$ as a "pivot" element in (9.24).

Section 9.10

J. Liouville was a French mathematician who lived from 1809 to 1882.

Chapter 10

GROUP ACTIONS

The importance of groups stems from the general way that they capture the concept of symmetry, as described in this chapter. In particular, each group contains various internal symmetries that constrain the structure of the group.

10.1 Actions

For a set X, recall that the set of all bijective maps from X to X forms a group $X!$ of permutations under composition, the symmetric group on X (compare Section 2.7, and Exercise 13 in Chapter 4). For a group G, Cayley's Theorem (page 110) shows that there is an injective group homomorphism

$$\Lambda : G \to G!; g \mapsto \lambda_g \qquad (10.1)$$

from G to $G!$. In order to study general symmetry, we replace the group $G!$ in (10.1) by the group $X!$ of permutations of a general set X.

DEFINITION 10.1 (Group action, permutation representation.)
Let X be a set, and let G be a group. Then the group G is said to have a (left) (group) action (X, G, λ) on the set X, or to act on the set X, if there is a homomorphism

$$\lambda : G \to X!; g \mapsto \lambda_g \qquad (10.2)$$

from G to $X!$. The homomorphism λ appearing in (10.2), or the full structure (X, G, λ), is said to be a permutation representation *of the group G.*

Example 10.2 (Regular action.)
As a permutation representation of a group $(G, \cdot, e)$, the homomorphism (10.1) from Cayley's Theorem, with $\lambda_g(x) = g \cdot x$ for g and x in G, is called the *(left) regular permutation representation*. ⬜

Example 10.3 (Natural action of a symmetric group.)
For each set X, the identity map $\mathrm{id}_{X!} : X! \to X!$ gives the *natural action* $(X, X!, \mathrm{id}_{X!})$ of the symmetric group $X!$ on X. ⬜

Example 10.4 (Matrix action.)
Left multiplication (2.6) by invertible real 2×2 matrices gives an action

$$L : \mathrm{GL}(2, \mathbb{R}) \to (\mathbb{R}_2^1)!; A \mapsto L_A$$

of the real general linear group $\mathrm{GL}(2, \mathbb{R})$ on the set $\mathbb{R}_2^1$ of 2-dimensional real column vectors. In this example, it is possible to replace the field $\mathbb{R}$ of real numbers by any other field F, giving an action

$$L : \mathrm{GL}(2, F) \to (F_2^1)!; A \mapsto L_A \tag{10.3}$$

of the *general linear group* $\mathrm{GL}(2, F)$ of invertible 2×2 matrices over F on the set F_2^1 of 2-dimensional column vectors over F. ⬚

A group G is defined by the multiplication, identity, and inversion. Given a set X, and maps $\lambda_g : X \to X$ for each element g of G, the group properties may be used directly to confirm the presence of an action.

PROPOSITION 10.5 (Conditions for a group action.)
Let X be a set and let $(G, \cdot, e)$ be a group. Suppose that a map $\lambda_g : X \to X$ is defined for each element g of G. Then there is a group action (10.2) of G on X if and only if

(a) $\lambda_e(x) = x$ *and*

(b) $\lambda_{g \cdot h}(x) = \lambda_g\big(\lambda_h(x)\big)$

for x in X and g, h in G.

PROOF   First, suppose that there is an action (10.2). Since $\lambda : G \to X!$ is a group homomorphism, λ_e is the identity element id_X of $X!$, so that (a) holds. Again, for elements g and h of G, we have $\lambda_{g \cdot h} = \lambda_g \circ \lambda_h$, so that (b) holds.

Conversely, suppose that (a) and (b) hold. In particular, (a) shows that $\lambda_e = \mathrm{id}_X$. Then for each element g of G, condition (b) gives

$$\lambda_g \circ \lambda_{g^{-1}} = \lambda_{g \cdot g^{-1}} = \lambda_e = \mathrm{id}_X$$

and

$$\lambda_{g^{-1}} \circ \lambda_g = \lambda_{g^{-1} \cdot g} = \lambda_e = \mathrm{id}_X \,,$$

so that λ_g is invertible. In particular, there is a map

$$\lambda : G \to X!; g \mapsto \lambda_g \,. \tag{10.4}$$

Condition (b), in the form $\lambda_{g \cdot h} = \lambda_g \circ \lambda_h$, shows that the map λ of (10.4) is a semigroup homomorphism. By Proposition 5.5 (page 96), it follows that λ is a group homomorphism, so that (10.4) becomes an instance of (10.2). ⬚

COROLLARY 10.6
Let x be an element of a set X with a group action (X, G, λ). Define

$$G_x = \{g \ \text{in} \ G \mid \lambda_g(x) = x\}. \tag{10.5}$$

Then G_x is a subgroup of G.

PROOF By Proposition 10.5(a), the identity element e of G lies in G_x. By Proposition 10.5(b), the set G_x is closed under multiplication. Finally, if g lies in G_x, then

$$\lambda_{g^{-1}}(x) = \lambda_{g^{-1}}\big(\lambda_g(x)\big) = \lambda_e(x) = x$$

by Proposition 10.5(b), so that G_x is closed under inversion. ⬜

DEFINITION 10.7 (Stabilizers.) *Suppose (X, G, λ) is a permutation representation. Let x be an element of the set X. Then the group G_x of (10.5) is called the stabilizer $\mathrm{Stab}_\lambda(x)$ of x in G.*

The notation G_x of (10.5) makes no specific mention of the representation λ. Should reference to λ be required, the notation $\mathrm{Stab}_\lambda(x)$ of Definition 10.7 may be used.

Example 10.8 (Trivial representations.)
Let X be a set, and let G be a group. For x in X and g in G, define $\epsilon_g(x) = x$. The conditions (a) and (b) of Proposition 10.5 are trivially satisfied. The representation (X, G, ϵ) is called the *trivial representation* of G on X. The whole group G is the stabilizer of each element of X. ⬜

Example 10.9 (Inner automorphisms.)
Let $(G, \cdot, e)$ be a group. For g in G, define

$$\tau_g : G \to G; x \mapsto g \cdot x \cdot g^{-1}. \tag{10.6}$$

Note that

$$\tau_e(x) = x$$

and

$$\tau_{g \cdot h}(x) = (g \cdot h) \cdot x \cdot (g \cdot h)^{-1} = g \cdot h \cdot x \cdot h^{-1} \cdot g^{-1} = \tau_g\big(\tau_h(x)\big)$$

for g, h and x in G. Thus by Proposition 10.5,

$$\tau : G \to G!; g \mapsto \tau_g \tag{10.7}$$

is an action of the group G on the set G, the *conjugation action*. In particular, each map $\tau_g : G \to G$ is bijective. Moreover,

$$\tau_g(x \cdot y) = g \cdot x \cdot y \cdot g^{-1} = g \cdot x \cdot g^{-1} \cdot g \cdot y \cdot g^{-1} = \tau_g(x) \cdot \tau_g(y)$$

for x and y in G, so that $\tau_g : G \to G$ is a group homomorphism. The maps τ_g are known as *inner automorphisms* of the group G.   ☐

DEFINITION 10.10 (Center, conjugation.)   *Let G be a group.*

(a) *The kernel of the group homomorphism* (10.7) *is the* center Z or $Z(G)$ *of the group G.*

(b) *The image $\tau(G)$ of the group homomorphism* (10.7) *is known as the* inner automorphism group $\mathrm{Inn}(G)$ *of the group G.*

(c) *For a specific element g of the group G, the map*

$$\tau_g : G \to G; x \mapsto g \cdot x \cdot g^{-1}$$

is known as conjugation *by g.*

(d) *For a specific element x of the set G, the stabilizer*

$$C_G(x) = \{g \ \text{in} \ G \mid \tau_g(x) = x\}$$

of x in the conjugation action is known as the centralizer *of x in G.*

10.2   Orbits

Let (X, G, λ) be an action of a group G on a set X. For an element x of X, the *orbit* of x is the set

$$\lambda_G(x) = \{\lambda_g(x) \mid g \ \text{in} \ G\}$$

of images of x under the actions λ_g of the elements g of G.

Example 10.11 (Plane rotations.)
For each real number θ, define a map

$$\kappa_\theta : \mathbb{R}_2^1 \to \mathbb{R}_2^1; \begin{bmatrix} x_1 \\ x_2 \end{bmatrix} \mapsto \begin{bmatrix} \cos\theta & -\sin\theta \\ \sin\theta & \cos\theta \end{bmatrix} \begin{bmatrix} x_1 \\ x_2 \end{bmatrix}.$$

Note that for an angle θ (in radians), the map $\kappa_\theta : \mathbb{R}_2^1 \to \mathbb{R}_2^1$ rotates the plane $\mathbb{R}_2^1$ counterclockwise by θ about the origin O. Now $(\mathbb{R}_2^1, \mathbb{R}, \kappa)$ is an action of the additive group $(\mathbb{R}, +, 0)$ of real numbers (Exercise 15). Then for a point P in the plane $\mathbb{R}^1_2$, the orbit $\kappa_{\mathbb{R}}(P)$ is the circle around the origin through the point P (Figure 10.1). If the Sun is located at the origin O, and P represents a planet, then the mathematical orbit $\kappa_{\mathbb{R}}(P)$ of P represents the astronomical orbit of the planet around the sun. ▯

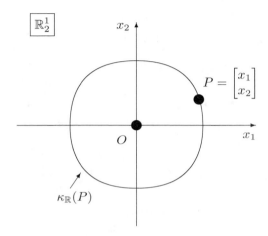

FIGURE 10.1: The orbit $\kappa_{\mathbb{R}}(P)$ of a point P in the plane $\mathbb{R}^1_2$.

A group action may be trimmed down to a smaller action in two ways: by passing to a subgroup of the group, or by discarding some of the orbits.

DEFINITION 10.12 *Let (X, G, λ) be an action of a group G on a set X.*

(a) *For a subgroup H of G, the action (X, H, λ) given by the restriction*

$$\lambda : H \to X!; h \mapsto \lambda_h$$

to the subgroup H of the group homomorphism

$$\lambda : G \to X!; g \mapsto \lambda_g$$

is called the restriction *of the action (X, G, λ) to the subgroup H.*

(b) *Let Y be a union of orbits in the action (X, G, λ). Then the action (Y, G, λ) given by*

$$\lambda_g : Y \to Y; y \mapsto \lambda_g(y)$$

for each g in G is called the restriction *of the action (X, G, λ) to the subset Y of X.*

Example 10.13 (**Right cosets.**)

Let H be a subgroup of a group G. Consider the restriction (G, H, λ) to H of the left regular representation of G (compare Example 10.2). Then for each element x of G, the orbit $\lambda_H(x)$ is the right coset Hx. Now if K is a subgroup of G that contains H, the action (G, H, λ) may be restricted to (K, H, λ). ⬚

For a subgroup H of a group G, the right cosets Hx of H in G are the classes of an equivalence relation on G (recall Proposition 4.49, page 83). The following proposition states that the orbits of a general group action (X, G, λ) are the classes of an equivalence relation. The proof, analogous to the proof of Proposition 4.49 (page 83), is relegated to Exercise 16.

PROPOSITION 10.14 (**Orbits as equivalence classes.**)

Let (X, G, λ) *be a permutation representation of a group G on a set X. Define a relation* $\xrightarrow{\lambda_G}$ *on X by*

$$x \xrightarrow{\lambda_G} y \quad \text{if and only if} \quad y = \lambda_g(x) \quad \text{for some} \quad g \quad \text{in} \quad G.$$

(a) *The relation* $\xrightarrow{\lambda_G}$ *is an equivalence relation on X.*

(b) *For x in X, the equivalence class of x is the orbit $\lambda_G(x)$ of x.*

Example 10.15 (**Conjugacy.**)

Let x be an element of a group G. Then the orbit $\tau_G(x)$ in the conjugacy action (G, G, τ) of G is the *conjugacy class*

$$\{g \cdot x \cdot g^{-1} \mid g \ \text{in} \ G\}$$

of x in G. The equivalence relation $\xrightarrow{\tau_G}$ on G given by Proposition 10.14 is known as *conjugacy*. In the group $\mathrm{GL}(2, F)$, the conjugacy relation is known as *matrix similarity*. ⬚

10.3 Transitive actions

Consider a permutation representation (X, G, λ) of a group G on a set X. According to Proposition 10.14, the set X is partitioned by the orbits. The representation is said to be *transitive* if there is just one orbit. Equivalently, the action (X, G, λ) is transitive if X is nonempty, and for each pair of elements x, y of X, there is at least one element g of G with $y = \lambda_g(x)$.

Example 10.16 (Regular action.)
For a group G, the regular representation is transitive. See Corollary 4.54 (page 86) and Exercise 42 in Chapter 4. □

Example 10.17 (Matrix action.)
Let F be a field. Consider the left multiplication action (10.3) of $GL(2, F)$ on F_1^2. This action is never transitive, since the zero column vector is always alone in its orbit. However, there is only one other orbit (Exercise 17). □

Example 10.18 (Restriction to an orbit.)
Let (X, G, λ) be a permutation representation. For an element x of X, let Y be the orbit $\lambda_G(x)$ of x. Then the restriction (Y, G, λ) to Y is transitive. □

The most important example of a transitive action is given by the following proposition.

PROPOSITION 10.19 (Action on cosets.)
Let H be a subgroup of a group G. Consider the set

$$G/H = \{xH \mid x \ \ in \ \ G\} \tag{10.8}$$

of left cosets of H. Then a transitive action of G on G/H is defined by setting

$$\lambda_g(xH) = gxH \tag{10.9}$$

for each g in G.

PROOF First, verify the conditions (a) and (b) of Proposition 10.5.

(a): For the identity element e of G, (10.9) gives

$$\lambda_e(xH) = exH = xH$$

for each element x of G.

(b): For elements g and h of G, (10.9) gives

$$\lambda_{gh}(xH) = (gh)xH = g(hxH) = \lambda_g(\lambda_h(xH))$$

for each element x of G.

Now consider elements x and y of G, with corresponding elements xH and yH of G/H. Then for $g = y \cdot x^{-1}$ in G, we have

$$\lambda_g(xH) = g \cdot xH = y \cdot x^{-1} \cdot xH = yH,$$

so the action of G on G/H is transitive. □

DEFINITION 10.20 (Homogeneous spaces.)   *Let H be a subgroup of a group G.*

(a) *The set G/H of* (10.8) *is called a* homogeneous space.

(b) *The action $(G/H, G, \lambda)$ of Proposition 10.19 is called the* homogeneous space action *of G on G/H. This action is often denoted simply by G/H.*

Example 10.21
Let G be a group, with identity element e.

(a) The homogeneous space G/G is trivial.

(b) For the trivial subgroup $\{e\}$ of G, the homogeneous space $G/\{e\}$ is essentially the left regular representation of G, elements x of G just being rewritten as singleton sets $\{x\}$.

□

Up to renaming, such as $x \mapsto \{x\}$ in Example 10.21(b), every transitive action is a homogeneous space.

THEOREM 10.22 (Transitive actions are homogeneous spaces.)
Let $\mu : G \to X!$ be a transitive action of a group G on a nonempty set X. For each element a of X, consider the stabilizer G_a of a under the action (X, G, μ). Let $(G/G_a, G, \lambda)$ be the homogeneous space action of G on G/G_a.

(a) *There is a well-defined bijection*

$$b : G/G_a \to X; h \cdot G_a \mapsto \mu_h(a).$$

(b) *For elements g and h of G,*

$$b\big(\lambda_g(h \cdot G_a)\big) = \mu_g\big(b(h \cdot G_a)\big).$$

PROOF   (a): The map b is a well-defined injection, since

$$h \cdot G_a = k \cdot G_a \quad \Leftrightarrow \quad k^{-1}h \text{ lies in } G_a$$
$$\Leftrightarrow \quad \mu_{k^{-1}h}(a) = a \quad \Leftrightarrow \quad \mu_h(a) = \mu_k(a)$$

for elements h and k of G. The map b surjects, since (X, G, μ) is transitive.

(b): We have

$$b\big(\lambda_g(h \cdot G_a)\big) = b(gh \cdot G_a) = \mu_{gh}(a) = \mu_g\big(\mu_h(a)\big) = \mu_g\big(b(h \cdot G_a)\big)$$

for elements g and h of G.

□

Theorem 10.22 reduces the analysis of any transitive action (X, G, μ) of a group G to computations entirely within the group itself. Pick a fixed element a of X, and use the bijection b of Theorem 10.22(a) to label the elements $\mu_h(a)$ of X by left cosets $h \cdot G_a$ of the stabilizer G_a. Then, to compute the effect of the action μ_g on an element labeled $h \cdot G_a$, Theorem 10.22(b) shows that it suffices to compute the effect of the corresponding homogeneous space action λ_g on the coset $h \cdot G$. With $\lambda_g(h \cdot G_a)$ as the coset $gh \cdot G_a$, we obtain $\mu_g(b(h \cdot G_a))$ as the element labeled by $gh \cdot G_a$. The procedure is illustrated in Figure 10.2.

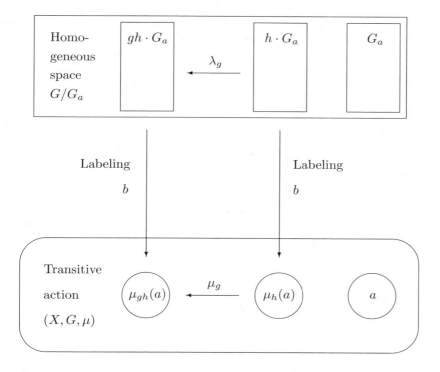

FIGURE 10.2: Tracking a transitive action in a homogeneous space.

Theorem 10.22 has a corollary that is very useful for counting.

COROLLARY 10.23 (Orbit sizes divide the group order.)

Let (X, G, μ) be a permutation representation of a finite group G. Then for each element a of X, the orbit $\mu_G(a)$ is finite, with size $|\mu_G(a)| = |G|/|G_a|$

equal to the index of the stabilizer G_a. In particular, the size of each orbit is a divisor of the order of G.

PROOF Apply Theorem 10.22(a) to the transitive action $\big(\mu_G(a), G, \mu\big)$ obtained by restriction of (X, G, μ) to the orbit $\mu_G(a)$ of a. The bijection b establishes an isomorphism between the orbit and the set of left cosets of the stabilizer G_a in G. ⬜

Example 10.24 (Matrix action.)

Let F be a finite field, with q elements. Consider the left multiplication action (10.3) of GL$(2, F)$ on F_2^1, with orbits as described in Example 10.17. The set F_2^1 of two-dimensional column vectors has q^2 elements, so the orbit of nonzero vectors has $q^2 - 1$ elements. Corollary 10.23 then shows that the size of GL$(2, F)$ is a multiple of $q^2 - 1$. In fact, each invertible matrix

$$A = \begin{bmatrix} a_{11} & a_{12} \\ a_{21} & a_{22} \end{bmatrix}$$

over the field F is constructed from two column vectors

$$\mathbf{a}_1 = \begin{bmatrix} a_{11} \\ a_{21} \end{bmatrix} \quad \text{and} \quad \mathbf{a}_2 = \begin{bmatrix} a_{12} \\ a_{22} \end{bmatrix}.$$

In setting up the matrix A, the first vector $\mathbf{a}_1$ may be chosen as any of the $q^2 - 1$ nonzero vectors in F_2^1. In order to guarantee that det $A \neq 0$, the second vector $\mathbf{a}_2$ must then avoid the q scalar multiples $l\mathbf{a}_1$ with l in F. There are $q^2 - q$ such choices for $\mathbf{a}_2$. In total, there are

$$(q^2 - 1)(q^2 - q) = (q + 1)q(q - 1)^2$$

elements in the group GL$(2, F)$. ⬜

10.4 Fixed points

Let (X, G, λ) be a permutation representation of a group $(G, \cdot, e)$ on a set X. Let x be an element of X, and let g be an element of G. Then x is said to be a *fixed point* of g (*under* λ) if

$$\lambda_g(x) = x. \tag{10.10}$$

Note that x is a fixed point of g if and only if g lies in the stabilizer G_x of x. By Proposition 10.5, each element of X is a fixed point of e.

Example 10.25 (Conjugation action.)
Let G be a group. Consider the conjugation action (G, G, τ) of G on itself
— compare (10.7). Then elements x and y of G commute if and only if x is
a fixed point of y under τ:

$$x \cdot y = y \cdot x \quad \text{if and only if} \quad x = y \cdot x \cdot y^{-1} = \tau_y(x).$$

Each element of G is a fixed point of a central element z. In other words, the
center $Z(G)$ is the set of elements z commuting with each element x of G. ⬜

Example 10.26 (Regular action.)
Let $(G, \cdot, e)$ be a group. Consider the regular representation (G, G, λ) of G
on itself — compare Example 10.2. As observed above, each element of G is
a fixed point of e under λ. For elements x and g of G, the element x is a
fixed point of g under λ if and only if $g \cdot x = x$, which happens if and only
if $g = e$. Thus a nonidentity element of G has no fixed points in the regular
representation. ⬜

Example 10.27 (Plane rotations.)
Consider the rotational representation $(\mathbb{R}_1^2, \mathbb{R}, \kappa)$ of the additive group $(\mathbb{R}, +, 0)$
of real numbers from Example 10.11. Note that the origin O is a fixed point
of every real number θ. However, a real number θ that is not a multiple of 2π
has no other fixed points. ⬜

Now let (X, G, λ) be a permutation representation of a finite group $(G, \cdot, e)$
on a finite set X. For each element g of G, let $\pi_\lambda(g)$ or $\pi(g)$ denote the
number of fixed points of g under λ. According to Example 10.26,

$$\pi(g) = \begin{cases} n & \text{if } g = e \\ 0 & \text{otherwise} \end{cases} \tag{10.11}$$

in the regular representation of a finite group $(G, \cdot, e)$ of order n.

The following result relates the number of orbits in the representation to
the number of points that are fixed by each element of G.

THEOREM 10.28 (Burnside's Lemma.)
*Let G be a group of finite order $|G|$. In a permutation representation (X, G, λ)
of G on a finite set X, the total number of orbits is equal to the average number*

$$\frac{1}{|G|} \sum_{g \in G} \pi(g) \tag{10.12}$$

of fixed points of an element of G. In particular, the sum

$$\sum_{g \in G} \pi(g)$$

of the number of fixed points $\pi(g)$ of each element g of G is a multiple of $|G|$.

PROOF Consider the subset

$$F = \{(g, x) \text{ in } G \times X \mid \lambda_g(x) = x\}$$

of $G \times X$. The number of elements in the finite set F will be computed in two different ways. Equating the two answers will then prove the theorem.

First, consider the pairs (g, x) in F for a particular element g of G. Each such pair comprises a fixed point x of g in its second slot, so there are $\pi(g)$ such pairs altogether. Summing over all the group elements g, we obtain

$$|F| = \sum_{g \in G} \pi(g) \tag{10.13}$$

as the first expression for $|F|$.

Now let $\mathrm{Orb}(\lambda)$ be the set of orbits in the action (X, G, λ). The set F may be described as the disjoint union

$$F = \bigcup_{x \in X} G_x \times \{x\}$$

of products of stabilizers with singletons containing points of X. With an application of Corollary 10.23, we obtain

$$|F| = \sum_{x \in X} |G_x| = \sum_{x \in X} |G|/|\lambda_G(x)| = |G| \sum_{x \in X} 1/|\lambda_G(x)|$$

$$= |G| \sum_{Y \in \mathrm{Orb}(\lambda)} \sum_{x \in Y} 1/|Y| = |G| \sum_{Y \in \mathrm{Orb}(\lambda)} 1$$

$$= |G| \cdot |\mathrm{Orb}(\lambda)| . \tag{10.14}$$

Equating the two quantities (10.13) and (10.14) for $|F|$ yields the desired expression (10.12) for the number $|\mathrm{Orb}(\lambda)|$ of orbits. ☐

Example 10.29 (Regular action.)
Consider the regular representation of a finite group G with n elements. Using (10.11), the expression (10.12) for the number of orbits becomes

$$\frac{1}{n}\left(n + (n-1) \cdot 0\right) = 1 ,$$

which of course is the correct count for a transitive representation. ☐

Example 10.30 (Matrix action.)
Let F be the 2-element field $\mathbb{Z}/_2$. Consider the left multiplication action $\left(F_2^1, \mathrm{GL}(2, F), L\right)$ from Example 10.4. As discussed in Example 10.17, the

Element g of $GL(2, F)$	Set of fixed points of g in F_2^1	Number $\pi(g)$ of fixed points
$\begin{bmatrix} 1 & 0 \\ 0 & 1 \end{bmatrix}$	F_2^1	4
$\begin{bmatrix} 0 & 1 \\ 1 & 0 \end{bmatrix}$	$\left\{ \begin{bmatrix} 0 \\ 0 \end{bmatrix}, \begin{bmatrix} 1 \\ 1 \end{bmatrix} \right\}$	2
$\begin{bmatrix} 1 & 0 \\ 1 & 1 \end{bmatrix}$	$\left\{ \begin{bmatrix} 0 \\ 0 \end{bmatrix}, \begin{bmatrix} 0 \\ 1 \end{bmatrix} \right\}$	2
$\begin{bmatrix} 1 & 1 \\ 0 & 1 \end{bmatrix}$	$\left\{ \begin{bmatrix} 0 \\ 0 \end{bmatrix}, \begin{bmatrix} 1 \\ 0 \end{bmatrix} \right\}$	2
$\begin{bmatrix} 1 & 1 \\ 1 & 0 \end{bmatrix}$	$\left\{ \begin{bmatrix} 0 \\ 0 \end{bmatrix} \right\}$	1
$\begin{bmatrix} 0 & 1 \\ 1 & 1 \end{bmatrix}$	$\left\{ \begin{bmatrix} 0 \\ 0 \end{bmatrix} \right\}$	1

FIGURE 10.3: Fixed points in the action of $GL(2, F)$ on F_2^1, for $F = \mathbb{Z}/_2$.

number of orbits in this representation is 2. The various elements of $GL(2, F)$, with their fixed point sets, are listed in Figure 10.3.

The expression (10.12) for the number of orbits becomes

$$\frac{1}{6}\left(4 + 3 \cdot 2 + 2 \cdot 1\right),$$

which reduces correctly to 2. In fact, this count may be used as proof that the set of nonzero column vectors forms a single orbit. ⬜

10.5 Faithful actions

Consider the rotational representation

$$\kappa : \mathbb{R} \to (\mathbb{R}_2^1)!; \theta \mapsto \kappa_\theta \tag{10.15}$$

of Example 10.11. Recall that for a real number θ, the action κ_θ rotates the plane $\mathbb{R}_2^1$ counterclockwise by an angle of θ radians. Thus if distinct real numbers θ and φ differ by a multiple of 2π, their actions κ_θ and κ_φ coincide.

DEFINITION 10.31 (Faithful action.) *A permutation action*

$$\lambda : G \to X!; g \mapsto \lambda_g$$

of a group G is said to be faithful *if the homomorphism λ is injective.*

Example 10.32 (Plane rotations.)
The rotational representation (10.15) is not faithful.   ⬜

Example 10.33 (Regular actions.)
By Cayley's Theorem, the regular representation of a group G is faithful. ⬜

Faithful permutation actions are very close to groups of permutations.

PROPOSITION 10.34
Let (X, G, λ) be a faithful permutation action of a group G on a set X. Then the abstract group G is isomorphic to a group of permutations of the set X.

PROOF The injective group homomorphism $\lambda : G \to X!; g \mapsto \lambda_g$ yields an isomorphism
$$G \to \lambda(G); g \mapsto \lambda_g$$
of G with the group $\lambda(G)$ of permutations of the set X.   ⬜

In the converse direction, let G be a group of permutations on a set X. Let $j : G \hookrightarrow X!; g \mapsto g$ be the inclusion of G as a subgroup of $X!$. Then the permutation action (X, G, j) of G is faithful. The following theorem provides a more general source of faithful actions.

THEOREM 10.35 (Every action induces a faithful action.)
Let (X, G, λ) be a permutation representation of a group G. Let $\operatorname{Ker} \lambda$ be the group kernel of the homomorphism $\lambda : G \to X!$. For each element g of G, let $\overline{g}$ denote the coset $(\operatorname{Ker} \lambda)g$. Let $\overline{G}$ denote the quotient group $G/\operatorname{Ker}\lambda$. Then there is a faithful action $(X, \overline{G}, \overline{\lambda})$ given by a well-defined homomorphism
$$\overline{\lambda} : \overline{G} \to X!; \overline{g} \mapsto \lambda_g .$$

PROOF Apply the First Isomorphism Theorem for groups to the group homomorphism
$$\lambda : G \to X!; g \mapsto \lambda_g ,$$
factorizing λ as the composite $j \circ b \circ s$ of the projection $G \to \overline{G}; g \mapsto \overline{g}$, the isomorphism $b : \overline{G} \to \lambda(G); \overline{g} \to \lambda_g$, and the insertion $j : \lambda(G) \hookrightarrow X!$. Then $\overline{\lambda}$, as the composite $b \circ s$, is injective.   ⬜

COROLLARY 10.36
The kernel of the representation λ is the intersection

$$\text{Ker}\,\lambda = \bigcap_{x \in X} G_x$$

of the stabilizers G_x of the elements x of X.

PROOF   An element g of G lies in the kernel $\text{Ker}\,\lambda$ if and only if

$$\lambda_g(x) = x$$

for each x in X. This happens if and only if g lies in G_x for each x in X.   □

DEFINITION 10.37   *Let (X, G, λ) be a permutation representation of a group G. Then the faithful representation $(X, \overline{G}, \overline{\lambda})$ of Theorem 10.35 is called the* faithful representation *(or* faithful action*) induced by (X, G, λ).*

Example 10.38
The rotational representation $\left(\mathbb{R}_2^1, \mathbb{R}, \kappa\right)$ of Example 10.11 induces a faithful permutation representation $\left(\mathbb{R}_2^1, \mathbb{R}/2\pi\mathbb{Z}, \overline{\kappa}\right)$ of the group $(\mathbb{R}/2\pi\mathbb{Z}, +)$.   □

Example 10.39
Let G be a group. The conjugacy action (G, G, τ) of G induces a faithful action $(G, \text{Inn}(G), \overline{\tau})$ of the inner automorphism group $\text{Inn}(G)$ of G.   □

10.6   Cores

Suppose that (X, G, λ) is an action of a group $(G, \cdot, e)$. Let x and y be elements of X that lie in the same orbit, say $\lambda_g(x) = y$ for some element g of the group G. Consider an element s of the stabilizer G_x of x in G, so

$$\lambda_s(x) = x.$$

Now

$$\lambda_{g \cdot s \cdot g^{-1}}(y) = \lambda_g \circ \lambda_s \circ \lambda_{g^{-1}}(y) = \lambda_g \circ \lambda_s(x) = \lambda_g(x) = y\,,$$

so that $g \cdot s \cdot g^{-1}$ lies in the stabilizer G_y of y. In other words, when $\lambda_g(x) = y$, the inner automorphism τ_g yields a map

$$\tau_g : G_x \to G_y; s \mapsto g \cdot s \cdot g^{-1}$$

from the stabilizer of x to the stabilizer of y. This map is a bijection, with two-sided inverse $\tau_{g^{-1}}$.

Two subgroups S and T of a group G are said to be *conjugate* if there is an inner automorphism τ_g of G with $\tau_g(S) = T$. We obtain the following:

PROPOSITION 10.40 (Conjugacy of stabilizers.)
Let (X, G, λ) be a permutation representation of a group G.

(a) *If elements x and y of X lie in the same orbit, then their respective stabilizers G_x and G_y are conjugate in G.*

(b) *The conjugate $\tau_g(G_x)$ of the stabilizer G_x of an element x of X is the stabilizer of the element $\lambda_g(x)$ of X.*

(c) *If the action (X, G, λ) is transitive, then the stabilizers G_x of elements of X form a full set of conjugate subgroups of G.*

Example 10.41
Consider the action of S_3 on $\{0, 1, 2\}$. The respective stabilizers of the points 0, 1, and 2 are the subgroups $\{(0), (1\ 2)\}$, $\{(0), (2\ 0)\}$, and $\{(0), (0\ 1)\}$. Then

$$\tau_{(0\ 1\ 2)} : (1\ 2) \mapsto (2\ 0) \mapsto (0\ 1) \mapsto (1\ 2)$$

— compare Exercise 38 in Chapter 2.   ⬚

Proposition 10.40 has an important application to homogeneous spaces.

DEFINITION 10.42 (The core of a subgroup.)   *Let H be a subgroup of a group G. Then the core of H in G is the intersection*

$$\mathrm{Core}_G(H) = \bigcap_{g \in G} \tau_g(H)$$

of the full set of conjugates of H in G.

THEOREM 10.43 (Stabilizers in homogeneous spaces.)
Let H be a subgroup of a group $(G, \cdot, e)$. Let $(G/H, G, \lambda)$ be the corresponding homogeneous space, with actions

$$\lambda_g : G/H \to G/H; x \cdot H \mapsto gx \cdot H$$

for elements g of G.

(a) *The stabilizer of H in G is H.*

(b) *For each element g of G, the stabilizer of $\lambda_g(H) = g \cdot H$ in G is $\tau_g(H)$.*

(c) *The group kernel of the representation*

$$\lambda : G \to (G/H)!$$

is the core of H in G. In particular, the core of H is a normal subgroup of G.

(d) If N is a normal subgroup of G that is contained in H, then N is also a subgroup of the core of H.

PROOF (a): An element g of G lies in the stabilizer of H if and only if $H = \lambda_g(H) = g \cdot H$, so if and only if g lies in H itself.

(b): Apply (a) and Proposition 10.40(b).

(c): Apply (b) and Corollary 10.36.

(d): If a normal subgroup N is contained in H, then $N = \tau_g(N)$ is contained in each conjugate $\tau_g(H)$ of H. ▯

Parts (c) and (d) of Theorem 10.43 are often summarized by saying that the core of H is the largest normal subgroup of G that is contained in H. Theorem 10.43 has a useful corollary.

COROLLARY 10.44 (The index of a core.)
Let G be a group, with a subgroup H of finite index r.

(a) *The index of $\mathrm{Core}_G(H)$ in G is finite.*

(b) *The index of $\mathrm{Core}_G(H)$ in G is a divisor of $r!$.*

(c) *The index of $\mathrm{Core}_G(H)$ in G is a multiple of r.*

PROOF (a): By Theorem 10.43(c) and Theorem 10.35, the homogeneous space representation $(G/H, G.\lambda)$ induces a faithful representation

$$(G/H, G/\mathrm{Core}_G(H), \overline{\lambda})$$

of the quotient group $G/\mathrm{Core}_G(H)$. Now Proposition 10.34 implies that the group $G/\mathrm{Core}_G(H)$ is isomorphic to a subgroup S of the finite group $(G/H)!$ of order $r!$.

(b): By Lagrange's Theorem, the order of S is a divisor of $r!$.

(c): By Lagrange's Theorem, $|G/\mathrm{Core}_G(H)| = r \cdot |H/\mathrm{Core}_G(H)|$. ▯

Example 10.45 (Subgroups of index 2.)
Let G be a group, and let H be a subgroup of index 2. Corollary 10.44 shows that the proper subgroup $\mathrm{Core}_G(H)$ of G has index 2, and therefore coincides with H. In other words, H is a normal subgroup of G. (Exercise 28 asks for a direct proof that a subgroup of index 2 is normal.) ▯

10.7   Alternating groups

Let n be a positive integer, and let

$$x_n > \cdots > x_1 > x_0$$

be a set of $n+1$ distinct real numbers. For α in $S_{n+1} = \{0, 1, \ldots, n\}!$, define the *sign* $\varepsilon(\alpha)$ of the permutation α to be the quotient

$$\varepsilon(\alpha) = \frac{\prod_{n \geq j > i \geq 0}(x_{\alpha(j)} - x_{\alpha(i)})}{\prod_{n \geq j > i \geq 0}(x_j - x_i)} = \prod_{n \geq j > i \geq 0} \frac{x_{\alpha(j)} - x_{\alpha(i)}}{x_j - x_i} . \tag{10.16}$$

For example, if α is the permutation $(0\ 1\ 2)$ in S_3, then

$$\varepsilon(\alpha) = \frac{(x_0 - x_2)(x_0 - x_1)(x_2 - x_1)}{(x_2 - x_1)(x_2 - x_0)(x_1 - x_0)} = 1 .$$

If α is the permutation $(0\ 1)$ in S_3, then

$$\varepsilon(\alpha) = \frac{(x_2 - x_0)(x_2 - x_1)(x_0 - x_1)}{(x_2 - x_1)(x_2 - x_0)(x_1 - x_0)} = -1 . \tag{10.17}$$

The numerator of the left-hand fraction in (10.16) includes all the factors from the denominator, either with the same or reversed order. Thus the sign of a permutation is either 1 or -1.

PROPOSITION 10.46

Let n be a positive integer. Then the sign map

$$\varepsilon : S_{n+1} \to \{\pm 1\}; \alpha \mapsto \varepsilon(\alpha) \tag{10.18}$$

is a group homomorphism from S_{n+1} onto the group of units $\{\pm 1\}$ of the monoid $(\mathbb{Z}, \cdot, 1)$ of integers.

PROOF   For permutations α and β in S_{n+1}, we have

$$\varepsilon(\alpha \circ \beta) = \prod_{n \geq j > i \geq 0} \frac{x_{\alpha \circ \beta(j)} - x_{\alpha \circ \beta(i)}}{x_j - x_i}$$

$$= \prod_{n \geq j > i \geq 0} \frac{x_{\alpha \circ \beta(j)} - x_{\alpha \circ \beta(i)}}{x_{\beta(j)} - x_{\beta(i)}} \prod_{n \geq j > i \geq 0} \frac{x_{\beta(j)} - x_{\beta(i)}}{x_j - x_i}$$

$$= \prod_{n \geq j > i \geq 0} \frac{x_{\alpha(j)} - x_{\alpha(i)}}{x_j - x_i} \prod_{n \geq j > i \geq 0} \frac{x_{\beta(j)} - x_{\beta(i)}}{x_j - x_i}$$

$$= \varepsilon(\alpha)\varepsilon(\beta) .$$

By (10.17), $\varepsilon(0\ 1) = -1$, so ε is surjective. $\Box$

DEFINITION 10.47 *Let n be a positive integer.*

(a) *The group kernel* $\mathrm{Ker}\,\varepsilon$ *of the sign homomorphism* (10.18) *is known as the* alternating group A_{n+1}.

(b) *A permutation is said to be* even *if its sign is* 1.

(c) *A permutation is said to be* odd *if its sign is* -1.

Recall that $|S_{n+1}| = (n+1)!$. By the First Isomorphism Theorem for groups applied to (10.18), it is apparent that

$$|A_{n+1}| = \frac{1}{2}(n+1)!$$

for each positive integer n.

Although the definition (10.16) of the sign is ideally suited to the proof of Proposition 10.46, it is not very practical for computation. If x and y are two distinct elements of a set X, then the cycle

$$(x\ y)$$

that interchanges x and y, while fixing all the other elements of X, is called a *transposition* (as an element of the permutation group $X!$). Transpositions in S_{n+1} are odd:

PROPOSITION 10.48 (Transpositions are odd.)
For a positive integer n, let σ be a transposition $(k\ l)$ in S_{n+1}. Then σ is an odd permutation.

PROOF Suppose that $l > k$. In order to determine $\varepsilon(\sigma)$, we examine the factor

$$\frac{x_{\sigma(j)} - x_{\sigma(i)}}{x_j - x_i} \tag{10.19}$$

from the right-hand side of (10.16) for each pair $j > i$ of integers between n and 0. If the set $\{j, i\}$ has no element in common with $\{l, k\}$, then the numerator and denominator of (10.19) are equal, canceling to 1. For the pairs $j > l$ and $j > k$, the product

$$\frac{(x_{\sigma(j)} - x_{\sigma(l)})(x_{\sigma(j)} - x_{\sigma(k)})}{(x_j - x_l)(x_j - x_k)} = \frac{(x_j - x_k)(x_j - x_l)}{(x_j - x_l)(x_j - x_k)}$$

cancels to 1. Similarly, for the pairs $l > i$ and $k > i$, the product

$$\frac{(x_{\sigma(l)} - x_{\sigma(i)})(x_{\sigma(k)} - x_{\sigma(i)})}{(x_l - x_i)(x_k - x_i)} = \frac{(x_k - x_i)(x_l - x_i)}{(x_l - x_i)(x_k - x_i)}$$

cancels to 1. For $l > j$ and $j > k$, the product

$$\frac{(x_{\sigma(l)} - x_{\sigma(j)})(x_{\sigma(j)} - x_{\sigma(k)})}{(x_l - x_j)(x_j - x_k)} = \frac{(x_k - x_j)(x_j - x_l)}{(x_l - x_j)(x_j - x_k)}$$

reduces to 1. Finally, for the pair $l > k$, the quotient

$$\frac{x_{\sigma(l)} - x_{\sigma(k)}}{x_l - x_k} = \frac{x_k - x_l}{x_l - x_k}$$

reduces to -1. Overall, $\varepsilon(\sigma) = -1$.   ⬜

COROLLARY 10.49

A permutation is even or odd if and only if it can be expressed as a product of an even or odd number of transpositions respectively.

PROOF   Suppose that a permutation α is the product of t transpositions. Then Proposition 10.48 and Proposition 10.46 show that $\varepsilon(\alpha) = (-1)^t$.   ⬜

Using Corollary 10.49, Proposition 10.46, and the fact that permutations in S_{n+1} decompose as products of disjoint cycles, it is possible to determine the sign of a given permutation more directly. Noting the equality

$$(a_0 \ a_1 \ a_2 \ \ldots \ a_r) = (a_0 \ a_r) \circ \cdots \circ (a_0 \ a_2) \circ (a_0 \ a_1) \qquad (10.20)$$

for distinct elements $a_0, a_1, \ldots, a_r$ of $\underline{\mathbf{n}}$ with $r+1 \le n$ (Exercise 31), we obtain

$$\varepsilon(a_0 \ a_1 \ \ldots \ a_r) = (-1)^r \ :$$

cycles of odd length are even, while cycles of even length are odd. If a general permutation α is the product of h cycles of respective lengths $l_1, \ldots, l_h$, the sign $\varepsilon(\alpha)$ is then given as the power

$$(-1)^{h+l_1+\cdots+l_h}$$

of -1.

Example 10.50

Consider the symmetries of the regular tetrahedron that are discussed in Study Project 2, Chapter 2. The 8 cycles of length 3 that appear in (2.33) are even, as are the 4 elements of the Klein 4-group V_4 (Example 2.31, page 38). Together, these 12 even permutations form the full alternating group A_4.   ⬜

10.8 Sylow Theorems

In the final section of this chapter, we use group actions to relate the order of a finite group with its structure.

PROPOSITION 10.51

Let p be a prime number. For positive integers a and m, let G be a group of finite order $p^a m$. (The prime p may be a divisor of m.) Let n_a be the number of subgroups of G of order p^a. Then

$$n_a \equiv 1 \mod p. \tag{10.21}$$

In particular, G does have a subgroup of order p^a.

PROOF Let X_a be the set of all the p^a-element subsets of G, subgroups or not. For each p^a-element subset S of G, define

$$\sigma_g(S) = g \cdot S$$

for g in G. Since the conditions of Proposition 10.5 are satisfied, we obtain an action (X_a, G, σ) of G on X_a. For a p^a-element subset S of G, Corollary 10.23 shows that the orbit $\sigma_G(S)$ contains $|G|/|G_S|$ elements, G_S being the stabilizer of S in the action. Now $G_S \cdot S = S$ implies that S is a union of right cosets of the subgroup G_S, so $|G_S| = p^b$ with $b \le a$.

Let x be an element of G. If s is an element of S, then x lies in $\sigma_{xs^{-1}}(S)$. In other words, for each p^a-element subset S of G, the subsets in the orbit $\sigma_G(S)$ cover all of G.

- If no two subsets in the orbit $\sigma_G(S)$ overlap, then there are exactly m subsets in the orbit. Orbits of this kind are called **nonoverlapping**. This case obtains precisely when G_S is a subgroup of G of size p^a. In particular, G_S is the unique member of the orbit that contains the identity element 1 of G. Thus the number of nonoverlapping orbits is n_a, the same as the number of subgroups of size p^a.

- Otherwise, the size $p^a m/p^b$ of the orbit $\sigma_G(S)$ exceeds m, and thus is some multiple of pm. Orbits of this kind are described as **overlapping**.

The sum of all the orbit sizes, the total number of p^a-element subsets of a $p^a m$-element set, is

$$\binom{p^a m}{p^a} \tag{10.22}$$

(compare Exercise 43 in Chapter 6). Let k be the number of p^a-element subsets that lie in overlapping orbits. The size of each overlapping orbit is congruent to 0 modulo pm, so $k \equiv 0 \mod pm$. There are n_a nonoverlapping orbits, each containing m subsets. Thus

$$\binom{p^a m}{p^a} = n_a \cdot m + k \equiv n_a \cdot m \mod pm. \tag{10.23}$$

If G is the cyclic group $\mathbb{Z}/p^a m \mathbb{Z}$, then there is a unique subgroup $\langle m + p^a m \mathbb{Z} \rangle$ of size p^a, so $n_a = 1$ in this case. Thus the binomial coefficient (10.22) is congruent to m modulo pm.

Now return to the general group G of order $p^a m$. The congruence (10.23) becomes

$$m \equiv \binom{p^a m}{p^a} \equiv n_a \cdot m + k \equiv n_a \cdot m \mod pm.$$

Since pm divides $n_a \cdot m - m = (n_a - 1) \cdot m$, we have $p \mid (n_a - 1)$. The desired congruence (10.21) follows.   □

Let x be an element of a finite group G. Lagrange's Theorem shows that the order $|\langle x \rangle|$ of x is a divisor of $|G|$. Conversely, for a divisor d of $|G|$, there may be no element x of order d. For example, the symmetric group S_3 has no element of order 6. The first consequence of Proposition 10.51 shows that for prime divisors, the situation is different.

COROLLARY 10.52 (Cauchy's Theorem.)
If p is a prime divisor of the order of a finite group G, then G contains an element of order p.

PROOF   Taking $a = 1$ in Proposition 10.51 shows that G has a subgroup H of order p. Each nonidentity element of H has order p.   □

We now study subgroups of maximal prime-power order.

DEFINITION 10.53 (Sylow subgroups.)   *Let G be a nontrivial finite group of order n. Suppose that n factorizes as a product*

$$n = p_1^{e_1} p_2^{e_2} \ldots p_r^{e_r}$$

of powers of distinct prime factors $p_1, p_2, \ldots, p_r$.

(a) *For each such factor p_i, a subgroup of G of order $p_i^{e_i}$ is called a Sylow p_i-subgroup of G.*

(b) *The set of all Sylow p_i-subgroups of G is written as $\mathrm{Syl}_{p_i}(G)$.*

THEOREM 10.54 (Sylow's First Theorem.)
Let G be a finite group, and let p be a prime factor of the order of G. Then

$$|\mathrm{Syl}_p(G)| \equiv 1 \mod p.$$

In particular, G has at least one Sylow p-subgroup.

PROOF In Proposition 10.51, take the case where m is coprime to p. ☐

Example 10.55
Consider the alternating group A_4, of order $12 = 2^2 \cdot 3$ (see Example 10.50). The Klein 4-group is the unique Sylow 2-subgroup. The 8 cycles of length 3 that appear in (2.33) pair up, together with the identity element, to form 4 Sylow 3-subgroups. Note that $4 \equiv 1 \mod 3$. ☐

THEOREM 10.56 (Sylow's Second Theorem.)
Let G be a finite group, and let p be a prime factor of the order of G.

(a) *All the Sylow p-subgroups of G are conjugate.*

(b) *If $|H| = p^a$ for a subgroup H of G, and some positive integer a, then H is contained in a Sylow p-subgroup of G.*

PROOF Let P be a particular Sylow p-subgroup of G. To prove (b), we will show that H is contained in some conjugate $\tau_g(P)$ of P. Statement (a) follows by taking H to be a Sylow p-subgroup of G. In this case, since $|H| = |P| = |\tau_g(P)|$, the containment of H in $\tau_g(P)$ implies the equality of H and $\tau_g(P)$.

To prove (b) by contradiction, suppose that H is not contained in any conjugate $\tau_g(P)$ of P. Then the intersection

$$H \cap \tau_g(P) = \{g \text{ in } G \mid g \text{ in } H \text{ and } g \text{ in } \tau_g(P)\}$$

is a proper subgroup of H, so the quotient $|H|/|H \cap \tau_g(P)|$ is divisible by p for each element g of G.

For h in H and p in P, define

$$\beta_{(h,p)}(x) = h \cdot x \cdot p^{-1}$$

for x in G. Since

$$\beta_{(k,q)}\big(\beta_{(h,p)}(x)\big) = k(hxp^{-1})q^{-1} = (kh)x(qp)^{-1} = \beta_{(k,q)(h,p)}(x)$$

for k in H and q in Q, Proposition 10.5 shows that we obtain an action $(G, H \times P, \beta)$ of $H \times P$ on G.

The orbit $\beta_{H\times P}(1)$ of the identity element 1 of G is the subset $H \cdot P$ of G. The stabilizer of 1 is

$$\{(h,p) \text{ in } H \times P \mid hp^{-1} = 1\} = \{(h,h^{-1}) \mid h \text{ in } H \cap P\}.$$

Thus

$$|H \cdot P| = |H| \cdot |P|/|H \cap P|. \tag{10.24}$$

The orbit $\beta_{H\times P}(g)$ of an element g of G is the set HgP (a *double coset* in the notation of Exercise 39 in Chapter 4). Apply Proposition 4.47, and (10.24) with $\tau_g(P)$ in place of P. The orbit $\beta_{H\times P}(g)$ is seen to have size

$$|HgP| = |HgPg^{-1}| = |H \cdot \tau_g(P)| = \frac{|H| \cdot |\tau_g(P)|}{|H \cap \tau_g(P)|},$$

a multiple of $p|P|$. Since $|G|$ is the sum of the sizes of the orbits, we obtain the contradiction that $|G|$ is a multiple of $p|P|$. ☐

COROLLARY 10.57
Let G be a finite group, and let p be a prime factor of the order of G. If P is a unique Sylow p-subgroup, then P is a normal subgroup of G.

PROOF For each element g of G, consider the conjugate $\tau_g(P)$. As a Sylow p-subgroup of G, it coincides with P. Thus P is normal. ☐

Sylow's Theorems may be used to obtain strong limitations on the number of Sylow subgroups.

PROPOSITION 10.58
Let G be a finite group, and let p be a prime factor of the order of G. Then

$$|\mathrm{Syl}_p(G)| \equiv 1 \mod p \tag{10.25}$$

and

$$|\mathrm{Syl}_p(G)| \,\Big|\, |G|. \tag{10.26}$$

PROOF The congruence (10.25) is just a restatement of Sylow's First Theorem. The divisibility relationship (10.26) follows by Corollary 10.23, since G acts transitively on $\mathrm{Syl}_p(G)$ by conjugation. ☐

Example 10.59
Let G be a group of order 33. Since 1 is the only divisor of $33 = 3 \cdot 11$ which is congruent to 1 modulo 11, Proposition 10.58 shows that there is a unique Sylow 11-subgroup of G. By Corollary 10.57, this subgroup is normal. A similar argument shows that there is a unique Sylow 3-subgroup, again a normal subgroup. ☐

10.9 Exercises

1. Let $(G, \cdot, e)$ be a group. Define a map

$$\rho_g : G \to G; x \mapsto x \cdot g^{-1}$$

for each element g of G. Show that there is a permutation representation

$$\rho : G \to G!; g \mapsto \rho_g. \tag{10.27}$$

The representation (10.27) is called the *right regular representation* of the group G.

2. Let G be a group. For elements g and h of G, show that

$$\lambda_g \circ \rho_h = \rho_h \circ \lambda_g$$

in the notation of Example 10.2 and Exercise 1.

3. Let G be a group. Show that

$$\tau_g = \lambda_g \circ \rho_g$$

for each element g of G.

4. Let $(G, \cdot, e)$ be a group. If

$$\lambda_g \circ \rho_h(e) = e$$

for elements g and h of G, show that $g = h$.

5. Let (X, G, λ) and (Y, G, μ) be actions of a group G.

   (a) Show that an action
   $$(X \times Y, G, \pi)$$
   is defined by
   $$\pi_g(x, y) = \big(\lambda_g(x), \mu_g(y)\big)$$
   for g in G, x in X, and y in Y. [The action
   $$(X \times Y, G, \pi)$$
   is known as the *product* of the actions (X, G, λ) and (Y, G, μ).]

(b) If the sets X and Y are disjoint, show that an action

$$(X \cup Y, G, \sigma)$$

is defined by

$$\sigma_g(z) = \begin{cases} \lambda_g(z) & \text{for } z \text{ in } X \\ \nu_g(z) & \text{for } z \text{ in } Y \end{cases}$$

for each g in G. [The action

$$(X \cup Y, G, \sigma)$$

is known as the *sum* of the disjoint actions (X, G, λ) and (Y, G, μ).]

6. Show that the center of a group is abelian.

7. Show that a group G is abelian if and only if $G = Z(G)$.

8. For an abelian group G, show that the conjugation action is the trivial representation of the group G on the set G.

9. Let N be a subgroup of the center of a group G.

   (a) Show that N is a normal subgroup of G.

   (b) If the quotient group G/N is cyclic, show that G is abelian.

10. Consider the symmetric group S_3.

   (a) Show that the center of S_3 is trivial.

   (b) Show that $S_3 \cong \operatorname{Inn}(S_3)$.

11. Show that the center of the general linear group $\mathrm{GL}(2, F)$ over a field F is the set

$$\left\{ \begin{bmatrix} x & 0 \\ 0 & x \end{bmatrix} \ \middle|\ 0 \neq x \text{ in } F \right\}$$

of nonzero multiples of the identity matrix.

12. Show that there is an isomorphism $G/Z(G) \cong \operatorname{Inn}(G)$ for each group G.

13. Let G be a group.

   (a) Using the notation of Example 10.2 and Exercise 1, show that the map

$$T : G \times G \to G!; (g, h) \mapsto \lambda_g \circ \rho_h$$

   is a group homomorphism.

   (b) Determine the kernel of the homomorphism T.

14. For each normal subgroup N of a group $(G, \cdot, e)$, define

$$\overline{N} = \{(g, n \cdot g) \mid g \text{ in } G, \ n \text{ in } N\}.$$

   (a) Show that $\overline{N}$ is a subgroup of $G \times G$.

   (b) Show that a normal subgroup N of G is contained in $Z(G)$ if and only if $\overline{\{e\}}$ is a normal subgroup of $\overline{N}$.

15. In Example 10.11, verify that κ gives an action of the additive group $(\mathbb{R}, +, 0)$ of real numbers on the plane $\mathbb{R}^1_2$. (Compare Exercise 34 in Chapter 4.)

16. Prove Proposition 10.14.

17. For a field F, consider the action (10.3) of $\mathrm{GL}(2, F)$ on F^1_2.

   (a) For each nonzero vector

$$\mathbf{x} = \begin{bmatrix} x_1 \\ x_2 \end{bmatrix}$$

   in F^1_2, show that there is an invertible matrix

$$A = \begin{bmatrix} a_{11} & a_{12} \\ a_{21} & a_{22} \end{bmatrix}$$

   in $\mathrm{GL}(2, F)$ such that

$$L_A \begin{bmatrix} 1 \\ 0 \end{bmatrix} = \mathbf{x}.$$

   (b) Conclude that the set of nonzero 2-dimensional column vectors forms a single orbit in the action (10.3).

18. Let H be a subgroup of a group G. Let (X, G, λ) be an action of G, with corresponding restriction (X, H, λ) to the subgroup H. Let x be an element of X.

   (a) Show that the orbit $\lambda_H(x)$ is a subset of the orbit $\lambda_G(x)$.

   (b) Show that the orbit $\lambda_G(x)$ is a disjoint union of orbits $\lambda_H(y)$.

19. Consider a permutation $\alpha = (x_1 \ x_2 \ \dots \ x_{r-1} \ x_r)$ of a finite set X. Show that a permutation γ of X is conjugate to α in the symmetric group $X!$ if and only if $\gamma = (y_1 \ y_2 \ \dots \ y_{r-1} \ y_r)$ for distinct elements $y_1, \dots, y_r$ of X. (See Exercise 38 in Chapter 2.)

20. Determine the conjugacy class of the matrix

$$\begin{bmatrix} 1 & 0 \\ 1 & 1 \end{bmatrix}$$

   in the group $\mathrm{GL}(2, \mathbb{Z}/_2)$ of invertible 2×2 matrices over $\mathbb{Z}/_2$.

21. For a field F, define the *trace* function

$$\text{tr} : F_2^2 \to F; \begin{bmatrix} a_{11} & a_{12} \\ a_{21} & a_{22} \end{bmatrix} \mapsto a_{11} + a_{22}$$

— compare (5.17). Let A and B be invertible 2×2 matrices over F.

(a) Show that $\text{tr}(AB) = \text{tr}(BA)$.

(b) If A and B are conjugate in the group $\text{GL}(2, F)$, show that $\text{tr}(A) = \text{tr}(B)$.

(c) Give an example to show that $\text{tr}(A) = \text{tr}(B)$ does not imply the conjugacy of A and B.

22. Let $(M, \cdot, e)$ be a monoid, with group of units M^*. For u in M^*, define

$$\lambda_u : M \to M; x \mapsto u \cdot m.   \tag{10.28}$$

(a) Show that (M, M^*, λ) is an action of M^* on M.

(b) Show that M is a group if and only if (M, M^*, λ) is transitive.

23. Consider the permutation representation $(\mathbb{Z}, \mathbb{Z}^*, \lambda)$ given by (10.28) for the monoid $(\mathbb{Z}, \cdot, 1)$ of integers under multiplication. Show that the orbits are the classes of the kernel relation **ker** sq of the squaring function (2.3).

24. Consider the permutation representation $(\mathbb{Z}/_8, (\mathbb{Z}/_8)^*, \lambda)$ given by (10.28) for the monoid $(\mathbb{Z}/_8, \cdot, 1)$ of integers modulo 8 under multiplication.

(a) Determine the orbits.

(b) Determine the number $\pi(u)$ of fixed points for each element u of $(\mathbb{Z}/_8)^*$.

(c) Verify (10.12) for this action.

25. Repeat Exercise 24 for the monoid $(\mathbb{Z}/_9, \cdot, 1)$ of integers modulo 9 under multiplication.

26. Let G be a finite group with just two conjugacy classes. Suppose that the order of G is n.

(a) Show that all the centralizers of nonidentity elements of G are conjugate.

(b) Show that the centralizer of a nonidentity element has order

$$n/(n - 1).$$

(c) Conclude that G has order 2.

27. Let G be a group of order 35, and let H be a subgroup of order 7. Show that H is a normal subgroup of G.

28. Let G be a group, and let H be a subgroup of index 2 in G. Give a direct proof that H is a normal subgroup of G (without invoking Corollary 10.44).

29. Let G be a group of finite order n. Suppose that n factorizes into a product

$$n = p_1^{e_1} p_2^{e_2} \dots p_k^{e_k}$$

of powers of prime numbers $p_1, \dots, p_k$, with $p_1 < p_2 < \dots < p_k$. If H is a subgroup of G of index p_1, show that H is a normal subgroup of G.

30. Show that a permutation α is odd if and only if its inverse α^{-1} is odd.

31. Verify (10.20) by showing that both sides have the same effect on each element j of $\mathbf{n}$.

32. Determine the signs of each of the following permutations:

   (a) $(0\ 1\ 2) \circ (3\ 4) \circ (5\ 6)$;

   (b) $(0\ 1\ 2\ 3\ 4\ 5) \circ (6\ 7\ 8\ 9)$;

   (c) $(0\ 1\ 2\ 3\ 4\ 5) \circ (2\ 4\ 6\ 7)$.

33. Show that the identity permutation cannot be expressed as a product

$$\sigma_1 \circ \sigma_2 \circ \dots \circ \sigma_h$$

of an odd number h of (not necessarily distinct) transpositions σ_1, σ_2, $\dots$, σ_h.

34. A nonidentity element x of a group $(G, \cdot, e)$ is an *involution* if

$$x^2 = e.$$

For a positive integer n, show that each involution in the symmetric group S_{n+1} is a product of disjoint transpositions.

35. Let G be a group in which each nonidentity element is an involution. Show that G is abelian.

36. Determine the conjugacy class of the 3-cycle $(0\ 1\ 2)$ in each of the following groups:

   (a) The symmetric group S_3;

   (b) The symmetric group S_4;

   (c) The alternating group A_4.

37. Consider an integer $n > 1$.

    (a) Show that each transposition in the symmetric group S_n may be expressed as a product of transpositions from the set

$$T_n = \{(k-1 \ k) \mid 0 < k < n\}.$$

    (b) Show that each element of the symmetric group S_n is a product of transpositions from the set T_n.

    (c) Write each element of the symmetric group S_3 as a minimal product of elements of the set

$$T_3 = \{(0 \ 1), (1 \ 2)\}.$$

    (d) Express the element $(0 \ 2)$ of the symmetric group S_3 in two different ways as a minimal product of elements of the set

$$T_3 = \{(0 \ 1), (1 \ 2)\}.$$

38. Let S be a subset of a finite group $(G, \cdot, e)$.

    (a) Show that if

$$2|S| > |G|, \qquad\qquad (10.29)$$

    then $S \cdot S = G$.

    (b) Show that the strictness of the inequality in (10.29) is essential.

39. Show that the Klein 4-group V_4 is a normal subgroup of the alternating group A_4.

40. Determine the Sylow subgroups of the symmetric group S_3.

41. Let G be a group of order 15. Show that the Sylow subgroups of G are normal.

42. Consider the set

$$G = \left\{ \begin{bmatrix} x & t \\ 0 & y \end{bmatrix} \;\middle|\; x, y, z \text{ in } \mathbb{Z}/_3, \ xy \neq 0 \right\}$$

of invertible upper triangular matrices over the ring $\mathbb{Z}/_3$ of integers modulo 3.

    (a) Show that G forms a group under matrix multiplication.

    (b) Show that $|G| = 12$.

    (c) Show that the diagonal matrices in the set G form a subgroup D of order 4.

    (d) Is D a normal subgroup of G?

    (e) Determine whether the group G is or is not isomorphic to the alternating group A_4. (Hint: Consider Sylow 2-subgroups.)

10.10   Study projects

1. **Even and odd functions.** Consider the set $C(\mathbb{R})$ of all continuous functions $f : \mathbb{R} \to \mathbb{R}$. For such a function, define

$$Tf : \mathbb{R} \to \mathbb{R}; x \mapsto f(-x).$$

Define f to be *even* if $Tf = f$, and *odd* if $Tf = -f$ (see Figure 10.4).

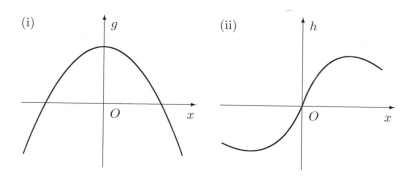

FIGURE 10.4:   (i) An even function g.   (ii) An odd function h.

(a) Show that the assignments $(0\ 1) \mapsto T$ and $(0) \mapsto \mathrm{id}_{C(\mathbb{R})}$ give an action of the symmetric group $S_2 = \{(0), (0\ 1)\}$ on $C(\mathbb{R})$.

(b) For each natural number n, show that the power function x^n is even if n is even, and odd if n is odd.

(c) Show that each continuous function f is the sum $g + h$ of an even function g and an odd function h.

(d) Suppose that a continuous function f is both even and odd. Show that f is identically zero.

(e) If $f : \mathbb{R} \to \mathbb{R}$ is a continuous odd function, show that

$$\int_{-r}^{r} f(x)dx = 0$$

for each positive real number r.

(f) If $f : \mathbb{R} \to \mathbb{R}$ is a continuous odd function, what is the value of $f(0)$? Justify your answer.

(g) If $f : \mathbb{R} \to \mathbb{R}$ is a differentiable even function, what is the value of $f'(0)$? Justify your answer.

2. **The projective line.** Let F be a field. In the plane F_2^1, each line through the origin is the set

$$X_{\mathbf{h}} = \left\{ \begin{bmatrix} x_1 \\ x_2 \end{bmatrix} \;\middle|\; h_1 x_1 + h_2 x_2 = 0 \right\}$$

of solutions

$$\mathbf{x} = \begin{bmatrix} x_1 \\ x_2 \end{bmatrix}$$

to the matrix equation $\mathbf{h}^T \mathbf{x} = 0$ or

$$\begin{bmatrix} h_1 & h_2 \end{bmatrix} \begin{bmatrix} x_1 \\ x_2 \end{bmatrix} = 0$$

for a nonzero row vector $\begin{bmatrix} h_1 & h_2 \end{bmatrix}$, the transpose $\mathbf{h}^T$ of a column vector

$$\mathbf{h} = \begin{bmatrix} h_1 \\ h_2 \end{bmatrix}.$$

The vertical line (or x_2-axis) is $X_{[1\ 0]}$. Each nonvertical line, of slope m, is $X_{[m\ -1]}$. Note that $X_{\mathbf{h}} = X_{\mathbf{k}}$ if and only if $\mathbf{h} = c\mathbf{k}$ for a nonzero constant c in F. The set of all the lines $X_{\mathbf{h}}$ through the origin is called the *projective line* or the 1-*dimensional projective space* $\mathrm{PG}(1, F)$ over the field F. In this context, a line $X_{\mathbf{h}}$ through the origin in the plane F_2^1 is called a *point* on the projective line $\mathrm{PG}(1, F)$.

(a) Consider the action $\left(F_2^1, \mathrm{GL}(2, F), L \right)$ of Example 10.4, with

$$L_A(\mathbf{x}) = A\mathbf{x}$$

for a matrix A in $\mathrm{GL}(2, F)$ and a column vector $\mathbf{x}$ in F_2^1. For each point $X_{\mathbf{h}}$ on the projective line $\mathrm{PG}(1, F)$, show that

$$\Lambda_A(X_{\mathbf{h}}) = \{ L_A(\mathbf{x}) \mid \mathbf{x} \text{ in } X_{\mathbf{h}} \} \tag{10.30}$$

is again a point on the projective line.

(b) Show that $\Lambda_A(X_{\mathbf{h}}) = X_{(A^{-1})^T \mathbf{h}}$.

(c) Show that there is an action

$$\left(\mathrm{PG}(1, F), \mathrm{GL}(2, F), \Lambda \right)$$

of $\mathrm{GL}(2, F)$ on $\mathrm{PG}(1, F)$ with Λ_A defined by (10.30).

(d) Show that the action $\left(\mathrm{PG}(1, F), \mathrm{GL}(2, F), \Lambda \right)$ is transitive.

(e) Show that a line $X_{\mathbf{h}}$ through the origin in F_2^1 is a fixed point of an invertible matrix A in the action $\left(\mathrm{PG}(1, F), \mathrm{GL}(2, F), \Lambda \right)$ if and only if the nonzero members of $X_{\mathbf{h}}$ are eigenvectors of A.

(f) For the 2-element field $F = \mathbb{Z}/_2$, determine the set of fixed points in $\mathrm{PG}(1, F)$ for each element A of $\mathrm{GL}(2, F)$.

(g) Show that $\mathrm{GL}(2, \mathbb{Z}/_2) \cong S_3$.

3. **The Class Equation.** Let $(G, \cdot, e)$ be a group of finite order n, with conjugacy classes

$$C_1 = \{e\}, C_2, \ldots, C_s.$$

Applying Proposition 10.40(b) to the conjugacy action (G, G, τ) of G, it is apparent that for $1 \le i \le s$, the various elements x of the class C_i all have conjugate centralizers $C_G(x)$. Let k_i denote the common order of the centralizers $C_G(x)$ of the elements x of C_i. In particular, note that $k_1 = n$.

(a) Show that $|C_i| = n/k_i$ for $1 \le i \le s$.

(b) Obtain the *Class Equation*

$$n = 1 + \sum_{i=2}^{s} \frac{n}{k_i} \tag{10.31}$$

or its equivalent form

$$1 = \sum_{i=1}^{s} \frac{1}{k_i}. \tag{10.32}$$

(c) Suppose that a finite group has prime power order p^e. By reading the Class Equation (10.31) modulo p, show that C_1 cannot be the only conjugacy class containing just one element. Conclude that a group of prime power order has a nontrivial center.

(d) If p is a prime number, show that a group of order p^2 is abelian.

(e) For a prime number p, consider the set

$$U_p = \left\{ \begin{bmatrix} 1 & x & z \\ 0 & 1 & y \\ 0 & 0 & 1 \end{bmatrix} \ \middle|\ x, y, z \text{ in } \mathbb{Z}/p \right\}$$

of 3×3 matrices over $\mathbb{Z}/p$. Show that under matrix multiplication, U_p forms a nonabelian group of order p^3. Determine $Z(U_p)$.

(f) Prove the following statement by induction on the positive integer s: For each positive real number r, there are only finitely many (and maybe no) decreasing sequences

$$k_1 \ge k_2 \ge \cdots \ge k_s$$

of s positive integers k_i such that

$$r = \sum_{i=1}^{s} \frac{1}{k_i}. \tag{10.33}$$

[Hint: If (10.33) holds, then $0 < k_s \le s/r$.]

(g) By considering the Class Equation in its alternative form (10.32), show that there are only finitely many possible orders n for a finite group with a fixed number s of conjugacy classes. (Hint: $k_1 = n$.)

4. **Simple groups.** A nonabelian group is said to be *simple* if it has no proper, nontrivial normal subgroups. Suppose that a group G has a normal subgroup N. Then many properties of G (such as the order) may be recovered from the smaller groups G/N and N. This kind of decomposition continues until we reach simple groups or abelian groups. Thus determination of the simple groups is an important basic step towards the study of general groups.

Consider the group A_5, of order 60.

(a) Show that A_5 contains 20 permutations of the form $(a\ b\ c)$, all of which are conjugate in A_5.

(b) Show that A_5 contains 15 permutations of the form $(a\ b)(c\ d)$, all of which are conjugate in A_5.

(c) Show that A_5 contains 24 permutations of the form $(a\ b\ c\ d\ e)$.

(d) Show that the 24 permutations of the form $(a\ b\ c\ d\ e)$ break up into two conjugacy classes in A_5, each of size 12.

(e) Show that the conjugacy classes of A_5 have respective sizes

$$1, 20, 15, 12, 12. \tag{10.34}$$

(f) Show that no sum of numbers from the list (10.34), except for 1 and 60, is a divisor of 60.

(g) Given that a normal subgroup N of A_5 is a union of conjugacy classes, with $|N| \mid 60$, show that A_5 is simple.

10.11 Notes

Section 10.1

In a permutation representation (X, G, λ), the stabilizer G_x of a point x of X is also known as the *isotropy subgroup* or *inertial subgroup*.

Section 10.4

W. Burnside was an English mathematician who lived from 1852 to 1927.

Section 10.8

A.L. Cauchy was a French mathematician who lived from 1789 to 1857. L. Sylow (or Sylov) was a Norwegian mathematician who lived from 1832 to 1918. (His name is pronounced "seal off," not "sigh low.")

Chapter 11

QUASIGROUPS

The multiplication in a group $(G, \cdot)$ satisfies two key properties. Along with associativity, there is the property discussed in Corollary 4.54 (page 86): If the equation

$$x \cdot y = z$$

holds in G, then knowledge of any two of the elements x, y, z of G specifies the third uniquely. In particular, the latter property implies that the body of the multiplication table of a finite group is a Latin square (Theorem 4.55, page 86). For many purposes, the combination of these two group properties is too strong, and it becomes necessary to consider sets that are closed under a multiplication satisfying just one of the two properties. Sets closed under an associative multiplication are semigroups. This chapter studies sets that are closed under a multiplication satisfying the second property. Such structures are known as quasigroups.

11.1 Quasigroups

DEFINITION 11.1 (Quasigroups.) *Let a given set Q be closed under a multiplication $x \cdot y$ or xy of its elements x, y. Suppose that when the equation*

$$x \cdot y = z$$

holds for elements x, y, z of Q, then knowledge of any two of x, y, z specifies the third uniquely. In this case, the structure $(Q, \cdot)$ or Q is said to be a quasigroup.

By Corollary 4.54 (page 86), each group $(G, \cdot)$ forms a quasigroup. The empty set also forms a quasigroup: There are no elements for which the closure and equation-solving conditions of Definition 11.1 have to be checked. This quasigroup is associative, since it contains no counterexamples to the associative law. The following examples give nonassociative quasigroups.

Example 11.2 (Subtraction modulo 4.)

Consider the set $\mathbb{Z}/_4$ of integers modulo 4, with the operation of subtraction. Consider the equation

$$x - y = z \tag{11.1}$$

between elements x, y, z of $\mathbb{Z}/_4$. If x and y are given, then (11.1) specifies z uniquely. If (11.1) holds, and y, z are given, then x is specified uniquely as $x = y + z$. If (11.1) holds, and x, z are given, then y is specified uniquely as $y = x - z$. Thus $(\mathbb{Z}/_4, -)$ is a quasigroup. Since

$$(1 - 1) - 1 = 3 \neq 1 = 1 - (1 - 1),$$

the associative law is not satisfied: The quasigroup is not a group. □

Arguing as in Example 11.2, it may be shown that the set of integers forms a nonassociative quasigroup under the operation of subtraction (Exercise 1).

Example 11.3 (Arithmetic means.)

Consider the set $\mathbb{R}$ of real numbers. The *arithmetic mean* of two real numbers x and y is

$$x \circ y = \frac{x + y}{2}.$$

Geometrically, the arithmetic mean represents the midpoint of the real line segment from x to y (Figure 11.1). Note that $x \circ y = y \circ x$, so $\circ$ gives a commutative multiplication on the set $\mathbb{R}$ of real numbers. Consider the equation

$$x \circ y = z \tag{11.2}$$

between real numbers x, y, and z. Certainly the arithmetic mean z is uniquely specified by x and y. If y and z are given, then x is uniquely specified as $x = 2z - y$. Similarly, y is uniquely specified by (11.2) in terms of x and z. Thus $(\mathbb{R}, \circ)$ is a quasigroup. Since

$$(0 \circ 4) \circ 8 = 2 \circ 8 = 6 \neq 3 = 0 \circ 6 = 0 \circ (4 \circ 8),$$

the associative law is not satisfied: The quasigroup is not a group. □

FIGURE 11.1: The arithmetic mean $x \circ y$ of real numbers x and y.

Example 11.4 (Nonzero bit strings.)

Fix a natural number r. The set $(\mathbb{Z}/_2)^{r+1}$ of bit strings of length $r+1$ carries the componentwise abelian group structure inherited from the group $(\mathbb{Z}/_2, +)$ of integers modulo 2 (Example 4.35, page 78). The zero element of the additive group $((\mathbb{Z}/_2)^{r+1}, +)$ is the string $000\ldots00$ of $r+1$ zeroes. Let P_r denote the subset of $(\mathbb{Z}/_2)^{r+1}$ consisting of nonzero bit strings. The set P_r has $2^{r+1} - 1$ elements, which may be considered as the binary expansions of the positive integers less than 2^{r+1}. It is often convenient to write the elements of P_r as such integers (using base ten expansions), rather than writing long bit strings. Thus P_1 becomes

$$\{01, 10, 11\} \quad \text{or} \quad \{1, 2, 3\},$$

while P_2 becomes

$$\{001, 010, 011, 100, 101, 110, 111\} \quad \text{or} \quad \{1, 2, 3, 4, 5, 6, 7\}.$$

Now define a multiplication $*$ on P_r by

$$x * y = \begin{cases} x & \text{if } x = y; \\ x + y & \text{if } x \neq y. \end{cases} \tag{11.3}$$

This means that in the equation

$$x * y = z$$

for elements x, y, z of P_r, either $x = y = z$ or

$$x + y + z = 0$$

in the group $((\mathbb{Z}/_2)^{r+1}, +)$. At any rate, the *bit string quasigroup* $(P_r, *)$ is a quasigroup for each natural number r, and is not a group for $r > 0$ (Exercise 4). □

11.2   Latin squares

By definition, quasigroups satisfy the property (Corollary 4.54, page 86) ensuring that the body of a finite group multiplication table is a Latin square. Thus the body of the multiplication table of a finite (nonempty) quasigroup will also be a Latin square. For illustration, the multiplication table of the quasigroup $(\mathbb{Z}/_4, -)$ of Example 11.2 is shown in Figure 11.2. Note that each of the four integers modulo 4 appears exactly once in each row and each column of the body of the table.

$-$	0	1	2	3
0	0	3	2	1
1	1	0	3	2
2	2	1	0	3
3	3	2	1	0

FIGURE 11.2:   Subtraction modulo 4.

In Figure 11.2, it is possible to change the column labels, making the table that results present the addition operation for the group of integers modulo 4 (Exercise 6). In general, however, there may be no way to label a Latin square so that it forms the body of a group multplication table. (In Section 11.7, it will be shown that the smallest such Latin square has size 5×5.) But as the following theorem indicates, a Latin square always gives a quasigroup table.

THEOREM 11.5 (Latin squares as quasigroup tables.)
A finite, nonempty set Q with a multiplication $\cdot$ forms a quasigroup $(Q, \cdot)$ if and only if the body of the multiplication table of $(Q, \cdot)$ forms a Latin square.

PROOF   If $(Q, \cdot)$ is a finite, nonempty quasigroup, then the argument showing that the body of the multiplication table of $(Q, \cdot)$ forms a Latin square is exactly the same as the argument used to prove Theorem 4.55 (page 86) — compare Exercise 7.

Conversely, suppose that Q has n elements, say $x_1, \ldots, x_n$ in a certain order. Consider the multiplication table of $(Q, \cdot)$, with the rows and columns each labeled by the elements $x_1, \ldots, x_n$ in order. Suppose that the body of the table forms a Latin square. Note that for $1 \le i, j \le n$, the table entry appearing in the row labeled x_i and the column labeled x_j is $x_i \cdot x_j$ (Figure 11.3). Suppose that the equation

$$x_i \cdot x_j = x_k \tag{11.4}$$

holds in $(Q, \cdot)$, with $1 \le i, j, k \le n$. If x_i and x_j are given, then x_k is specified uniquely in (11.4) by the multiplication $\cdot$ defined on Q. Now suppose that x_i and x_k are given in Q, so that (11.4) holds. Consider the table row labeled by the element x_i. Since the body of the table is a Latin square, this row contains the element x_k exactly once. Let x_j be the label of the column in which this table entry x_k appears. Then x_j is the unique solution of (11.4) for the given x_i and x_k. In similar fashion, it may be shown that (11.4) has a unique solution x_i for given x_j and x_k (Exercise 8). Thus $(Q, \cdot)$ is a quasigroup.   □

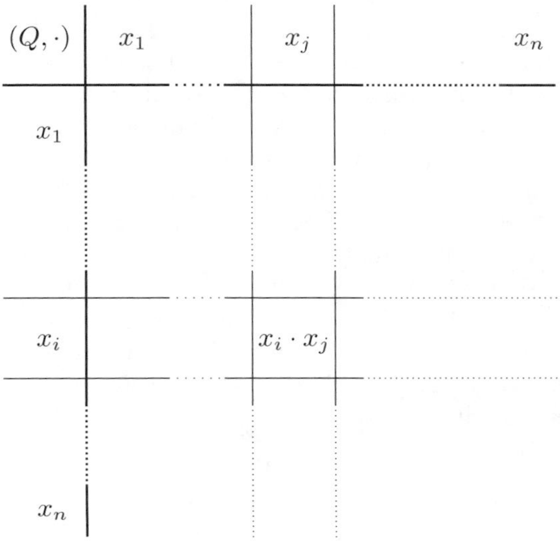

FIGURE 11.3: A multiplication table.

Theorem 11.5 provides a direct way to construct finite quasigroups. We first build a Latin square, and then make it into a quasigroup multiplication table by providing it with row and column labels taken from the set of entries of the square (Figure 11.4). Although it is customary and convenient to use these labels in a specified order (for example, increasing numerical order as in Figure 11.4), the labels may be applied in any order to give a quasigroup multiplication table.

1	3	2	5	6	4
3	2	1	6	4	5
2	1	3	4	5	6
4	5	6	1	2	3
5	6	4	2	3	1
6	4	5	3	1	2

$\cdot$	1	2	3	4	5	6
1	1	3	2	5	6	4
2	3	2	1	6	4	5
3	2	1	3	4	5	6
4	4	5	6	1	2	3
5	5	6	4	2	3	1
6	6	4	5	3	1	2

FIGURE 11.4: From a Latin square to a quasigroup multiplication table.

There are various approaches to the construction of Latin squares. We may certainly take the body of any finite, nonempty group or quasigroup multiplication table. Another approach is to build up the Latin square by gradually adding entries until the square is complete. At each stage, we must make sure that no entries are repeated in any row or column of the partial table. At the same time, we may take advantage of the fact that each element has to appear somewhere in each row and column. Sometimes, this procedure will stall: There will be no way to complete the partial table to a Latin square. If this happens, we have to withdraw a step or two, and then try again.

As an example, consider the problem of building up a 4×4 Latin square using the set $\{0, 1, 2, 3\}$ of integers modulo 4. Suppose that we have reached the following partial square, which seems to be acceptable since there are no repeated elements in any row or column:

0		1	
	2		
			2

The elements 2 and 3 have to appear somewhere in the first row. The element 2 cannot appear in the second column of the first row, since 2 already appears in the second row of the second column. This forces 3 to be the second entry in the first row, leaving 2 to appear in the final entry of the first row:

0	3	1	2
	2		
			2

Now we encounter the problem that the element 2 is appearing twice in the final column. To avoid this problem, we may go back to the original square and withdraw the element 2 from the third row of the final column:

0		1	
	2		

Arguing as before, we reach the square

0	3	1	2
2			

which may then be completed to a full Latin square, say

0	3	1	2
1	2	3	0
2	1	0	3
3	0	2	1

for instance.

11.3   Division

In semigroups, such as the multiplicative structures of fields and rings, division is not always possible. For example, we cannot divide by the element 2 within the ring $\mathbb{Z}$ of integers, nor by 0 in the field $\mathbb{R}$ of real numbers. On the other hand, quasigroups are defined so that division is always possible. In fact, there are two forms of division in a quasigroup: from the left, and from the right.

DEFINITION 11.6 (Quasigroup divisions.) *Let $(Q, \cdot)$ be a quasigroup. Consider elements x and y of Q.*

(a) *The element $x \backslash y$ of Q is defined as the unique solution z of the equation $x \cdot z = y$. In other words,*

$$x \cdot (x \backslash y) = y . \qquad (11.5)$$

The element $x \backslash y$ may be read as "x dividing y" or "x backslash y." The operation $\backslash$ on the set Q is known as left division *in the quasigroup $(Q, \cdot)$.*

(b) *The element x/y of Q is defined as the unique solution z of the equation $z \cdot y = x$. In other words,*

$$(x/y) \cdot y = x . \qquad (11.6)$$

The element x/y may be read as "x divided by y" or "x slash y." The operation $/$ on the set Q is known as right division *in the quasigroup $(Q, \cdot)$.*

Introduction to Abstract Algebra

Example 11.7 (Subtraction as a right division.)
Consider the abelian group $(\mathbb{Z}, +)$ of integers as a quasigroup, in which the "quasigroup multiplication" is given by addition. Equation (11.6) becomes

$$(x/y) + y = x \,.$$

Thus the right division x/y in the quasigroup $(\mathbb{Z}, +)$ is just the subtraction $x - y$. More generally, subtraction is right division in any additive group. □

Example 11.8 (Division in bit string quasigroups.)
In the bit string quasigroups $(P_r, *)$ of Example 11.4,

$$x * (x * y) = y \quad \text{and} \quad (x * y) * y = x \tag{11.7}$$

(Exercise 12). Thus in this case, $x \backslash y = x * y = x/y$: The two division operations are the same as the quasigroup multiplication. □

Example 11.9 (Divisions in groups.)
Let $(G, \cdot)$ be a group, considered as a quasigroup. Then

$$x \backslash y = x^{-1}y \quad \text{and} \quad x/y = xy^{-1} \tag{11.8}$$

(Exercise 13). It is worth recalling the role of the right division in a group, as the single operation that is used in the subgroup test (Proposition 4.43, page 80; compare Remark 4.44). □

Example 11.10 (Reflection as a quasigroup division.)
Consider the arithmetic mean quasigroup structure $(\mathbb{R}, \circ)$ on the real line, as given in Example 11.3. In solving (11.2) for x in terms of y and z, it was shown there that

$$z/y = 2z - y \,.$$

This operation of right division in the arithmetic mean quasigroup has a geometrical interpretation, as the reflection of y in a mirror located at z (Figure 11.5). □

FIGURE 11.5: The reflection z/y of y in a mirror at z.

PROPOSITION 11.11 (Properties of divisions.)
Let $(Q, \cdot)$ be a quasigroup. Consider elements x and y of Q.

(a) $x\backslash(x \cdot y) = y$.

(c) $(y \cdot x)/x = y$.

(b) $x = y/(x\backslash y)$.

(d) $x = (y/x)\backslash y$.

PROOF Suppose $x \cdot y = z$. Recall that since $(Q, \cdot)$ is a quasigroup, the element y is the unique solution t in Q to the equation

$$x \cdot t = z . \tag{11.9}$$

On the other hand, (11.5) — with z written in place of y — shows that $x\backslash z$ is also a solution t to (11.9). Thus $x\backslash z = y$, as required for (a). Moreover, since $x \cdot (x\backslash z) = z$, the equation $x = z/(x\backslash z)$ follows by (11.6). Replacing z by y yields (b). For (c) and (d), which are similar, see Exercise 14. $\quad\square$

Proposition 11.11 yields a new characterization of quasigroups, often more convenient than Definition 11.1. (The standard labeling of the equations in the following proposition is justified in Study Project 3.)

PROPOSITION 11.12 (Characterization of quasigroups.)
A set Q forms a quasigroup $(Q, \cdot)$ under a multiplication $\cdot$ if and only if it is equipped with a left division $\backslash$ and a right division $/$ such that

(SL) $x \cdot (x\backslash y) = y$.

(SR) $(x/y) \cdot y = x$.

(IL) $x\backslash(x \cdot y) = y$.

(IR) $(y \cdot x)/x = y$.

for all x, y in Q.

PROOF If $(Q, \cdot)$ is a quasigroup, then (SL) and (SR) are the respective defining equations (11.5) and (11.6) for the left and right divisions, while Proposition 11.11(a) and (c) provide (IL) and (IR).

Conversely, suppose that Q is equipped with operations $\cdot$, $\backslash$, and $/$ satisfying the identities (SL), (IL), (SR), and (IR) of the proposition. Consider the equation

$$x \cdot y = z \tag{11.10}$$

for elements x, y, z of Q. If y and z are fixed, the equation (SR) gives $(z/y) \cdot y = z$, so that $x = z/y$ is a solution to (11.10). If s and t are solutions, (IR) gives $s = (s \cdot y)/y = z/y = (t \cdot y)/y = t$, so the solution is unique. Similar use of (SL) and (IL) shows that (11.10) has a unique solution for y when x and z are fixed (Exercise 15). Thus $(Q, \cdot)$ is a quasigroup. $\quad\square$

One further consequence of Proposition 11.11 is that the left and right divisions provide new quasigroup multiplications.

THEOREM 11.13 (Divisions as quasigroup multiplications.)
Let $(Q, \cdot)$ be a quasigroup, with left division $\backslash$ and right division $/$. Then $(Q, \backslash)$ and $(Q, /)$ are quasigroups.

PROOF   Consider the equation

$$x \backslash y = z \qquad\qquad (11.11)$$

involving elements x, y, z of Q. If x and y are given, then z is determined directly and uniquely by (11.11). Now suppose that y and z are given. By Proposition 11.11(d), the equation (11.11) has a solution $x = y/z$. But if $t \backslash y = z$ as well as $x \backslash y = z$, then Proposition 11.11(b) gives

$$x = y/z = y/(t \backslash y) = t \, ,$$

so the solution is unique. Similar arguments show that (11.11) has a unique solution for y in terms of x and z, so that $(Q, \backslash)$ becomes a quasigroup. Again, similar arguments show that $(Q, /)$ is a quasigroup (Exercise 16). ☐

Example 11.14 (Subtraction modulo 4, revisited.)
Theorem 11.13 gives an immediate proof for the content of Example 11.2, showing that the set $\mathbb{Z}/4$ of integers modulo 4 forms a quasigroup under subtraction. As noted in Example 11.7, subtraction is the right division for the addition in any additive group like $(\mathbb{Z}/4, +)$. ☐

Another way to obtain new quasigroups is by reversing given quasigroup multiplications. Suppose that a multiplication $\cdot$ is given on a set Q. Then the *opposite* multiplication $\circ$ is defined by

$$x \circ y = y \cdot x$$

for elements x and y of Q. The following result is readily checked (Exercise 18).

PROPOSITION 11.15
If $(Q, \cdot)$ is a quasigroup, then so is its opposite $(Q, \circ)$.

DEFINITION 11.16 (Conjugates of a quasigroup.)   *Let $(Q, \cdot)$ be a quasigroup. Then the conjugates of $(Q, \cdot)$ are the quasigroups $(Q, \cdot)$ itself, its opposite $(Q, \circ)$, the quasigroups $(Q, \backslash)$ and $(Q, /)$ of Theorem 11.13, and their respective opposites $(Q, \backslash\backslash)$ and $(Q, //)$.*

11.4   Quasigroup homomorphisms

Just like rings, groups, and semigroups, quasigroups are abstract algebras in their own right. As such, they come equipped with their corresponding concepts of substructure, homomorphism, and product.

DEFINITION 11.17 (Subquasigroups.)   *Let $(Q, \cdot)$ be a quasigroup. Then a* subquasigroup S *of* $(Q, \cdot)$ *is a subset of Q which forms a quasigroup* $(S, \cdot)$ *under the multiplication $\cdot$ of* $(Q, \cdot)$.

Example 11.18 (Subquasigroups of bit string quasigroups.)
The bit string quasigroup P_1 of Example 11.4, taken on the subset

$$\{1, 2, 3\} = \{001, 010, 011\}$$

of the underlying set

$$\{1, 2, 3, 4, 5, 6, 7\} = \{001, 010, 011, 100, 101, 110, 111\}$$

of the quasigroup $(P_2, *)$, forms a subquasigroup of $(P_2, *)$. The various 3-element subquasigroups of $(P_2, *)$ are displayed by the straight lines and curved line in Figure 11.6.   ⬜

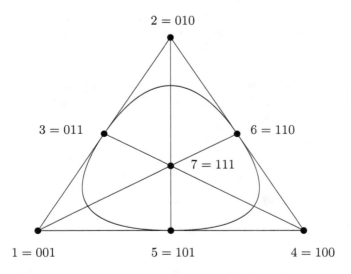

FIGURE 11.6:   Subquasigroups of the bit string quasigroup P_2.

Subquasigroups of a quasigroup are characterized by closure under the three operations of multiplication, left division, and right division.

PROPOSITION 11.19 (The subquasigroup test.)
Let S be a subset of a quasigroup $(Q, \cdot)$. Then S is a subquasigroup of $(Q, \cdot)$ if and only if

$$x \cdot y, \quad x \backslash y, \quad and \quad x/y \quad lie \ in \quad S \tag{11.12}$$

for all x, y in S.

PROOF   If S is closed under the three operations of multiplication, left division, and right division, then Proposition 11.12 shows that S forms a quasigroup under the multiplication of $(Q, \cdot)$. Conversely, suppose that $(S, \cdot)$ is a quasigroup. In particular, $x \cdot y$ lies in S for x, y in S. Moreover, for given y and z in S, the unique solution x to

$$x \cdot y = z \tag{11.13}$$

in S must agree with the unique solution x to (11.13) in Q. This solution is z/y. Thus S is closed under the right division. A similar argument shows that S is also closed under the left division (Exercise 23).   ◻

In a quasigroup $(Q, \cdot)$, closure of a subset S under the multiplication alone is generally insufficient to make S a subquasigroup. For example, the set $\mathbb{N}$ of natural numbers is closed under the addition operation in the group $(\mathbb{Z}, +)$ of integers, but $\mathbb{N}$ does not form a subquasigroup there. For a natural "nonassociative" example, see Exercise 24.

DEFINITION 11.20 (Quasigroup homomorphism, isomorphism.)
Suppose that $(P, *)$ and $(Q, \circ)$ are quasigroups.

(a) *A function $f : P \rightarrow Q$ is a* quasigroup homomorphism, *denoted by $f : (P, *) \rightarrow (Q, \circ)$, if*

$$f(x) \circ f(y) = f(x * y) \tag{11.14}$$

   for all x, y in P.

(b) *If a quasigroup homomorphism $f : (P, *) \rightarrow (Q, \circ)$ is bijective, it is called an* isomorphism.

(c) *Quasigroups P and Q are* isomorphic *(written $P \cong Q$) if there is an isomorphism between them.*

Example 11.21 (Inclusion of a subquasigroup.)
Let P be a subquasigroup of a quasigroup $(Q, \cdot)$. Then the inclusion function

$$j : P \hookrightarrow Q; x \mapsto x$$

is a quasigroup homomorphism $j : (P, \cdot) \to (Q, \cdot)$. ▯

The following result is an analogue of Proposition 5.5 (page 96).

PROPOSITION 11.22 (Homomorphisms respect division.)
Let $f : (P, \cdot) \to (Q, \cdot)$ be a quasigroup homomorphism, between quasigroups
with $\backslash$ and $/$ as the respective divisions. Then

$$f(x)\backslash f(y) = f(x\backslash y) \quad and \quad f(x)/f(y) = f(x/y)$$

for x and y in P.

PROOF Since $x \cdot (x\backslash y) = y$ in $(P, \cdot)$, the fact that f is a quasigroup
homomorphism implies

$$f(x) \cdot f(x\backslash y) = f(y)$$

in Q. However, the unique solution z to the equation $f(x) \cdot z = f(y)$ in Q
is $z = f(x)\backslash f(y)$. Thus $f(x)\backslash f(y) = f(x\backslash y)$, as required. The proof that
$f : P \to Q$ preserves right divisions is similar (Exercise 26). ▯

PROPOSITION 11.23 (Componentwise quasigroup structure.)
Let $(P, \cdot)$ and $(Q, \cdot)$ be quasigroups. Then the product set $P \times Q$, equipped
with the componentwise multiplication

$$(x_1, x_2) \cdot (y_1, y_2) = (x_1 \cdot y_1, x_2 \cdot y_2),$$

forms a quasigroup $(P \times Q, \cdot)$. In this quasigroup, the left division

$$(x_1, x_2)\backslash(y_1, y_2) = (x_1\backslash y_1, x_2\backslash y_2)$$

and right division

$$(x_1, x_2)/(y_1, y_2) = (x_1/y_1, x_2/y_2)$$

are given componentwise in terms of the left and right divisions on the factors
P and Q.

PROOF Consider the equation

$$(x_1, x_2) \cdot (y_1, y_2) = (z_1, z_2)$$

in $P \times Q$. For given (y_1, y_2) and (z_1, z_2) in $P \times Q$, i.e., for given y_1, z_1 in P and y_2, z_2 in Q, there is a unique solution

$$(x_1, x_2) = (z_1/y_1, z_2/y_2)$$

in $P \times Q$. The other verifications are similar (Exercise 29).   □

DEFINITION 11.24 (Product quasigroups.) *For quasigroups $(P, \cdot)$ and $(Q, \cdot)$, the quasigroup $(P \times Q, \cdot)$ of Proposition 11.23 is known as the product of the quasigroups $(P, \cdot)$ and $(Q, \cdot)$.*

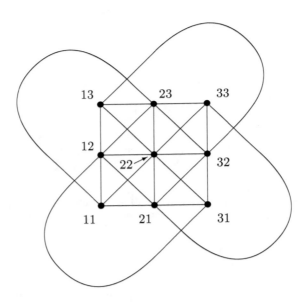

FIGURE 11.7:   Subquasigroups of the product quasigroup $P_1 \times P_1$.

Example 11.25 (A product of bit string quasigroups.)
Consider the bit string quasigroup $P_1 = \{1, 2, 3\}$. The product quasigroup $P_1 \times P_1$ is displayed in Figure 11.7. Note that ordered pairs such as $(3, 1)$ are written simply as juxtaposed digits like 31. The straight and curved lines denote the 3-element subquasigroups, e.g., $\{11, 23, 32\}$ curving round the top left-hand corner. Compare with Figure 11.6.   □

11.5 Quasigroup homotopies

The previous section discussed various algebraic concepts associated with quasigroups: substructures, homomorphisms, and products. These concepts are very similar to the analogous concepts for semigroups, groups, rings, and the other kinds of algebra encountered in this book. For quasigroups, however, it turns out that the concepts of homomorphism and isomorphism do not tell the full story. Consider the quasigroup $(\mathbb{Z}/_3, -)$ of integers modulo 3 under subtraction. Its full multiplication table is presented in Figure 11.8.

$-$	0	1	2
0	0	2	1
1	1	0	2
2	2	1	0

FIGURE 11.8: Subtraction modulo 3.

The quasigroup is not associative, for example

$$(0-2)-1 = 1-1 = 0 \neq 2 = 0-1 = 0-(2-1).$$

Now consider the bijective function

$$g : \mathbb{Z}/_3 \to \mathbb{Z}/_3; x \mapsto -x \qquad (11.15)$$

of negation modulo 3. Applying this permutation to the column labels in Figure 11.8 yields the addition table for the group $(\mathbb{Z}/_3, +)$ of integers modulo 3, displayed (with an unconventional ordering of the columns) in Figure 11.9.

$+$	0	2	1
0	0	2	1
1	1	0	2
2	2	1	0

FIGURE 11.9: Addition modulo 3.

Note that the bodies of the multiplication tables in Figures 11.8 and 11.9 consist of the same 3×3 Latin square, even though the second quasigroup is associative, while the first is not. Since the associativity property is preserved by quasigroup isomorphisms, it is apparent that the same Latin square has furnished multiplication tables for quasigroups which are not isomorphic.

Phenomena such as these are captured by the following definition, which may be contrasted with Definition 11.20.

DEFINITION 11.26 (Quasigroup homotopy, isotopy.) *Suppose that* $(P, *)$ *and* $(Q, \circ)$ *are quasigroups.*

(a) *A triple* (f, g, h) *of functions* $f : P \to Q$, $g : P \to Q$, *and* $h : P \to Q$ *is a quasigroup* homotopy, *denoted by*

$$(f, g, h) : (P, *) \to (Q, \circ),$$

if

$$f(x) \circ g(y) = h(x * y) \tag{11.16}$$

for all x, y *in* P.

(b) *The functions* f, g, h *of* (a) *appearing in the homotopy* (f, g, h) *are known as the* components *of the homotopy.*

(c) *If the components of a quasigroup homotopy*

$$(f, g, h) : (P, *) \to (Q, \circ)$$

are bijective, the homotopy is described as an isotopy.

(d) *Quasigroups* P *and* Q *are said to be* isotopic *(written* $P \sim Q$) *if there is an isotopy between them.*

Example 11.27 (Integers modulo 3.)
Consider the set $\mathbb{Z}/3$ of integers modulo 3. Define functions f and h to be the identity function on the set $\mathbb{Z}/3$. With $g : \mathbb{Z}/3 \to \mathbb{Z}/3$ as the negation (11.15) modulo 3, we have

$$f(x) + g(y) = h(x - y)$$

for integers x, y modulo 3. Thus

$$(f, g, h) : (\mathbb{Z}/3, -) \to (\mathbb{Z}/3, +) \tag{11.17}$$

is an isotopy.

The distinction between homotopy and homomorphism may be seen on comparing (11.16) with (11.14). In particular:

PROPOSITION 11.28 (When homotopies are homomorphisms.)
Suppose that $(P, *)$ and $(Q, \circ)$ are quasigroups.

(a) *A quasigroup homotopy $(f, g, h) : (P, *) \to (Q, \circ)$ is a homomorphism if its three components agree: $f = g = h$.*

(b) *Each quasigroup homomorphism $f : (P, *) \to (Q, \circ)$ forms a homotopy $(f, f, f) : (P, *) \to (Q, \circ)$.*

Composites of homotopies are homotopies.

PROPOSITION 11.29 (Composites of homotopies.)
Suppose that $(N, \cdot)$, $(P, *)$, and $(Q, \circ)$ are quasigroups, with homotopies $(f, g, h) : (P, *) \to (Q, \circ)$ and $(f', g', h') : (N, \cdot) \to (P, *)$. Then

$$(f \circ f', g \circ g', h \circ h') : (N, \cdot) \to (Q, \circ)$$

is again a quasigroup homomorphism.

PROOF By respective use of the homotopy property (11.16) of (f, g, h) and (f', g', h'), we have

$$(f \circ f')(x) \circ (g \circ g')(y) = f\big(f'(x)\big) \circ g\big(g'(y)\big)$$
$$= h\big(f'(x) * g'(y)\big)$$
$$= h\big(h'(x \cdot y)\big) = (h \circ h')(x \cdot y)$$

for x, y in N. ⬚

COROLLARY 11.30 (Isotopy as an equivalence relation.)
Isotopy forms an equivalence relation on any set of quasigroups.

PROOF By Proposition 11.28(b), the identity map id_Q on a quasigroup Q forms an isotopy. Thus the relation of isotopy is reflexive. Now suppose that $N \sim P$ and $P \sim Q$ for quasigroups N, P, and Q. By Proposition 11.29, the composite of respective isotopies from N to P and from P to Q is an isotopy from N to Q, so the relation of isotopy is transitive. Finally, suppose that $(f, g, h) : (P, *) \to (Q, \circ)$ is an isotopy. Consider elements x and y in Q, say with $f(x') = x$ and $g(y') = y$ for unique elements x', y' of P. By (11.16), we have

$$h(x' * y') = f(x') \circ g(y') = x \circ y.$$

Thus $h^{-1}(x \circ y) = x' * y' = f^{-1}(x) * g^{-1}(y)$, as required to show that $\big(f^{-1}, g^{-1}, h^{-1}\big) : (Q, \circ) \to (P, *)$ is an isotopy. It follows that the relation of isotopy is symmetric. ⬚

11.6 Principal isotopy

In order to simplify the concept of isotopy as much as possible, the following definition is useful.

DEFINITION 11.31 (Principal isotopy.)

(a) *A quasigroup isotopy*

$$(f, g, h) : (P, *) \to (Q, \circ)$$

between quasigroups $(P, *)$ and $(Q, \circ)$ is said to be a principal isotopy *if its third component h is the identity map $\mathrm{id}_P : P \to P$ on the set P (and thus in particular, if the domain set P and codomain set Q coincide).*

(a) *Two quasigroup structures $(Q, *)$ and $(Q, \circ)$ on a common underlying set Q are said to be* principally isotopic *if there is a principal isotopy $(f, g, \mathrm{id}_Q) : (Q, *) \to (Q, \circ)$.*

Example 11.32
The isotopy (11.17) of Example 11.27 is a principal isotopy. □

To within isomorphism, every isotopy is principal:

PROPOSITION 11.33 (Factorizing an isotopy.)
Consider a quasigroup isotopy $(f, g, h) : (P, *) \to (Q, \circ)$. Use the bijection $h : P \to Q$ to induce a multiplication

$$x \circ y = h^{-1}\big(h(x) \circ h(y)\big)$$

for x, y in P.

(a) *The structure $(P, \circ)$ is a quasigroup.*

(b) *There is an isomorphism $h : (P, \circ) \to (Q, \circ)$.*

(c) *The isotopy (f, g, h) factorizes as the composite*

$$(f, g, h) = (h, h, h) \circ \big(h^{-1} \circ f, h^{-1} \circ g, \mathrm{id}_P\big)$$

of the principal isotopy $\big(h^{-1} \circ f, h^{-1} \circ g, \mathrm{id}_P\big) : (P, *) \to (P, \circ)$ together with the isomorphism $h : (P, \circ) \to (Q, \circ)$.

Verification of the straightforward details in Proposition 11.33 is assigned as Exercise 39. The composite isotopy in Proposition 11.33(c) may be expressed symbolically as $(P, *) \sim (P, \circ) \cong (Q, \circ)$.

Principal isotopy clarifies the relationship between the various quasigroups obtained with a given Latin square as the body of their multiplication table.

THEOREM 11.34 (Bordering a Latin square.)
Let Q be a finite set. Then two quasigroups $(Q, *)$ and $(Q, \cdot)$ share a Latin square $L(Q)$ built on Q as the common body of their multiplication tables if and only if they are related by a principal isotopy $(f, g, \mathrm{id}_Q) : (Q, *) \to (Q, \cdot)$.

PROOF Suppose that Q has n elements $x_1, \ldots, x_n$. First, suppose that $(Q, *)$ and $(Q, \cdot)$ share a Latin square $L(Q)$ built on Q as the common body of their multiplication tables. In other words, there are permutations r', c', r, and c of the set Q such that

$*$	$c(x_1)$	$\cdots$	$c(x_n)$
$r(x_1)$			
$\vdots$		$L(Q)$	
$r(x_n)$			

is a multiplication table of $(Q, *)$ and

$\cdot$	$c'(x_1)$	$\cdots$	$c'(x_n)$
$r'(x_1)$			
$\vdots$		$L(Q)$	
$r'(x_n)$			

is a multiplication table of $(Q, \cdot)$. Then for ξ, η in Q, we have

$$r'(\xi) \cdot c'(\eta) = r(\xi) * c(\eta).$$

Substituting $\xi = r^{-1}(x)$ and $\eta = c^{-1}(y)$, we obtain

$$r'(r^{-1}(x)) \cdot c'(c^{-1}(y)) = x * y \qquad (11.18)$$

for x, y in Q. Define new permutations $f = r' \circ r^{-1}$ and $g = c' \circ c^{-1}$ of Q. The equation (11.18) becomes $f(x) \cdot g(y) = x * y$ for x, y in Q, yielding the principal isotopy $(f, g, \mathrm{id}_Q) : (Q, *) \to (Q, \cdot)$.

Conversely, suppose there is a principal isotopy $(f, g, \mathrm{id}_Q) : (Q, *) \to (Q, \cdot)$. Thus

$$f(x) \cdot g(y) = x * y \qquad (11.19)$$

for elements x, y of Q. Let $L(Q)$ be the Latin square on Q which forms the body of the multiplication table

*	x_1		x_n
x_1			
$\vdots$		$L(Q)$	
x_n			

of $(Q, *)$. Then by (11.19), the multiplication table of $(Q, \cdot)$ is

$\cdot$	$g(x_1)$		$g(x_n)$
$f(x_1)$			
$\vdots$		$L(Q)$	
$f(x_n)$			

Thus $(Q, *)$ and $(Q, \cdot)$ share the Latin square $L(Q)$ as the common body of their multiplication tables. $\quad\square$

11.7 Loops

Semigroups with an identity element are monoids. Quasigroups with an identity element are called loops.

DEFINITION 11.35 (Loops, identity element.) *A quasigroup $(Q, \cdot)$ is said to be a* loop *if it contains an element e such that*

$$e \cdot x = x = x \cdot e$$

for all elements x of Q. The element e of Q is called the identity element of *the loop $(Q, \cdot, e)$.*

Groups are certainly loops. Although it is not easy to find natural examples of nonassociative loops, each Latin square is the body of a multiplication table of a loop.

PROPOSITION 11.36 (Latin squares are loop tables.)

Let Q be a finite, nonempty set. Let $L(Q)$ be a Latin square built from the elements of Q. Then $L(Q)$ is the body of the multiplication table of a loop $(Q, \cdot, e)$ on the underlying set Q.

PROOF Suppose the Latin square is

$$L(Q) = \begin{bmatrix} x_{11} & x_{12} & \cdots & x_{1n} \\ x_{21} & x_{22} & \cdots & x_{2n} \\ \vdots & \vdots & & \vdots \\ x_{n1} & x_{n2} & \cdots & x_{nn} \end{bmatrix},$$

so that the n-element set Q is given as

$$Q = \{x_{11}, x_{21}, \ldots, x_{n1}\} = \{x_{11}, x_{12}, \ldots, x_{1n}\}.$$

Then the bordered version

$\cdot$	x_{11}		x_{1n}
x_{11}	x_{11}	$\cdots$	x_{1n}
$\vdots$	$\vdots$	$L(Q)$	$\vdots$
x_{n1}	x_{n1}	$\cdots$	x_{nn}

of the Latin square $L(Q)$ is the multiplication table of a loop $(Q, \cdot, x_{11})$ on the underlying set Q, with x_{11} as the identity element. ☐

Example 11.37 (Subtraction and addition modulo 3.)

Take Q to be the set of integers modulo 3, and take $L(Q)$ to be the body of the table of $(\mathbb{Z}/3, -)$, as illustrated in Figure 11.8. Then the addition table modulo 3, as displayed in Figure 11.9, exhibits the construction of the proof of Proposition 11.36. ☐

Proposition 11.36 shows that each finite, nonempty quasigroup is principally isotopic to a loop (Exercise 40). However, there is a more direct and general argument.

THEOREM 11.38 (Quasigroups are isotopic to loops.)
Let $(Q, \cdot)$ be a nonempty quasigroup, with left division $\backslash$ and right division $/$. Let a and b be elements of Q. Define a new multiplication $\circ$ on the set Q by

$$x \circ y = (x/b) \cdot (a\backslash y) \qquad (11.20)$$

for x, y in Q. Then $(Q, \circ, a \cdot b)$ is a loop that is principally isotopic to the quasigroup $(Q, \cdot)$.

PROOF By Proposition 11.11(a) and (11.6), we have

$$x \circ (a \cdot b) = (x/b) \cdot \big(a\backslash(a \cdot b)\big) = (x/b) \cdot b = x$$

for x in Q. Similarly, by Proposition 11.11(c) and (11.5), we have

$$(a \cdot b) \circ x = \big((a \cdot b)/b\big) \cdot (a\backslash x) = a \cdot (a\backslash x) = x$$

for x in Q. Thus $(Q, \circ, a \cdot b)$ is a loop.

Now define

$$\alpha : Q \to Q; y \mapsto a \cdot y$$

and

$$\beta : Q \to Q; x \mapsto x \cdot b.$$

The maps α and β are bijective, with corresponding inverses

$$\alpha^{-1} : Q \to Q; y \mapsto a\backslash y$$

and

$$\beta^{-1} : Q \to Q; x \mapsto x/b$$

(Exercise 41). By (11.20), the triple

$$(\beta^{-1}, \alpha^{-1}, \mathrm{id}_Q) : (Q, \circ) \to (Q, \cdot)$$

is an isotopy. Thus

$$(\beta, \alpha, \mathrm{id}_Q) : (Q, \cdot) \to (Q, \circ)$$

is the required principal isotopy from $(Q, \cdot)$ to the loop $(Q, \circ, a \cdot b)$. ☐

It is natural to ask why the concept of isotopy does not arise in the study of groups. The following theorem and its corollary provide an answer.

THEOREM 11.39 (Loop isotopes of groups are groups.)
If a loop is isotopic to a group, then it is isomorphic to that group.

PROOF It suffices to consider the case of a principal isotopy

$$(f, g, \mathrm{id}_Q) : (Q, *, e_*) \to (Q, \circ, e_\circ)$$

from a loop structure $(Q, *, e_*)$ on a set Q to a group structure $(Q, \circ, e_\circ)$ on Q. Thus

$$f(x) \circ g(y) = x * y \qquad (11.21)$$

for elements x, y of Q. Setting $y = e_*$ in (11.21) yields

$$f(x) \circ g(e_*) = x * e_* = x,$$

so that

$$f(x) = x \circ g(e_*)^{-1}$$

in the group $(Q, \circ, e_\circ)$. Similarly, setting $x = e_*$ in (11.21) yields

$$f(e_*) \circ g(y) = e_* * y = y,$$

so that

$$g(y) = f(e_*)^{-1} \circ y$$

in $(Q, \circ, e_\circ)$. Equation (11.21) may now be rewritten in the form

$$x \circ g(e_*)^{-1} \circ f(e_*)^{-1} \circ y = (x * y)$$

within the group $(Q, \circ, e_\circ)$. Multiplying from the left by $f(e_*)^{-1}$, and from the right by $g(e_*)^{-1}$, we obtain

$$f(e_*)^{-1} \circ x \circ g(e_*)^{-1} \circ f(e_*)^{-1} \circ y \circ g(e_*)^{-1}$$
$$= f(e_*)^{-1} \circ (x * y) \circ g(e_*)^{-1}. \qquad (11.22)$$

Consider the invertible map

$$\theta : Q \rightarrow Q; x \mapsto f(e_*)^{-1} \circ x \circ g(e_*)^{-1}$$

(compare Exercise 42). Written in terms of θ, the equation (11.22) becomes

$$\theta(x) \circ \theta(y) = \theta(x * y).$$

Thus $\theta : (Q, *, e_*) \rightarrow (Q, \circ, e_\circ)$ is the required isomorphism.   ▯

COROLLARY 11.40 (Isotopic groups.)
If two groups are isotopic, then they are isomorphic.

The final concern of this chapter is to resolve a critical issue that arose in Section 11.2:

> Can each Latin square be given suitable row and column labels so that it becomes the body of a group multiplication table?

(A positive answer would suggest that the study of quasigroups could be reduced to a study of groups.) By Theorem 11.34, the question is equivalent to asking whether each finite quasigroup is principally isotopic to a group. By Theorem 11.38 and the transitivity of the isotopy relation, the question reduces to asking whether each finite loop is principally isotopic to a group. Finally, by Theorem 11.39, the question becomes: are there any finite loops that are not associative?

There is a unique loop with identity 0 on the set $\{0,1\}$, the group $(\mathbb{Z}/_2, +)$. Now consider a loop of order 3, on the set $\{0,1,2\}$. With the natural ordering of the row and column labels, the body of the multiplication table becomes the incomplete Latin square

$$0\ 1\ 2$$
$$1\ a$$
$$2$$

There are apparently two choices for the element a, namely 0 or 2. However, in the former case, there is no way to complete the Latin square, since the completion procedure stalls at

$$0\ 1\ 2$$
$$1\ 0\ 2$$
$$2$$

On the other hand, choosing $a = 2$ forces a unique completion to the Latin square

$$0\ 1\ 2$$
$$1\ 2\ 0$$
$$2\ 0\ 1$$

that gives the multiplication table of the group $(\mathbb{Z}/_3, +)$. So loops of order 3 are associative. In Exercise 43, you are asked to apply similar techniques to show that each loop of order 4 is associative. However, the loop whose multiplication table is displayed in Figure 11.10 is not associative, since any group of order 5 is commutative.

·	0	1	2	3	4
0	0	1	2	3	4
1	1	4	3	0	2
2	2	0	4	1	3
3	3	2	0	4	1
4	4	3	1	2	0

FIGURE 11.10: A nonassociative loop of order 5.

11.8   Exercises

1. Show that the integers form a nonassociative quasigroup $(\mathbb{Z}, -)$ under subtraction.

2. The *geometric mean* of two positive real numbers x and y is

$$x * y = \sqrt{xy}. \qquad (11.23)$$

   Show that under the multiplication $*$ of (11.23), the set of positive real numbers forms a nonassociative quasigroup.

3. Let Q be the set of negative real numbers.

   (a) Show that (11.23) is defined for x, y in Q.

   (b) Show that (11.23) does not give a quasigroup multiplication on Q.

4. (a) Show that for a natural number r, the set P_r of nonzero bit strings of length $r + 1$ is closed under the multiplication $*$ of (11.3).

   (b) Show that for a positive integer r, the quasigroup P_r is not a group. (Hint: Consider the properties of the identity element of a group.)

5. Write out the multiplication tables for the quasigroups P_1 and P_2 of Example 11.4.

6. Find values for the unknowns a, b, c, d from the integers modulo 4 so that

	a	b	c	d
0	0	3	2	1
1	1	0	3	2
2	2	1	0	3
3	3	2	1	0

   becomes the addition table for the group $(\mathbb{Z}/_4, +)$.

7. Let $(Q, \cdot)$ be a finite, nonempty quasigroup. Show that the body of the multiplication table of $(Q, \cdot)$ forms a Latin square.

8. Complete the proof of Theorem 11.5.

9. Without directly using tables of group addition or subtraction, construct a 5×5 Latin square.

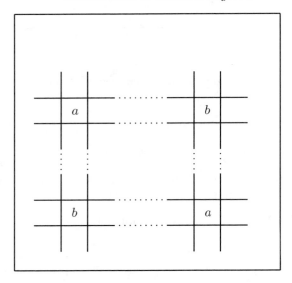

FIGURE 11.11:   An intercalate in a Latin square.

10. In a Latin square, an *intercalate* is a configuration of four entries at the intersection of two rows and two columns, containing just two distinct elements a and b (Figure 11.11).

    (a) Show that interchanging the entries a and b of an intercalate within one Latin square creates a new Latin square.

    (b) Let t be an element of a finite group G, with $t^2 = 1 \neq t$. Show that the multiplication table body of G contains an intercalate with entries 1 and t.

    (c) Use intercalates to create new quasigroups of order 6 from each group of order 6. Which of the new quasigroups are not associative?

11. Let $(A, +)$ be an additive group, considered as a quasigroup with $+$ as the quasigroup multiplication. Write the corresponding left division in terms of addition, subtraction, and negation.

12. Verify that (11.7) holds in each bit string quasigroup P_r. (Hint: There are two cases to consider, $x = y$ and $x \neq y$.)

13. Verify that the equations (11.8) hold in a group.

14. Prove Proposition 11.11(c),(d).

15. Complete the proof of Proposition 11.12.

16. Complete the proof of Theorem 11.13.

17. Suppose that $(Q, \cdot)$ is the quasigroup with multiplication table

·	0	1	2	3
0	0	3	1	2
1	1	2	3	0
2	2	1	0	3
3	3	0	2	1

Determine the multiplication tables for the quasigroups $(Q, \backslash)$ and $(Q, /)$.

18. Verify Proposition 11.15.

19. A quasigroup $(Q, \cdot)$ is said to be *commutative* if $x \cdot y = y \cdot x$ for all x, y in Q. Show that a quasigroup is commutative if and only if right division is the opposite of left division.

20. Show that a bit string quasigroup P_r (compare Example 11.4) coincides with each of its conjugates.

21. How many distinct conjugates does the group $(\mathbb{Z}, +)$ of integers possess?

22. Consider Figure 11.6.

    (a) How many distinct 3-element subquasigroups are displayed in the figure?

    (b) Show that knowledge of the 3-element subquasigroups, along with the observation that $x * x = x$ for each element x, specifies the multiplication $*$ in $(P_2, *)$ completely.

23. Complete the proof of Proposition 11.19.

24. Consider the closed unit interval $I = [0, 1]$, the set of real numbers from 0 to 1.

    (a) Show that I is closed under the multiplication $\circ$ of the arithmetic mean quasigroup $(\mathbb{R}, \circ)$ of Example 11.3.

    (b) Show that I does not form a subquasigroup of the arithmetic mean quasigroup $(\mathbb{R}, \circ)$.

25. Let $(G, \cdot)$ be a group, and let S be a nonempty subset of G. Show that S forms a subgroup of $(G, \cdot)$ if and only if it forms a subquasigroup of the quasigroup $(G, \cdot)$.

26. In the context of Proposition 11.22, show that the quasigroup homomorphism $f : (P, \cdot) \to (Q, \cdot)$ preserves right divisions.

27. Show that the arithmetic mean quasigroup $(\mathbb{R}, \circ)$ of Example 11.3 is isomorphic to the geometric mean quasigroup of Exercise 2.

28. Suppose that a quasigroup Q is isomorphic to a group G. Show that Q is associative.

29. Complete the proof of Proposition 11.23.

30. Let P and Q be quasigroups. Show that the projections

$$p_1 : P \times Q; (x_1, x_2) \mapsto x_1$$

and

$$p_2 : P \times Q; (x_1, x_2) \mapsto x_2$$

are quasigroup homomorphisms.

31. Let X be a set, and let $(Q, \cdot)$ be a quasigroup. Show that the set Q^X of functions $f : X \to Q$ from X to Q carries a componentwise quasigroup structure $(Q^X, \cdot)$, with

$$(f \cdot g)(x) = f(x) \cdot g(x)$$

for f, g in Q^X and x in X.

32. Consider Figure 11.7.

    (a) How many distinct 3-element subquasigroups are displayed in the figure?

    (b) Show that knowledge of the 3-element subquasigroups, along with the observation that $x * x = x$ for each element x, specifies the multiplication $*$ in $(P_1 \times P_1, *)$ completely.

33. Consider the product of the arithmetic mean quasigroup $(\mathbb{R}, \circ)$ with itself — compare Example 11.3.

    (a) Give a geometric interpretation of the product quasigroup structure $(\mathbb{R}^2, \circ)$ on the Cartesian plane $\mathbb{R}^2$.

    (b) Give a geometric interpretation of right division in the product quasigroup structure $(\mathbb{R}^2, \circ)$ on the Cartesian plane $\mathbb{R}^2$. (Hints: Compare Example 11.10. Recall the two types of reflection in the plane, point reflections and line reflections.)

34. Consider the quasigroup $(\mathbb{Z}/_3, -)$ of integers modulo 3, with subtraction as the quasigroup multiplication. Show that the product

$$(\mathbb{Z}/_3, -) \times (\mathbb{Z}/_2, +)$$

of $(\mathbb{Z}/_3, -)$ with the cyclic group $(\mathbb{Z}/_2, +)$ is isomorphic to the quasigroup $(\mathbb{Z}/_6, -)$ of integers modulo 6 under subtraction.

35. Show that the quasigroup of integers under subtraction is isotopic to the group of integers under addition.

36. Show that the arithmetic mean quasigroup $(\mathbb{R}, \circ)$ of Example 11.3 is isotopic to the additive group $(\mathbb{R}, +)$ of real numbers.

37. Show that the conjugates of a group are isotopic.

38. Let Q be a set. Show that principal isotopy forms an equivalence relation on the set of quasigroup structures $(Q, \cdot)$ on Q.

39. Verify the details of Proposition 11.33.

40. Use Proposition 11.36 to show that each finite, nonempty quasigroup is principally isotopic to a loop.

41. In the proof of Theorem 11.38, show that the maps α and β are bijective.

42. Show that the map $\theta : Q \to Q$ used in the proof of Theorem 11.39 is invertible.

43. Show that each 4×4 Latin square is the body of the multiplication table of a group.

44. Use intercalates (compare Exercise 10) to show that there are other nonassociative loops on the set $\{0, 1, 2, 3, 4\}$, besides the one displayed in Figure 11.10.

11.9   Study projects

1. **Quasigroups and Latin squares as experimental designs.**

   (a) A housing association is conducting an experiment to determine the best kind of wall siding to use for its houses: concrete, metal, plastic, or wood. For the experiment, it has houses, numbered 1, 2, 3, 4, at four different locations, with different climates and atmospheric conditions. Each house has walls facing in each of the cardinal directions: north, south, east, and west. How should the different kinds of siding be applied for the experiment, so that each kind of siding is tested on each house, and on each direction of wall?

   Set up a bordered 4×4 Latin square to plan how the experiment should be conducted. The house addresses from 1 to 4 should label the rows. The four directions should label the columns. The body should be a Latin square on the 4-element set

$$\{\text{concrete}, \text{metal}, \text{plastic}, \text{wood}\}$$

of siding types. The table entry in the row labeled i and column labeled d should indicate which type of siding is to be applied to the wall facing direction d on house number i.

(b) There are 7 students enrolled in a one-quarter algebra class. In each of 7 weeks of the course, the instructor wishes to designate a group of 3 students to prepare a special presentation. In order to assess everybody fairly, each student should be grouped exactly once with each other student. Use the bit-string quasigroup P_2 of Figure 11.6 to prepare an assignment plan for the instructor. Note that each student is involved in 2 different group presentations.

(c) Repeat the exercise of (b) for the case of 9 students in 12 groups of 3, during a one-semester course. Which quasigroup should be used in this case? In how many presentations is each student involved?

(d) If n students are to be assigned to t groups of 3, with each pair of students appearing in exactly one group as in (b) and (c) above, show that

$$\frac{n-1}{2} = t$$

and

$$\frac{n(n-1)}{2} = 3t.$$

(e) Conclude that an assignment plan for n students is only possible if n is congruent to 1 or 3 modulo 6.

2. **Orthogonal Latin squares.**

Let Q be a nonempty set with a finite number n of elements. Two Latin squares $L_1(Q)$ and $L_2(Q)$ on the set Q are said to be (*mutually*) *orthogonal*, if for each ordered pair (x_1, x_2) of elements of Q, there are unique integers $1 \leq i, j \leq n$ such that for $k = 1, 2$, the element x_k appears in the i-th row and j-th column of $L_k(Q)$. A pair of orthogonal Latin squares is displayed in Figure 11.12.

(a) Let $(Q, *)$ and $(Q, \circ)$ be quasigroup structures on the given set Q. Suppose that $L_*(Q)$ and $L_\circ(Q)$ are the respective bodies of the multiplication tables of $(Q, *)$ and $(Q, \circ)$, presented with a row and column labeling that is the same for each table. Show that the Latin squares $L_*(Q)$ and $L_\circ(Q)$ are orthogonal if and only if the function

$$Q \times Q \to Q \times Q; (x, y) \mapsto (x * y, x \circ y)$$

is bijective.

L_1	0	1	2
0	0	1	2
1	1	2	0
2	2	0	1

L_2	0	1	2
0	0	2	1
1	1	0	2
2	2	1	0

FIGURE 11.12:   A pair of orthogonal Latin squares.

(b) What quasigroup structures on the 3-element set $\mathbb{Z}/_3$ of residues modulo 3 correspond to the orthogonal Latin squares displayed in Figure 11.12?

(c) Let p be a prime number. Let l and m be distinct nonzero residues modulo p. Show that the quasigroups $(\mathbb{Z}/_p, *)$ with

$$x * y = x + ly$$

and $(\mathbb{Z}/_p, \circ)$ with

$$x \circ y = x + my$$

yield mutually orthogonal Latin squares on $\mathbb{Z}/_p$.

(d) Let F be a finite field. Let l and m be distinct nonzero elements of F. Show that the quasigroups $(F, *)$ with

$$x * y = x + ly$$

and $(F, \circ)$ with

$$x \circ y = x + my$$

yield mutually orthogonal Latin squares on F.

3. **Left and right multiplications.**

Let $(Q, \cdot)$ be a quasigroup. By analogy with (5.16), define a map

$$\lambda_q : Q \to Q; x \mapsto q \cdot x \qquad (11.24)$$

for each element q of Q. This map is known as *left multiplication* by the element q. Similarly, define the *right multiplication*

$$\rho_q : Q \to Q; x \mapsto x \cdot q \qquad (11.25)$$

for each element q of Q.

(a) Show that the respective identities (IL), (IR) of Proposition 11.12 imply the injectivity of the left and right multiplications λ_x, ρ_x.

(b) Show that the respective identities (SL), (SR) of Proposition 11.12 imply the surjectivity of the left and right multiplications λ_x, ρ_y.

(c) Conclude that in the quasigroup $(Q, \cdot)$, each left multiplication (11.24) and right multiplication (11.25) is a permutation of Q.

(d) Consider the case where $(Q, \cdot)$ is the quasigroup $(\mathbb{Z}/_n, -)$ of integers modulo a positive integer n, under subtraction. Show that the set

$$\{\lambda_x, \rho_x \mid x \text{ in } \mathbb{Z}/_n\}$$

forms a group of permutations on $\mathbb{Z}/_n$, isomorphic to the dihedral group D_n of Study Project 3 in Chapter 4.

(e) Show that a quasigroup $(Q, \cdot)$ is associative if and only if the map

$$\lambda : Q \to Q!; q \mapsto \lambda_q$$

is a quasigroup homomorphism.

11.10   Notes

Section 11.3

The symbols $/$ and $\backslash$ are often used within mathematical software in the same sense as in Definition 11.6. Thus if A and B are invertible (square) matrices, A/B may denote the matrix AB^{-1}, while $A\backslash B$ is used for $A^{-1}B$ — compare (11.8). The notation is extended to denote solutions to equations. For example, the solution $\mathbf{x}$ of the vector equation $A\mathbf{x} = \mathbf{y}$ is written as $\mathbf{x} = A\backslash\mathbf{y}$.

Conjugates of a quasigroup are sometimes described as "parastrophes."

Index